Collins Pocket Guide to

Stars and Planets

Ian Ridpath

Illustrated by
Wil Tirion

HarperCollins*Publishers*

Acknowledgements

This book is the result of an unusual international collaboration between an author and an illustrator in different countries, united by their common fascination with the sky. International cooperation of another kind is apparent in the photographic illustrations in this book, which are the result of the efforts of astronomers from many nations working with the world's finest telescopes. These awesome images rank with the most enduring works of art produced by mankind. It is fitting that the observatories concerned have made their results so readily available. The urge to study the sky transcends national boundaries, and so it should be. The skies are open to us all.

The authors would like to thank those whose photographs are used in this book, in particular the following who supplied specific images to our request: David Malin, Anglo-Australian Observatory; Brian Hadley, Royal Observatory Edinburgh; Walter Bonsack, University of Hawaii; the Royal Astronomical Society. Pictures are credited individually where they appear. Dr Peter Adams of the Institute of Geological Sciences provided prints of the US Air Force Lunar Reference Mosaic for making the Moon maps; the US Defense Mapping Agency gave permission for their use.

The following publications have been consulted extensively during the preparation of this edition: *Sky Atlas 2000.0* by Wil Tirion (Sky Publishing Corp., Cambridge, Mass.; and Cambridge University Press, England); *Sky Catalogue 2000.0*, Vols. 1 and 2 (Sky Publishing Corp. and Cambridge University Press); *The Bright Star Catalogue*, 4th edition, by Dorrit Hoffleit (Yale University Observatory); *Burnham's Celestial Handbook* by Robert Burnham, Jr (Dover, New York; Constable, London); *Astronomy* magazine (Milwaukee); *Sky & Telescope* magazine (Cambridge, Mass.). Further details about the origin and mythology of the constellations may be found in *Star Tales* by Ian Ridpath (Lutterworth, Cambridge; Universe, New York).

The Index was compiled for us by Enid Lake, former librarian of the Royal Astronomical Society. For this second edition we are grateful for the attentive editorial skills of John Woodruff.

I.R.
W.T.

HarperCollins*Publishers*
77–85 Fulham Palace Road, London W6 8JB

10 9 8 7
02 01 00 99 98

ISBN 0 00 219979 3

First published 1984
Second edition 1993

Typesetting and layout by Ian Ridpath
Printed and bound in Hong Kong

Contents

Section I

Section II

SECTION I

Introduction

The night sky is one of the most beautiful sights in nature. Yet many people remain lost among the jostling crowd of stars, and are baffled by the progressively changing appearance of the sky from hour to hour and season to season. The charts and descriptions in this book will guide you to the most splendid celestial sights, many of them within the range of simple optical equipment such as binoculars, and all accessible with an average-sized telescope as used by amateur astronomers.

It must be emphasized that you do not need a telescope to take up stargazing. Use the charts in this book to find your way among the stars first with your own eyes, and then with the aid of binoculars, which bring the stars more readily into view. Binoculars are a worthwhile investment, being relatively cheap, easy to carry and useful for many other purposes than stargazing.

In the night sky, stars appear to the naked eye as spiky, twinkling lights. Those stars near the horizon seem to flash and change colour. The twinkling and flashing effects are due not to the stars themselves but to the Earth's atmosphere: turbulent air currents cause the star's light to dance around. The steadiness of the atmosphere is referred to as the *seeing*. Steady air means good seeing. The spikiness of star images is due to optical effects in the observer's eye. In reality, stars are spheres of gas similar to our own Sun, emitting their own heat and light.

Stars come in various sizes, from giants to dwarfs, and in a range of colours according to their temperature. At first glance all stars appear white, but more careful inspection reveals that certain stars appear orange-red, notably Betelgeuse, Antares, Aldebaran and Arcturus, while others such as Rigel, Spica and Vega have a bluish tinge. Binoculars bring out the colours more readily than the naked eye does. Section II of this book explains more fully the different types of star that exist.

By contrast, planets are cold bodies that shine by reflecting the Sun's light. The planets are constantly on the move as they orbit the Sun, so they cannot be shown on the maps in this book. The planets are described in Section II.

About 2000 stars are visible to the naked eye on a clear, dark night, but you will not need to learn all of them. Start by learning the positions of the brightest stars and major constellations, and use these as signposts to the fainter, less prominent stars and constellations. Once you know the main features of the night sky, you will never again be lost among the stars.

Constellations The sky is divided into 88 sections, known as constellations, which astronomers use as a convenient way of locating and naming celestial objects. Each constellation is given a separate chart and description in this book. The main constellations of the sky were devised at the dawn of history, by Middle Eastern peoples who fancied that they could see a likeness to certain fabled creatures and mythological heroes among the stars. In particular, the 12 constellations of the zodiac, whose names are familiar to us from the astrological columns in newspapers and magazines, were of importance in ancient times. The zodiacal constellations are those that the Sun passes in front of in its yearly path around the heavens. However, it should be realized that the astrological 'signs' of the zodiac are not the same as the modern astronomical constellations, even though they share the same names.

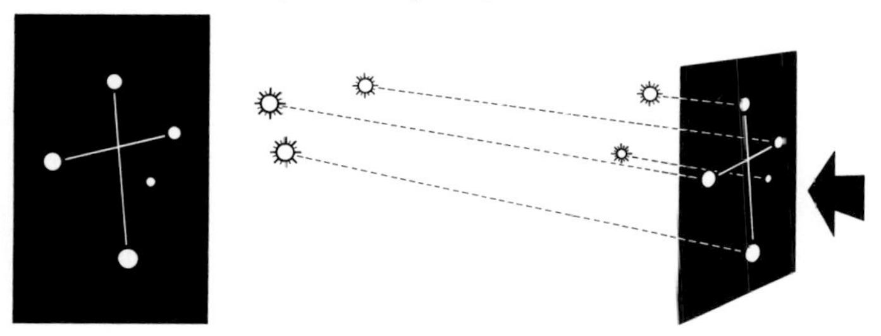

Stars in a constellation are usually unrelated to each other. Here the stars of Crux, the Southern Cross, are shown as they appear from Earth (left) *and as they actually lie in space* (right). Wil Tirion.

Most of the stars in a constellation have no real connection with each other at all; they may all lie at vastly differing distances from Earth, and simply form a pattern by chance. Some of the constellation patterns are easier to recognize than others, such as the magnificent Orion or the distinctive Cassiopeia. Others are faint and obscure, such as Lynx and Telescopium.

Our modern constellations derive from a list of 48, recognized by the Greek astronomer Ptolemy in AD 150. This list was expanded by navigators and celestial mapmakers, notably the Dutchmen Pieter Dirkszoon Keyser and Frederick de Houtman, the Pole Johannes Hevelius (1611–1687) and the Frenchman Nicolas Louis de Lacaille (1713–1762). Keyser and de Houtman introduced 12 new constellations, and Lacaille 14, in parts of the southern sky not visible from Mediterranean regions; Hevelius and others invented constellations to fill in the gaps between the figures recognized by the Greeks. The whole process sounds rather arbitrary, and indeed it was. A number of the newly devised patterns fell into disuse, leaving a total of 88 constellations that were officially adopted by the International Astronomical Union, astronomy's governing body, in 1930.

Star names The main stars in each constellation are labelled with a letter of the Greek alphabet, the brightest star usually (though not always!) being termed α (alpha). Notable exceptions include the constellations Orion and Gemini, in which the stars marked β (beta) are in fact the brightest. (For reference, the entire Greek alphabet is given in the table.) Particularly confusing are the constellations Vela and Puppis, which were once joined with Carina to make the larger figure of Argo Navis, the Ship. As a result of Argo's trisection, neither Vela nor Puppis possess stars labelled α or β, and there are gaps in the sequence of Greek letters in Carina as well.

Greek alphabet

α	alpha	ι	iota	ρ	rho
β	beta	$\varkappa$ or κ	kappa	σ	sigma
γ	gamma	λ	lambda	τ	tau
δ	delta	μ	mu	υ	upsilon
ε or ϵ	epsilon	ν	nu	φ or ϕ	phi
ζ	zeta	ξ	xi	χ	chi
η	eta	o	omicron	ψ	psi
ϑ or θ	theta	π	pi	ω	omega

The system of naming stars by Greek letters was introduced by Johann Bayer, so these designations are often known as Bayer letters. The genitive case of the constellation's name is always used when referring to a star within it; hence Canis Major, for instance, becomes Canis Majoris, and the name α Canis Majoris means 'the star α in Canis Major'. All constellation names have standard abbreviations. For instance, in abbreviated form Canis Major becomes CMa.

In some constellations, fainter stars are assigned Roman letters, such as L Puppis and P Cygni. An additional system of identifying stars is that of Flamsteed numbers, from their number in a star catalogue drawn up by England's first Astronomer Royal, John Flamsteed (1646–1719). Examples are 61 Cygni and 70 Ophiuchi.

Before the International Astronomical Union's ruling in 1930, there were no officially defined constellation boundaries; stars of one constellation could overlap parts of another constellation. Since 1930 some stars allocated by the Flamsteed system to one constellation have found themselves transferred to a neighbouring constellation, while retaining their old Flamsteed numbers. One example is 41 Lyncis (formerly in Lynx), which under the 1930 boundary rules now lies in Ursa Major.

Prominent stars also have proper names by which they are commonly known. α Canis Majoris is better known as Sirius, the brightest star in the sky. Stars' proper names originate from many sources. Some, such as Sirius, Castor and Pollux, date back to Greek times. Many others, such as Aldebaran, are of Arabic origin. Still others were added more recently by European astronomers who borrowed Arabic words in corrupted form; an example is Betelgeuse, which in its current form is meaningless in Arabic. To add to the confusion, spelling of names can vary from list to list, and some stars have more than one proper name.

Star clusters, nebulae and galaxies have a different system of identification. The most prominent of them are given numbers prefixed by the letter M from a catalogue compiled by the Frenchman Charles Messier. For example, M1 is the Crab Nebula and M31 the Andromeda Galaxy. Messier's catalogue contained 103 objects (a few more were added later by other astronomers). A far more comprehensive listing, containing many thousands of objects, is the *New General Catalogue* (NGC) compiled by J. L. E. Dreyer, with two supplements called the *Index Catalogues* (IC). Both the Messier numbers (where they exist) and NGC numbers remain in use by astronomers, and both are used in this book. On the charts, such objects are labelled with their Messier number if they have one, or by their NGC number (without the 'NGC' prefix) or IC number (prefix 'I').

Star brightnesses Stars appear of different brightness in the sky, for two reasons. Firstly, they do not all give out the same amount of light. But also, and just as importantly, they all lie at vastly differing distances. Hence, a modest star that is quite close to us can appear brighter than a tremendously powerful star that is a long way away.

Astronomers call a star's brightness its *magnitude*. The magnitude scale was introduced by the Greek astronomer Hipparchus in 129 BC. Hipparchus divided the naked-eye stars into six classes of brightness, from 1st magnitude (the brightest stars) to 6th magnitude (the faintest visible to the naked eye). In his day there was no means of measuring star brightness precisely, so this rough classification sufficed. But with the coming of technology it was possible to measure star brightnesses to an exact fraction of a magnitude.

In 1856 the English astronomer Norman Pogson put the magnitude scale on a precise mathematical footing, by defining a star of magnitude 1 as being exactly 100 times brighter than a star of magnitude 6. Since, on this scale, a difference of five magnitudes corresponds to a brightness difference of 100 times, a step of one magnitude is equal to a brightness difference of just over 2.5 times (the fifth root of 100).

Objects more than 250 times brighter than 6th magnitude are given negative (minus) magnitudes. For example, Sirius, the brightest star in the sky, is of magnitude –1.47. Stars fainter than magnitude 6 are given progressively larger positive magnitudes. The faintest objects detected with telescopes on Earth are of around magnitude 24. The system may sound confusing, but it works well and has the advantage that it can be extended infinitely, to both the very bright and the very faint.

When used without further qualification, the term 'magnitude' refers to how bright the star appears in the sky; strictly, this is the star's *apparent magnitude*. But because the distance of a star affects how bright it appears, the apparent magnitude bears little relation to its actual light output, or *absolute magnitude*. A star's absolute magnitude is defined as the brightness it would appear to have if it were at a standard distance of 10 parsecs from us (the parsec is explained below); the absolute magnitude is calculated by astronomers from knowledge of the star's nature and its distance.

Absolute magnitude is a good way of comparing the intrinsic brightness of stars. For instance, our daytime star the Sun has an apparent magnitude of –26.7, but an absolute magnitude of 4.8. Deneb (α (alpha) Cygni) has an apparent magnitude of 1.3, but an absolute magnitude of –7.5. From this comparison we deduce that Deneb gives out over 80,000 times as much light as the Sun and is hence one of the most luminous stars known, even though there is nothing at first sight to mark it out as extraordinary.

Magnitudes listed in various star catalogues may sometimes not agree with one another; this is because of differences in the instruments used to measure the brightness of stars. The values contained in this book come from *Sky Catalogue 2000.0*, 2nd edition (Sky Publishing Corp.). In addition, a number of stars actually vary in their light output, for various reasons. The nature of such so-called variable stars is discussed in Section II of this book.

Star distances In the Universe, distances are so huge that astronomers have abandoned the tiny kilometre (km) and have invented their own units. Most familiar of these is the *light year* (l.y.), the distance a beam of light travels in one year. Light moves at the fastest known speed in the Universe, 299,792.5 km per second. A light year is equivalent to 9.46 million million km. On average, stars are several light years apart. For instance, the closest star to the Sun, α (alpha) Centauri (actually a family of three stars), is 4.3 light years away.

The distance of the nearest stars can be found directly in the following way. A star's position is measured accurately when the Earth is on one side of the Sun, and then remeasured six months later when the Earth has moved around its orbit to the other side of the Sun. When viewed from two widely differing points in space in this way, a nearby star will appear to have shifted slightly in position with respect to more distant stars. This effect is known as *parallax*, and applies to any object viewed from two vantage points against a fixed background, such as a tree against the horizon. A star's parallax shift is so small as to be unnoticeable for all normal purposes – in the case of α Centauri, which has the greatest parallax shift of any star, the amount is about the same as the width of a small coin seen at a distance of 2 km. Once the star's

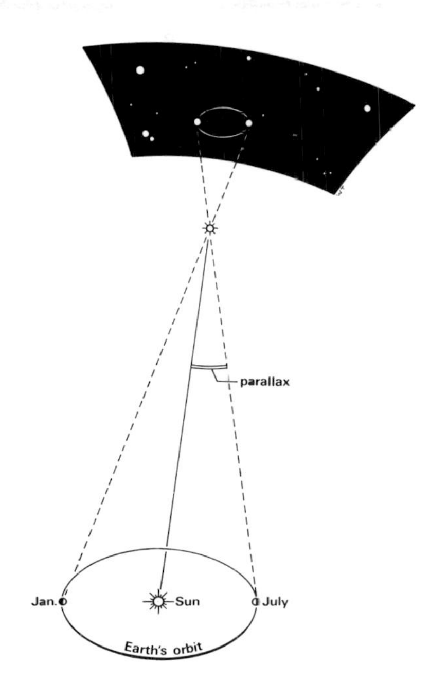

Parallax: as the Earth moves around its orbit, so a nearby star appears to change in position against the celestial background. The shift in position is known as the star's parallax. The nearer the star is to us, the greater its parallax. In this diagram, the amount of parallax is exaggerated for clarity. Wil Tirion.

parallax shift has been measured, a simple calculation reveals how far away it is.

An object close enough to us to show a parallax shift of 1″ (one second of arc – see page 10) would, in the jargon of professional astronomers, be said to lie at a distance of one *parsec*. In practice, no star is this close; the parallax of α Centauri is 0.75″. One parsec is equivalent to 3.26 light years. Astronomers frequently use parsecs in preference to light years because of the ease of converting parallax into distance: a star's distance in parsecs is simply the inverse of its parallax in seconds of arc. For example, a star 2 parsecs away has a parallax of 0.5″, 4 parsecs away it has a parallax of 0.25″, and so on.

The farther away a star is, the smaller its parallax. Beyond about 50 light years, a star's parallax becomes too small to be measured accurately by telescopes on Earth. Reliable parallaxes have been established for fewer than 1000 stars from Earth, although the astrometry satellite Hipparcos is in the process of increasing this to over 100,000 from space. For more distant stars, astronomers first of all estimate the star's absolute magnitude by studying the spectrum of its light. They then compare this estimated absolute magnitude with the observed apparent magnitude to determine the star's distance. The distance obtained in this way is open to considerable error, and the values quoted in various books and catalogues often differ widely because of this.

Distances of stars and galaxies given in this book are all expressed in light years. They are taken from *Sky Catalogue 2000.0*, supplemented where necessary by the *Bright Star Catalogue*. Most distances are given to two significant figures. Apart from the closest stars, whose distances have been determined from precise parallax measurements, not too much reliance should be placed on the second digit. Such is the uncertainty in estimating the distance of the more remote stars that in some cases even the first digit may be wrong. Nevertheless, it seemed better to give at least a rough figure rather than nothing at all.

Star positions To determine positions of objects in the sky, astronomers use a system of coordinates similar to latitude and longitude on Earth. The celestial equivalent of latitude is called *declination*, and the equivalent of longitude is called *right ascension*. Declination is measured in degrees, minutes and seconds (abbreviated °, ′ and ″) of arc from 0° on the celestial equator to 90° at the celestial poles. The celestial equator is the projection onto the sky of the Earth's equator, and the celestial poles lie exactly above the Earth's poles. Right ascension is measured in hours, minutes and seconds (abbreviated h, m and s), from 0 to 24 h. The 0 h line of right ascension, the celestial equivalent of the Greenwich meridian, is defined as the point where the Sun crosses the

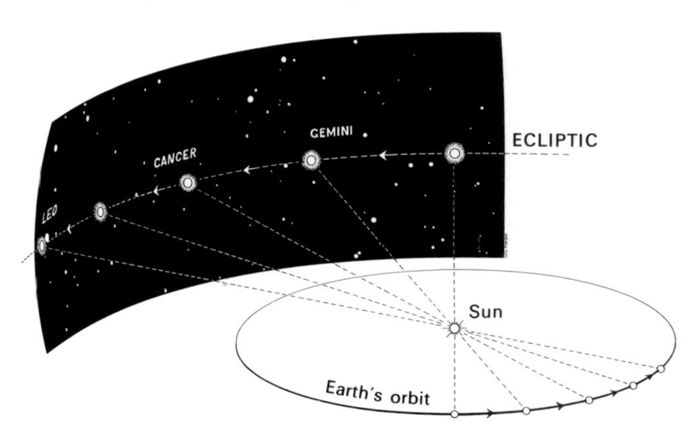

The ecliptic, the Sun's yearly path around the sky, is actually a result of the Earth's motion in orbit around the Sun. As the Earth moves along its orbit, the Sun is seen in different directions against the star background. The constellations that the Sun passes in front of during the year are known as the constellations of the zodiac. Wil Tirion.

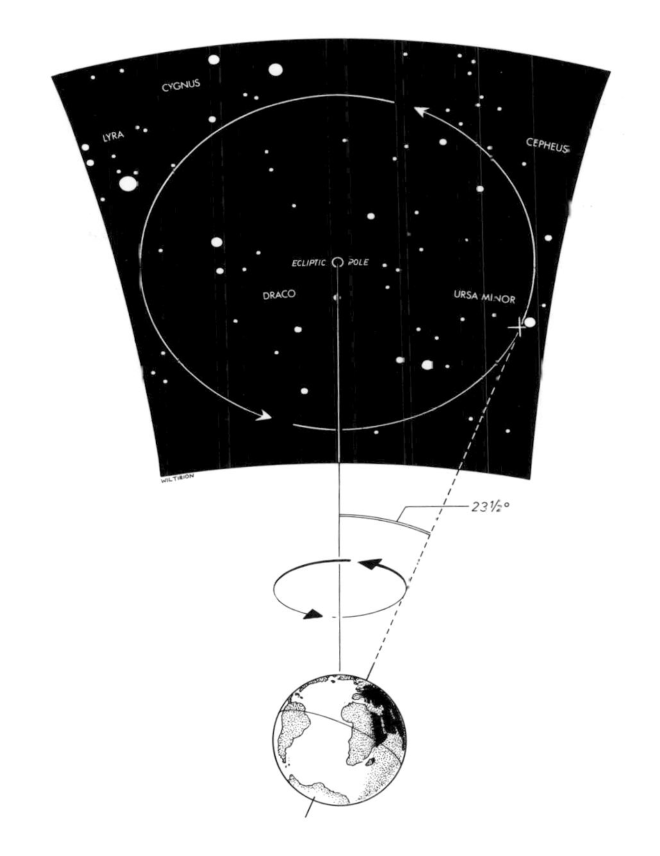

The Earth is very slowly wobbling in space like a tilted spinning top, an effect known as precession. As a result, the position of the celestial poles is constantly changing. The celestial poles trace out a complete circle on the sky every 26,000 years. Only the north celestial pole is shown here, but the effect applies to both poles. Wil Tirion.

celestial equator on its way north each year. Technically, this point is known as the spring (or vernal) equinox. In the constellation descriptions in this book, the positions of objects are given to the nearest minute of right ascension and the nearest degree of declination.

The Sun's path around the sky each year is known as the *ecliptic*. This path is inclined at 23½° to the celestial equator, because the Earth's axis is inclined at 23½° to the vertical. The most northerly point that the Sun reaches each year is called the *summer solstice*, 23½° north of the equator, and the most southerly point is the *winter solstice*, 23½° south of the equator. If the Earth's axis were directly upright with respect to its orbit around the Sun, then the equator and ecliptic would coincide. One result would be that we would have no seasons on Earth, for the Sun would always remain directly above the equator.

One additional effect that becomes important over long periods of time is that the Earth is slowly wobbling on its axis, like a spinning top. The axis remains inclined at an angle of 23½°, but the position in the sky to which the north and south poles of the Earth are pointing moves slowly. The Earth's poles describe a large circle on the sky, taking 26,000 years to return to their starting places. Hence the position of the celestial pole is always changing, albeit imperceptibly, as are the two points at which the Sun's path (the ecliptic) cuts the celestial equator. This wobbling of the Earth in space is termed *precession*. As an example of the effects caused by precession, whereas Polaris is the Pole Star today, in 12,000 years the north celestial pole will lie near Vega. And the vernal equinox which lay in Aries 2000 years ago now lies in Pisces.

The effect of precession means that the coordinates of celestial objects – the catalogued positions of stars, galaxies, and even constellation boundaries – are continually drifting. Astronomers usually draw up catalogues and star charts for a standard reference date, or *epoch*. The epoch of the star positions in this book is the year 2000. For most general purposes, precession does not introduce a noticeable error until after about 50 years, so the charts in this book will be usable without amendment until halfway through the 21st century.

Proper motions All the stars visible in the sky are members of a vast wheeling mass of stars called the Galaxy. The stars visible to the naked eye are among the nearest to us in the Galaxy. The more distant stars in the Galaxy crowd together in a hazy band called the Milky Way, which can be seen crossing the night sky.

The Sun and the other stars are all orbiting the centre of the Galaxy; it is so huge that the Sun takes about 250 million years to complete one orbit. Other stars move at different speeds, like cars in different lanes on a highway. The result is that stars are all very slowly changing their positions relative to one another. Such stellar movement is termed *proper motion*. Astronomers can detect proper motions with precision measuring techniques, but the motions are undetectable to the naked eye even over a human lifetime. If Ancient Greek astronomers were catapulted 2000 years forward to the present day, they would notice little difference in the sky, with the notable exception of Arcturus, a fast-moving bright star, which has moved more than two Moon diameters from its position in Ancient Greek times. Over very long periods of time the proper motions of stars considerably distort the shapes of all constellations. The diagrams on the facing page show some examples.

Appearance of the sky Three factors affect the appearance of the sky: the time of night, the time of year and your latitude on Earth.

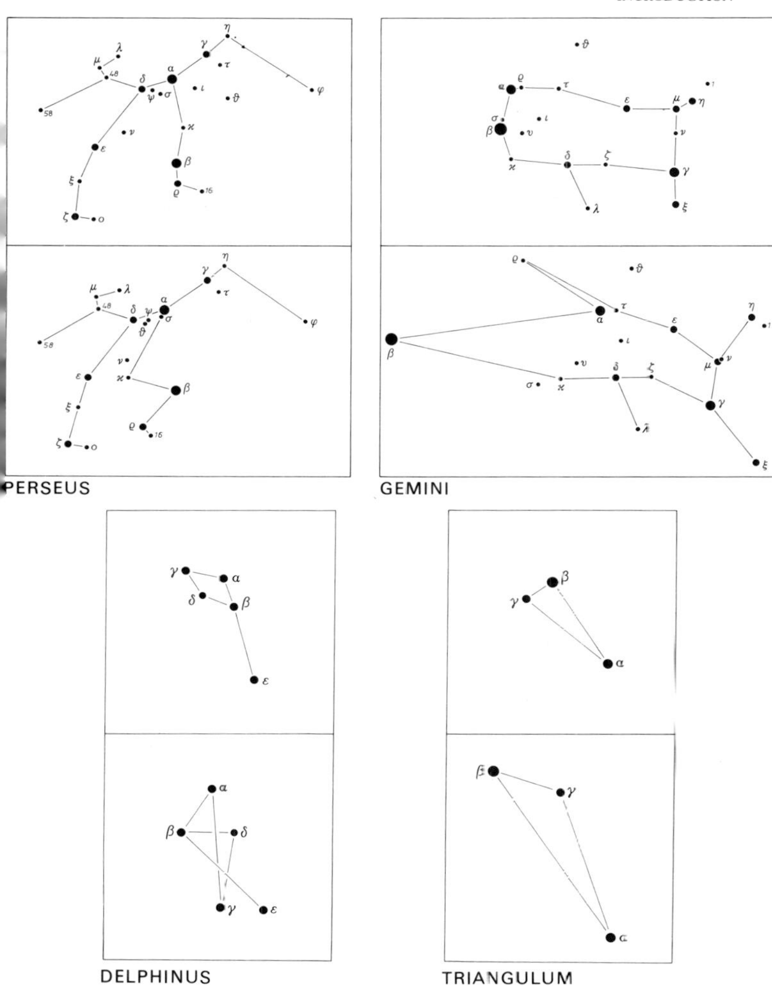

Four constellations as they appear today (top half of each diagram) and as they will appear in 100,000 years' time (bottom). Wil Tirion.

Firstly, let's consider the effect of latitude. An observer at one of the Earth's poles (latitude 90°) would see the celestial pole directly overhead, and as the Earth turned all stars would circle around the celestial pole without rising or setting. At the other extreme, an observer stationed exactly on the Earth's equator, latitude 0°, would see the

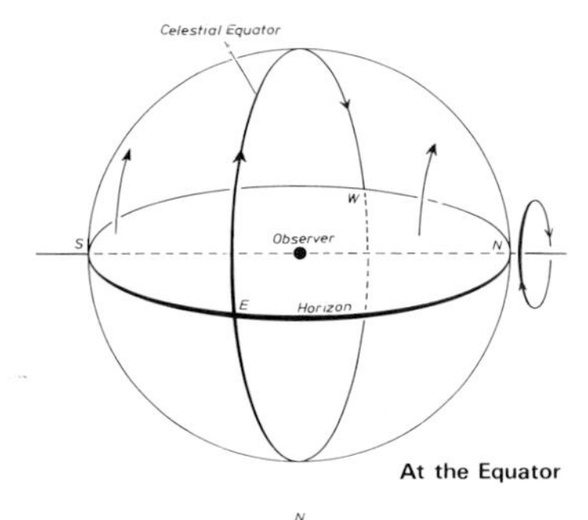

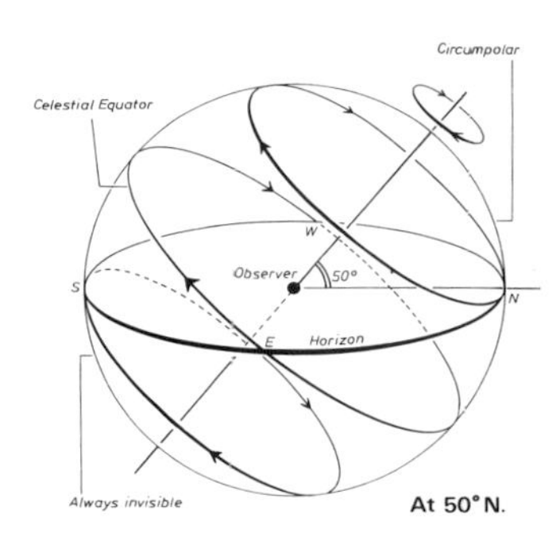

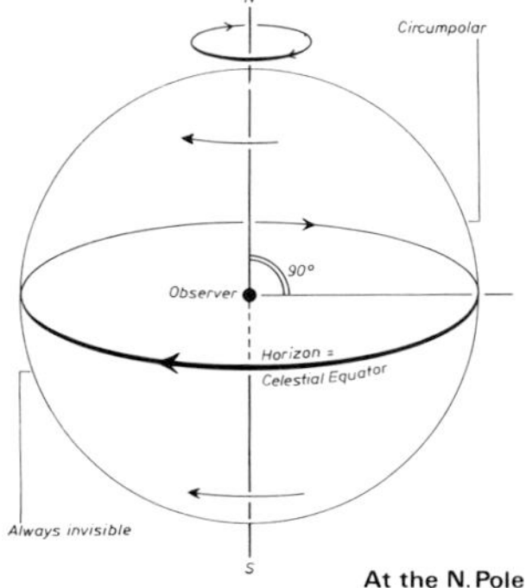

The changing appearance of the sky as seen from different locations on Earth.
At the equator, all the sky is visible at one time or another; as the Earth rotates, stars appear to rise in the east and set in the west.
At either pole, by contrast, only one half of the sky is ever visible, the other half being permanently below the horizon.
At intermediate latitudes, the situation is between the two extremes. Part of the sky is always above the horizon (the part marked 'circumpolar'), but an equal part is always below the horizon and hence is invisible.
Stars between these two regions rise and set during the night. Wil Tirion.

celestial equator directly overhead. The north and south celestial poles would lie on the north and south horizons respectively, and all parts of the sky would be visible at one time or another. All stars would rise in the east and set in the west as the Earth rotated.

For most observers, the sky appears somewhere between these two extremes: the celestial pole is at some intermediate altitude between horizon and zenith, and the stars closest to it circle around it without setting (they are said to be *circumpolar*) while the rest of the stars rise and set. The exact angle from the horizon to the celestial pole depends on the observer's latitude. For someone at latitude 40° north, for instance, the north celestial pole is 40° above the northern horizon. If you were at latitude 30° south, the south celestial pole would be 30° above the southern horizon. In other words, the altitude of the celestial pole above the horizon is exactly equal to your latitude, a fact long appreciated by navigators.

As the Earth turns, the stars march across the heavens at the rate of 15° per hour (the Earth rotates through 360° in 24 hours). Therefore the appearance of the sky changes with the time of night. An added complication is that the Earth is also orbiting the Sun each year, so the constellations visible will change with the seasons. For example, a brilliant constellation such as Orion, splendidly seen in December and January, will be in the daytime sky six months later and hence be invisible. The maps in this book will help you find out which stars are on view whenever and wherever you want to observe.

Using the star charts On pages 16–19 are charts showing the complete northern and southern hemispheres of the sky. In addition to the main stars of each hemisphere, the charts depict the hazy band of the Milky Way; the dashed line is the ecliptic, the Sun's path in the heavens. When planets are visible, they will be found near the ecliptic.

Around the rim of each chart are listed the months of the year to help you find which constellations are best placed at about 10 pm local time each month (11 pm when daylight-saving time is in operation). Observers in mid-northern latitudes should take the northern hemisphere chart and turn it so that the month of observation is at the bottom. The chart will show the sky as visible when you face due south that evening. Rotate the chart 15° anticlockwise for each hour after 10 pm, and turn it clockwise for each hour before 10 pm. Observers in mid-southern latitudes should take the southern hemisphere chart and turn it so that the month of observation is at the bottom. The chart will show the stars as they appear when you are facing due north. Turn the chart 15° clockwise for each hour after 10 pm, and 15° anticlockwise for each hour before. (In all cases above, for 10 pm read 11 pm when daylight-saving time is in operation.)

Next comes a series of maps showing the sky as seen when facing north or south at 10 pm (11 pm daylight-saving time) in mid-month from various latitudes. The first set of maps is usable from latitude 60° north to 10° north; the second set ranges from the equator to 50° south. (They will also be usable for 10° either side of this range without significant error.) Curved lines on each map show the horizon from each latitude. Taking these in conjunction with the maps of the celestial hemispheres, you should be able to identify the stars in the sky no matter where you are on Earth.

The centrepiece of this book consists of detailed charts and descriptions of each constellation. All stars down to magnitude 5.5 are shown. Some fainter stars have been added in areas of particular interest. The total number of stars shown on these maps is about 3000. All maps are to the same scale, with the exceptions of Hercules, Ursa Major and Serpens, which are slightly smaller in scale, and the rambling constellation Hydra, which is drawn to a significantly smaller scale.

We hope that the charts and descriptions in this book will serve as trusty companions for many nights of exploration under the stars. Good stargazing!

Northern Hemisphere
Stereographic projection
October
0h
September
AQUARIUS
21h
PISCES
PEGASUS
EQUULEUS
August
AQUILA
DELPHINUS
ANDROMEDA
SAGITTA
LACERTA
VULPECULA
CYGNUS
CEPHEUS
CASSIOPEIA
SCUTUM
SERPENS CAUDA
LYRA
CAMELOPARDALIS
18h
+90°
July
+80°
URSA MINOR
+70°
HERCULES
DRACO
OPHIUCHUS
+60°
URSA MAJOR
+50°
CORONA BOREALIS
+40°
CANES VENATICI
SERPENS CAPUT
BOÖTES
+30°
COMA BERENICES
+20°
June
LIBRA
15h
+10°
ECLIPTIC
VIRGO
0°
May
-10°
12h
April

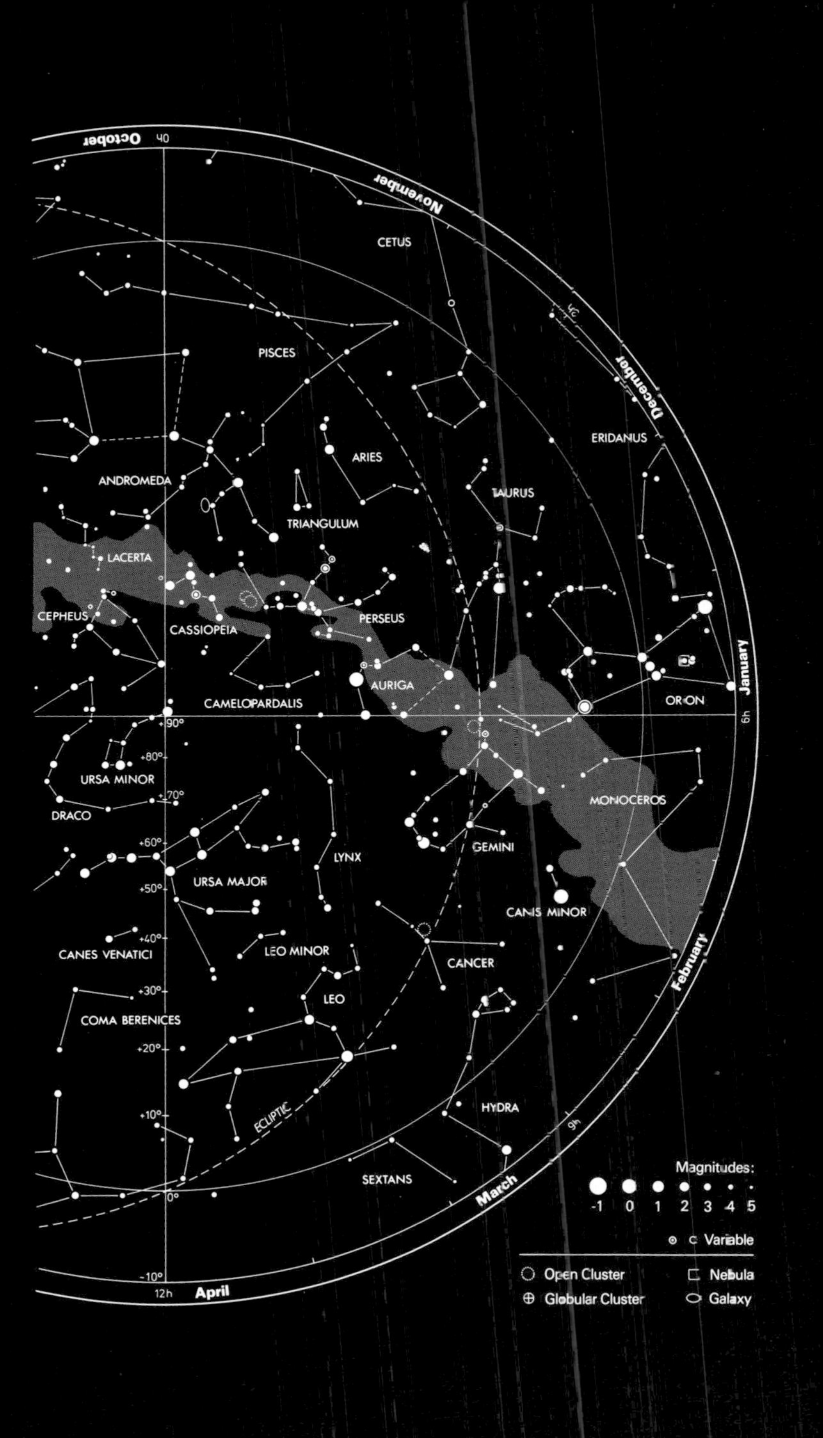

October
0h
November
December
January
6h
February
March
April
12h
9h
CETUS
PISCES
ARIES
ERIDANUS
ANDROMEDA
TAURUS
TRIANGULUM
LACERTA
CEPHEUS
CASSIOPEIA
PERSEUS
AURIGA
CAMELOPARDALIS
ORION
URSA MINOR
DRACO
MONOCEROS
GEMINI
LYNX
URSA MAJOR
CANIS MINOR
CANES VENATICI
LEO MINOR
CANCER
LEO
COMA BERENICES
HYDRA
ECLIPTIC
SEXTANS
+90°
+80°
+70°
+60°
+50°
+40°
+30°
+20°
+10°
0°
-10°
Magnitudes:
-1 0 1 2 3 4 5
Variable
Open Cluster
Nebula
Globular Cluster
Galaxy

Southern Hemisphere

Stereographic projection

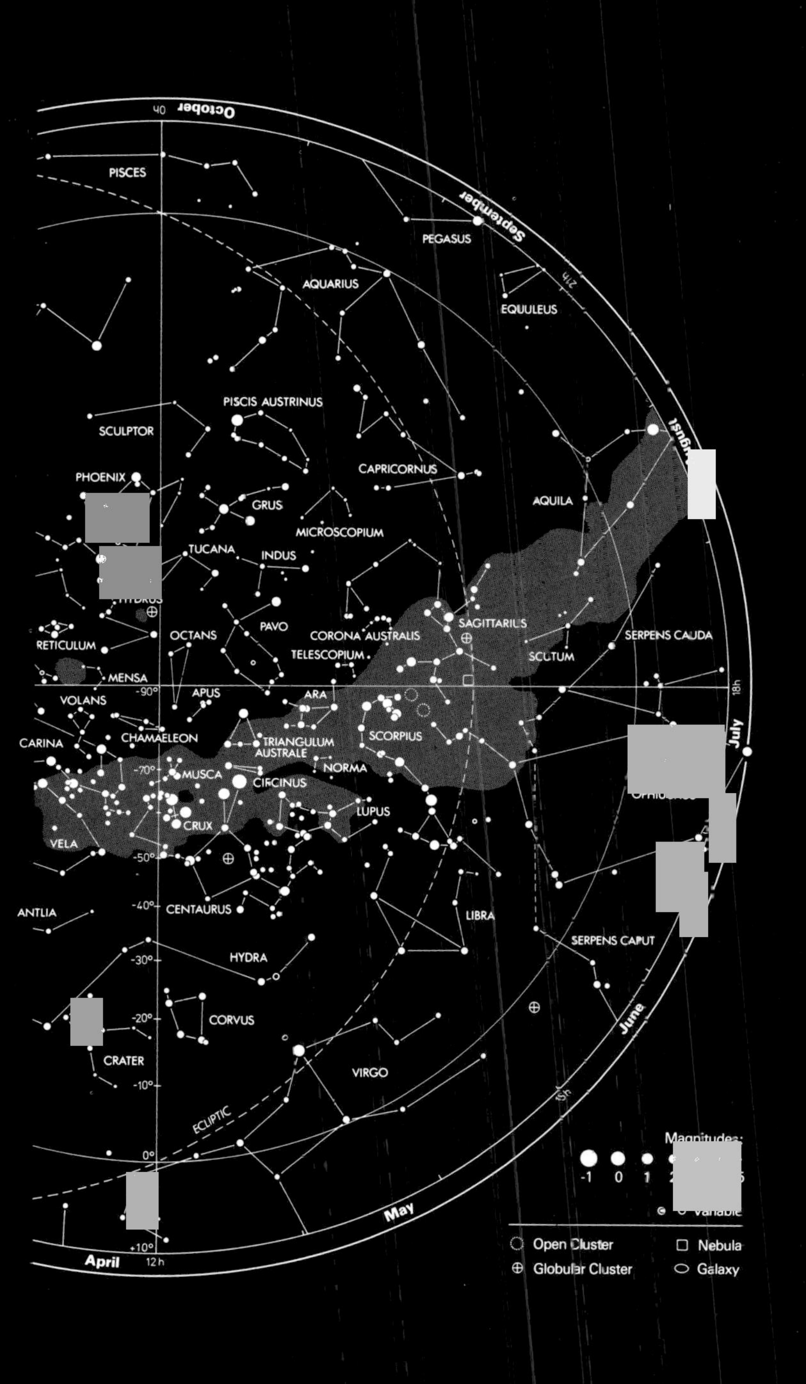

October
0h
PISCES
September
PEGASUS
21h
AQUARIUS
EQUULEUS
PISCIS AUSTRINUS
SCULPTOR
PHOENIX
CAPRICORNUS
AQUILA
GRUS
MICROSCOPIUM
TUCANA
INDUS
PAVO
SAGITTARIUS
SERPENS CAUDA
RETICULUM
OCTANS
CORONA AUSTRALIS
TELESCOPIUM
SCUTUM
MENSA
-90°
APUS
ARA
18h
VOLANS
CHAMAELEON
TRIANGULUM AUSTRALE
SCORPIUS
July
CARINA
NORMA
-70°
MUSCA
CIRCINUS
LUPUS
CRUX
VELA
-50°
ANTLIA
-40°
CENTAURUS
LIBRA
SERPENS CAPUT
-30°
HYDRA
-20°
CORVUS
June
CRATER
VIRGO
-10°
ECLIPTIC
Magnitudes:
-1
0
1
0°
May
+10°
April
12h
Open Cluster
Nebula
Globular Cluster
Galaxy

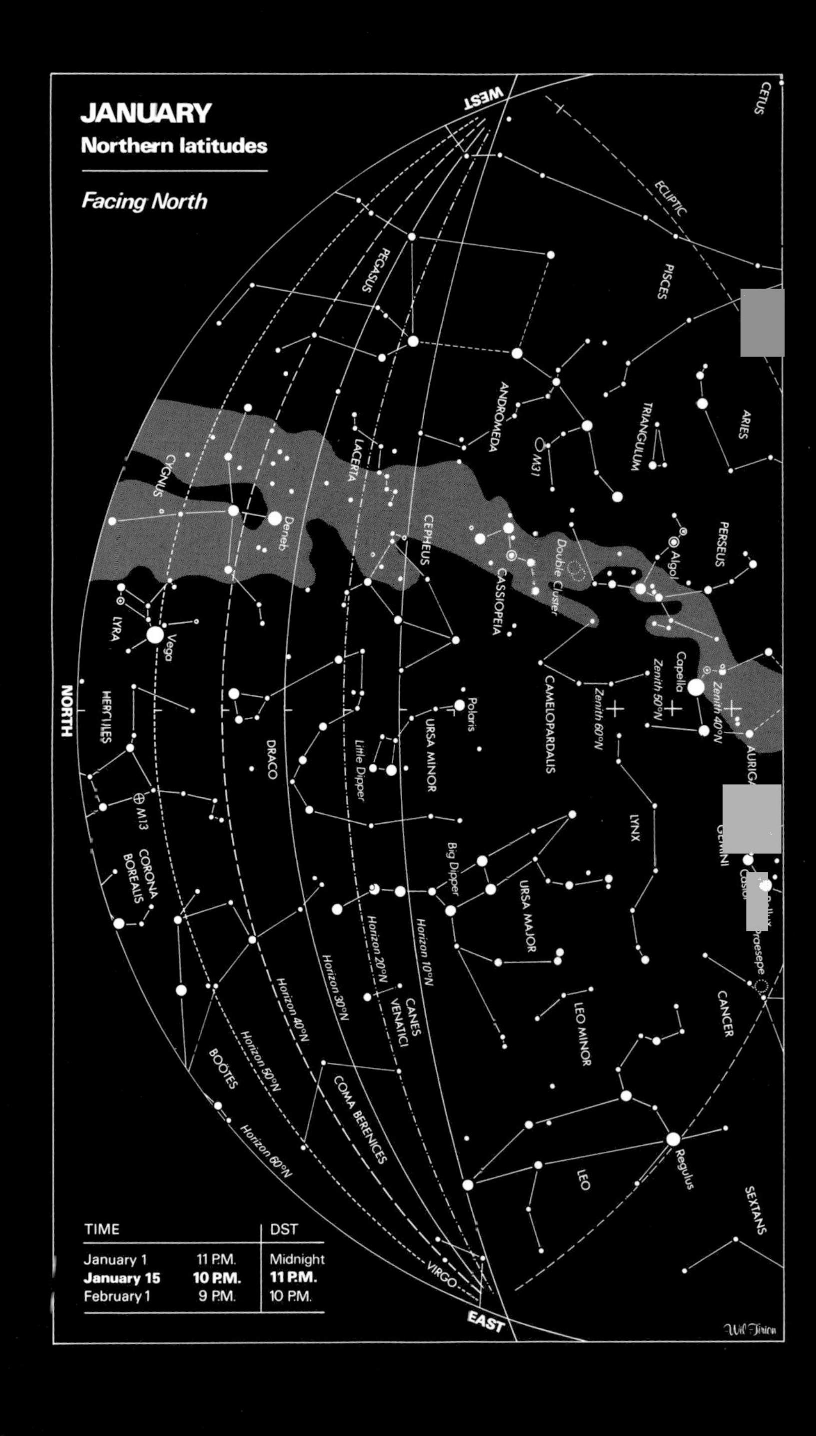

TIME		DST
January 1	11 P.M.	Midnight
January 15	**10 P.M.**	**11 P.M.**
February 1	9 P.M.	10 P.M.

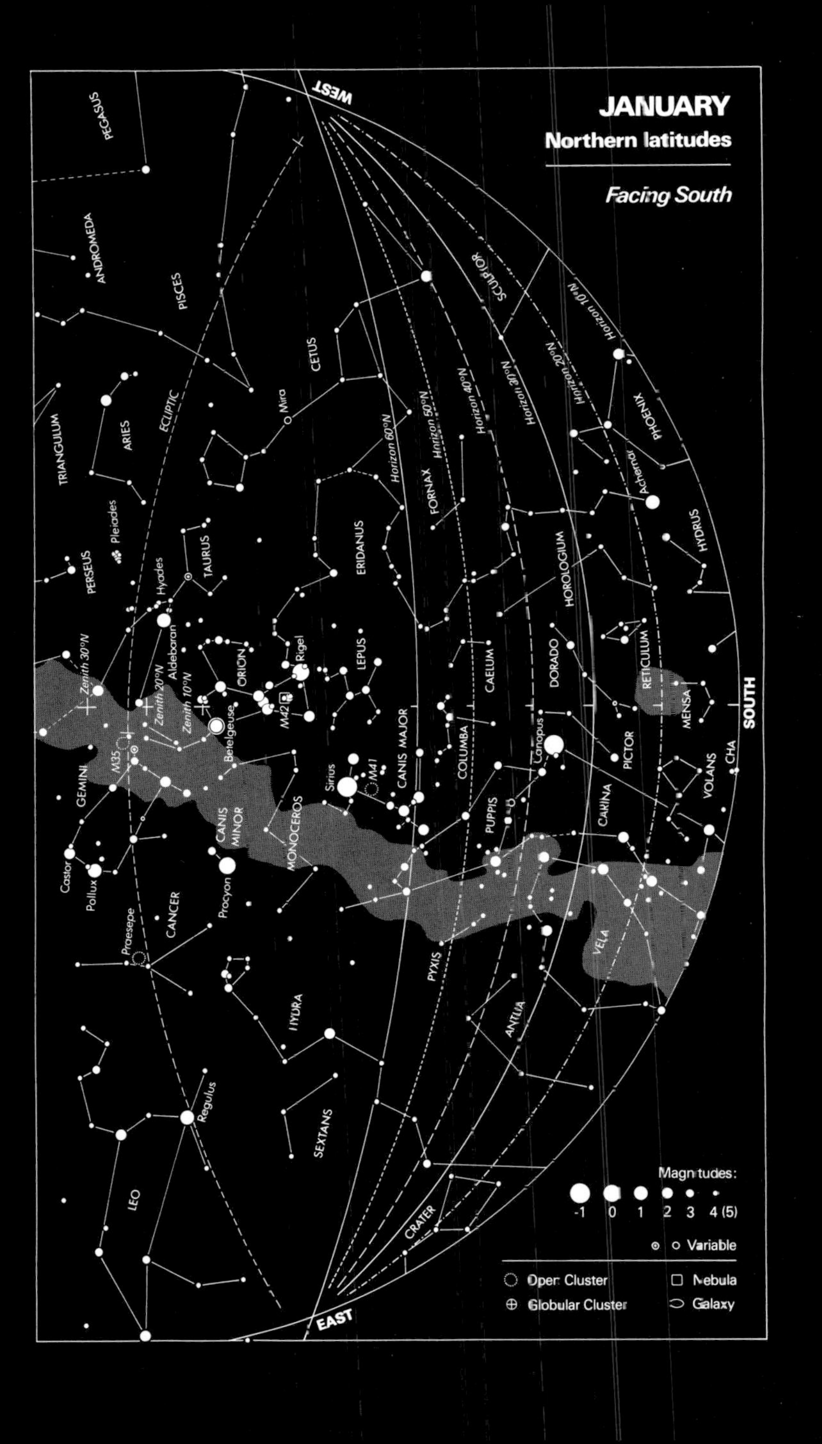

JANUARY
Northern latitudes
Facing South
WEST
EAST
SOUTH
PEGASUS
ANDROMEDA
PISCES
CETUS
SCULPTOR
TRIANGULUM
ARIES
ECLIPTIC
Mira
FORNAX
PHOENIX
Achernar
HYDRUS
Horizon 10°N
Horizon 20°N
Horizon 30°N
Horizon 40°N
Horizon 50°N
Horizon 60°N
Pleiades
PERSEUS
TAURUS
Hyades
ERIDANUS
HOROLOGIUM
Aldebaran
ORION
Rigel
LEPUS
CAELUM
DORADO
RETICULUM
MENSA
Zenith 30°N
Zenith 20°N
Zenith 10°N
Betelgeuse
M42
CANIS MAJOR
COLUMBA
Canopus
PICTOR
VOLANS
CHA
GEMINI
M35
M41
Sirius
CANIS MINOR
MONOCEROS
PUPPIS
CARINA
Castor
Pollux
Procyon
CANCER
Praesepe
VELA
PYXIS
HYDRA
ANTLIA
Regulus
SEXTANS
LEO
CRATER
Magnitudes:
-1 0 1 2 3 4 (5)
Variable
Open Cluster
Globular Cluster
Nebula
Galaxy

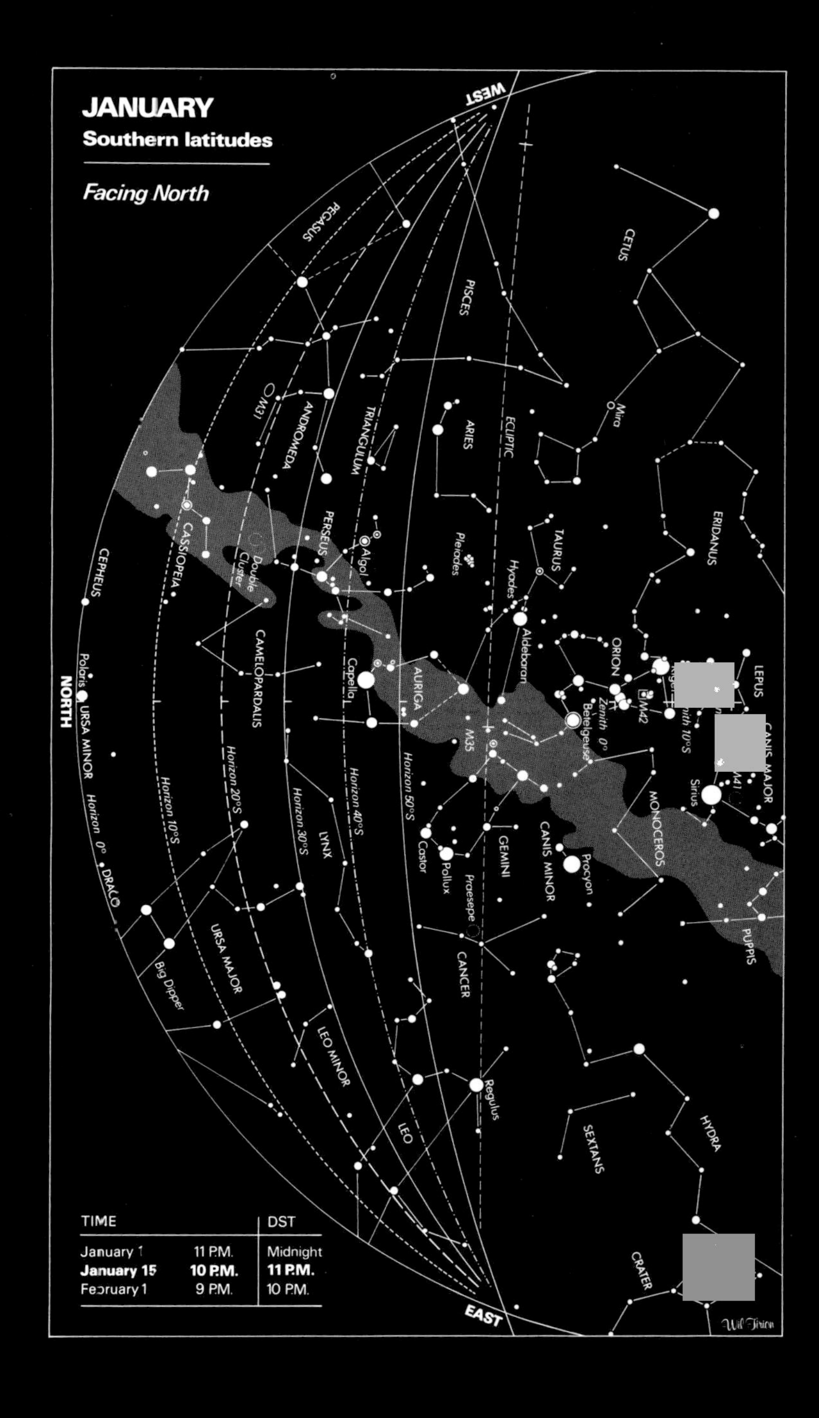
JANUARY
Southern latitudes
Facing North
WEST
EAST
NORTH
PEGASUS
PISCES
CETUS
ANDROMEDA
M31
TRIANGULUM
ARIES
ECLIPTIC
Mira
CASSIOPEIA
CEPHEUS
Double Cluster
PERSEUS
Algol
Pleiades
Hyades
TAURUS
ERIDANUS
CAMELOPARDALIS
Capella
AURIGA
Aldebaran
ORION
Rigel
LEPUS
M42
Zenith 0°
Betelgeuse
M35
CANIS MAJOR
M41
Sirius
Polaris
URSA MINOR
Horizon 0°
DRACO
Horizon 10°S
Horizon 20°S
Horizon 30°S
Horizon 40°S
Horizon 50°S
LYNX
Castor
Pollux
GEMINI
CANIS MINOR
Procyon
MONOCEROS
PUPPIS
Praesepe
CANCER
URSA MAJOR
Big Dipper
LEO MINOR
LEO
Regulus
SEXTANS
HYDRA
CRATER
Wil Tirion
TIME
DST
January 1 11 P.M. Midnight
January 15 10 P.M. 11 P.M.
February 1 9 P.M. 10 P.M.

JANUARY
Southern latitudes
Facing South
WEST
EAST
SOUTH
ECLIPTIC
PISCES
CETUS
SCULPTOR
AQUARIUS
PISCIS AUSTRINUS
Fomalhaut
MICROSCOPIUM
GRUS
FORNAX
PHOENIX
INDUS
SAGITTARIUS
ERIDANUS
HOROLOGIUM
Achernar
HYDRUS
47 Tuc
TUCANA
PAVO
TELESCOPIUM
CORONA AUSTRALIS
LEPUS
Zenith 30°S
Zenith 40°S
CAELUM
Zenith 50°S
RETICULUM
DORADO
MENSA
CHAMAELEON
OCTANS
TRIANGULUM AUSTRALE
ARA
SCORPIUS
Canopus
PICTOR
VOLANS
APUS
CIRCINUS
NORMA
M41
CANIS MAJOR
COLUMBA
CARINA
MUSCA
Acrux
Hadar
Rigil Kentaurus
PUPPIS
CRUX
Mimosa
VELA
LUPUS
PYXIS
ANTLIA
CENTAURUS
Horizon 0°
Horizon 10°S
Horizon 20°S
Horizon 30°S
Horizon 50°S
HYDRA
SEXTANS
CRATER
CORVUS
LEO
Magnitudes:
-1 0 1 2 3 4 (5)
Variable
Open Cluster
Globular Cluster
Nebula
Galaxy

FEBRUARY
Northern latitudes
Facing North

TIME		DST
February 1	11 P.M.	Midnight
February 15	**10 P.M.**	**11 P.M.**
March 1	9 P.M.	10 P.M.

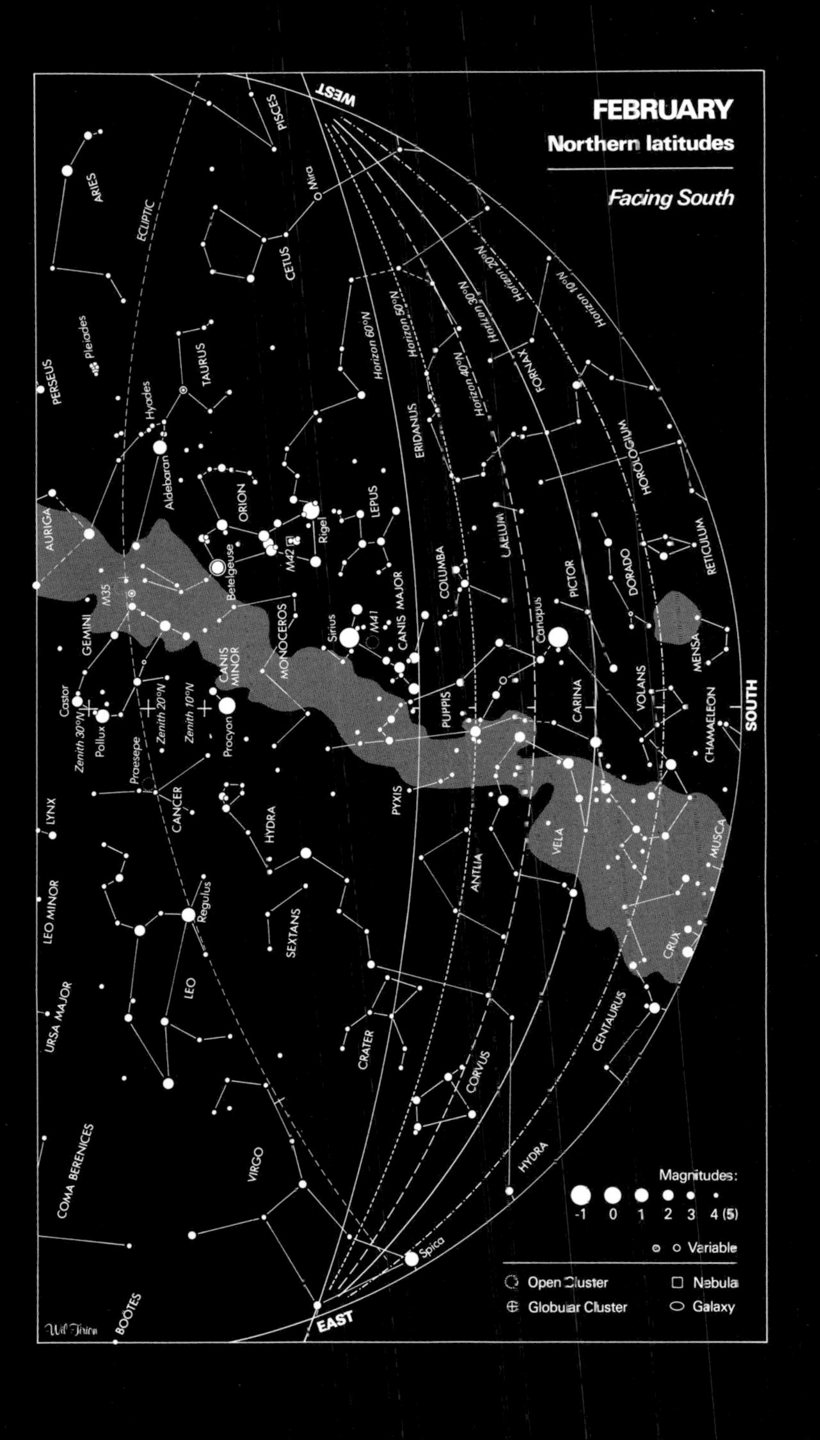
FEBRUARY
Northern latitudes
Facing South
WEST
EAST
SOUTH
PISCES
ARIES
ECLIPTIC
Mira
CETUS
Pleiades
PERSEUS
TAURUS
Hyades
Aldebaran
ORION
AURIGA
Rigel
M42
LEPUS
Betelgeuse
M35
GEMINI
MONOCEROS
CANIS MINOR
Sirius
M41
CANIS MAJOR
Castor
Pollux
Zenith 30°N
Zenith 20°N
Zenith 10°N
Praesepe
Procyon
CANCER
LYNX
HYDRA
PYXIS
PUPPIS
COLUMBA
ERIDANUS
Horizon 60°N
Horizon 50°N
Horizon 40°N
Horizon 30°N
Horizon 20°N
Horizon 10°N
FORNAX
CAELUM
HOROLOGIUM
PICTOR
DORADO
RETICULUM
Canopus
CARINA
VOLANS
MENSA
CHAMAELEON
VELA
ANTLIA
MUSCA
CRUX
CENTAURUS
LEO MINOR
Regulus
SEXTANS
LEO
URSA MAJOR
CRATER
CORVUS
HYDRA
VIRGO
COMA BERENICES
Spica
BOÖTES
Wil Tirion
Magnitudes:
-1 0 1 2 3 4 (5)
Variable
Open Cluster
Globular Cluster
Nebula
Galaxy

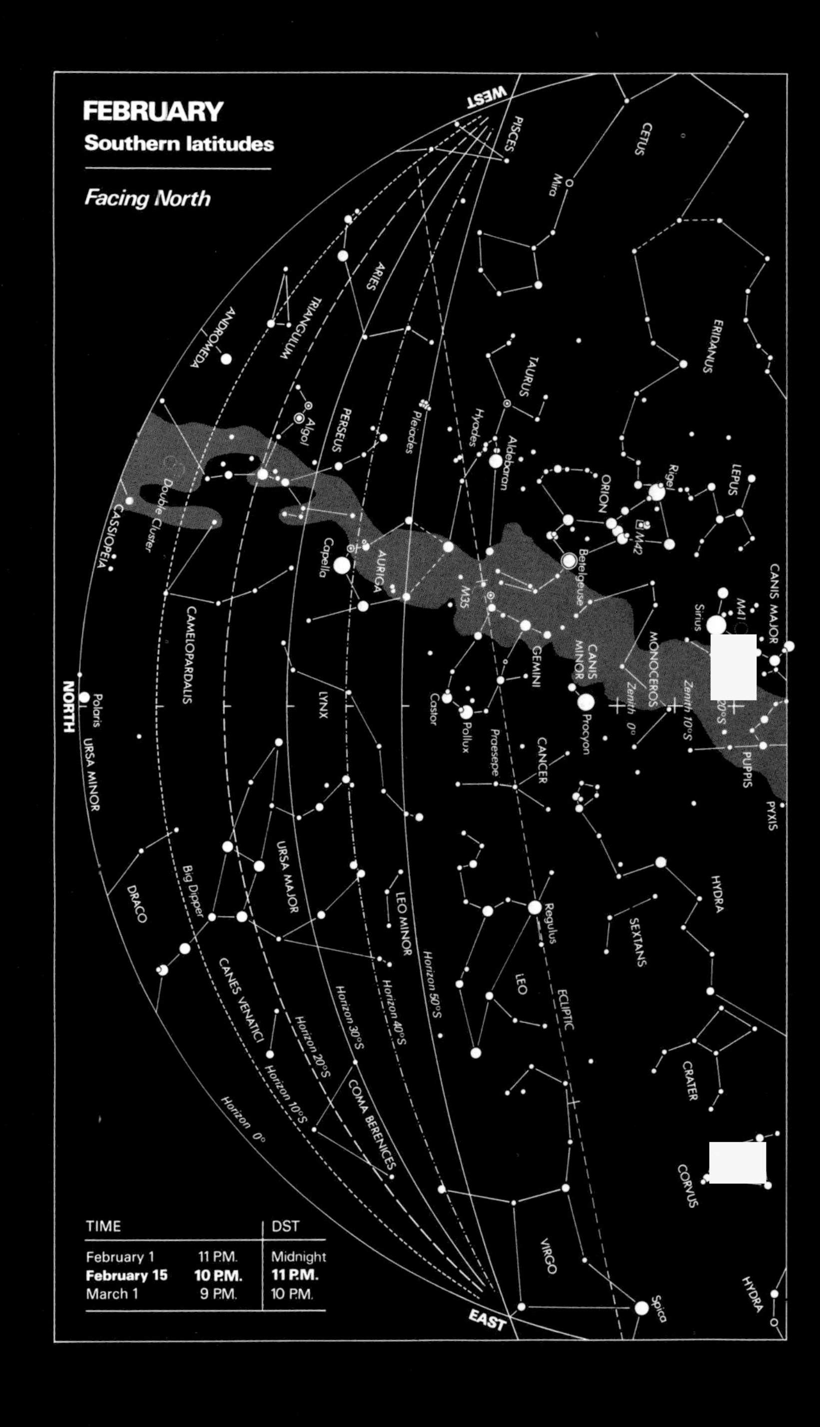
FEBRUARY
Southern latitudes
Facing North
WEST
EAST
NORTH
PISCES
CETUS
Mira
ARIES
TRIANGULUM
ANDROMEDA
ERIDANUS
TAURUS
Pleiades
Hyades
Aldebaran
PERSEUS
Algol
Double Cluster
CASSIOPEIA
Capella
AURIGA
ORION
Rigel
LEPUS
M42
Betelgeuse
CANIS MAJOR
M41
Sirius
M35
GEMINI
CANIS MINOR
MONOCEROS
CAMELOPARDALIS
LYNX
Castor
Pollux
Procyon
Zenith 0°
Zenith 10°S
20°S
Polaris
URSA MINOR
Praesepe
CANCER
PUPPIS
PYXIS
URSA MAJOR
LEO MINOR
HYDRA
SEXTANS
DRACO
Big Dipper
Regulus
LEO
CANES VENATICI
Horizon 30°S
Horizon 40°S
Horizon 50°S
ECLIPTIC
Horizon 20°S
Horizon 10°S
Horizon 0°
COMA BERENICES
CRATER
CORVUS
VIRGO
Spica
HYDRA
TIME
DST
February 1
11 P.M.
Midnight
February 15
10 P.M.
11 P.M.
March 1
9 P.M.
10 P.M.

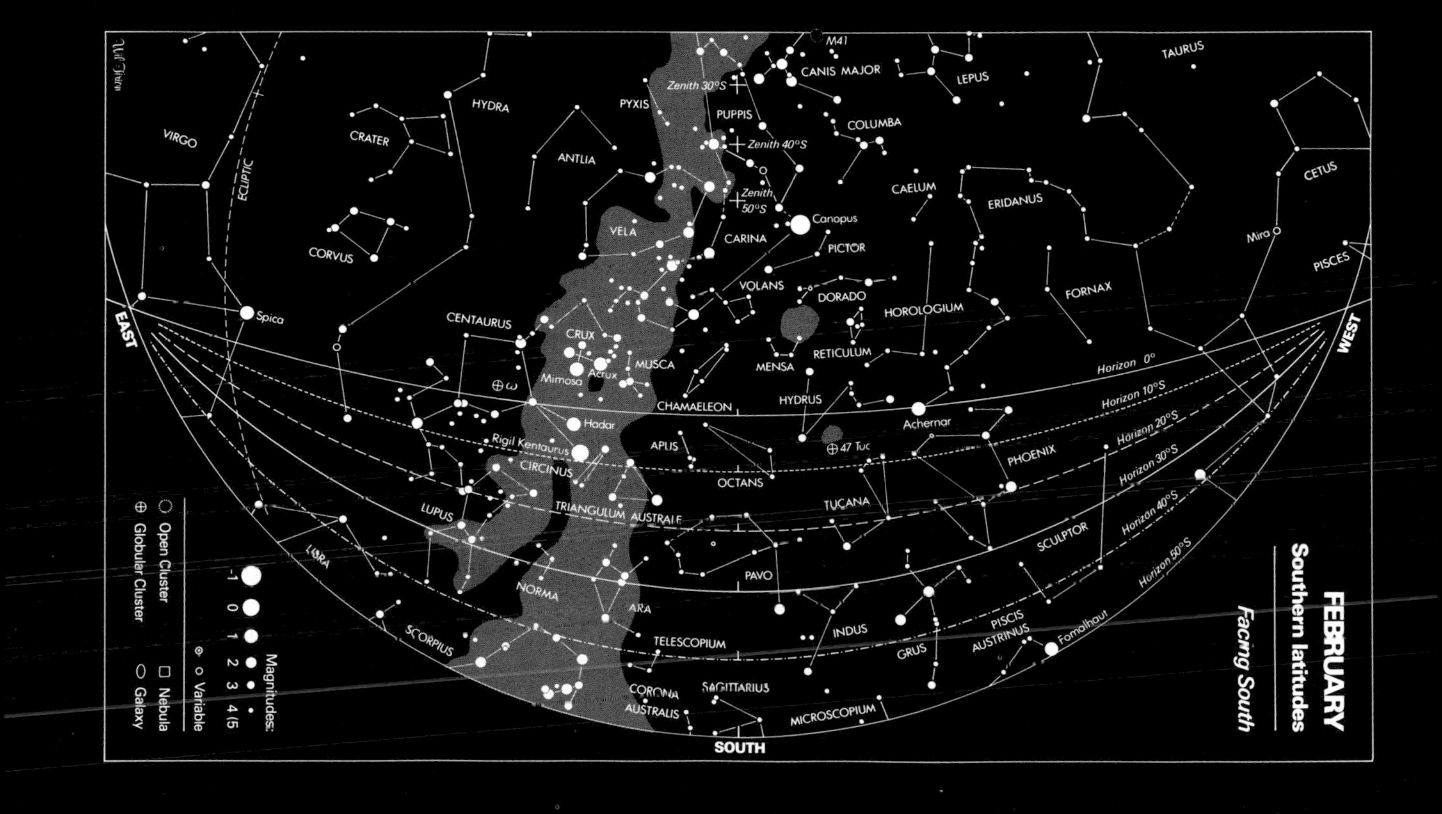
FEBRUARY
Southern latitudes
Facing South
WEST
EAST
SOUTH
Horizon 0°
Horizon 10°S
Horizon 20°S
Horizon 30°S
Horizon 40°S
Horizon 50°S
Zenith 30°S
Zenith 40°S
Zenith 50°S
ECLIPTIC
TAURUS
CETUS
PISCES
Mira
FORNAX
ERIDANUS
LEPUS
CAELUM
COLUMBA
CANIS MAJOR
M41
PUPPIS
PICTOR
Canopus
DORADO
HOROLOGIUM
RETICULUM
Achernar
PHOENIX
SCULPTOR
Fomalhaut
PISCIS AUSTRINUS
GRUS
TUCANA
47 Tuc
HYDRUS
MENSA
VOLANS
CARINA
VELA
PYXIS
ANTLIA
HYDRA
CRATER
CORVUS
VIRGO
Spica
CENTAURUS
CRUX
Acrux
Mimosa
Hadar
Rigil Kentaurus
MUSCA
CHAMAELEON
APUS
OCTANS
PAVO
INDUS
MICROSCOPIUM
SAGITTARIUS
TELESCOPIUM
CORONA AUSTRALIS
ARA
TRIANGULUM AUSTRALE
CIRCINUS
NORMA
LUPUS
SCORPIUS
LIBRA
Magnitudes:
-1 0 1 2 3 4 (5
Variable
Open Cluster
Globular Cluster
Nebula
Galaxy
Wil Tirion

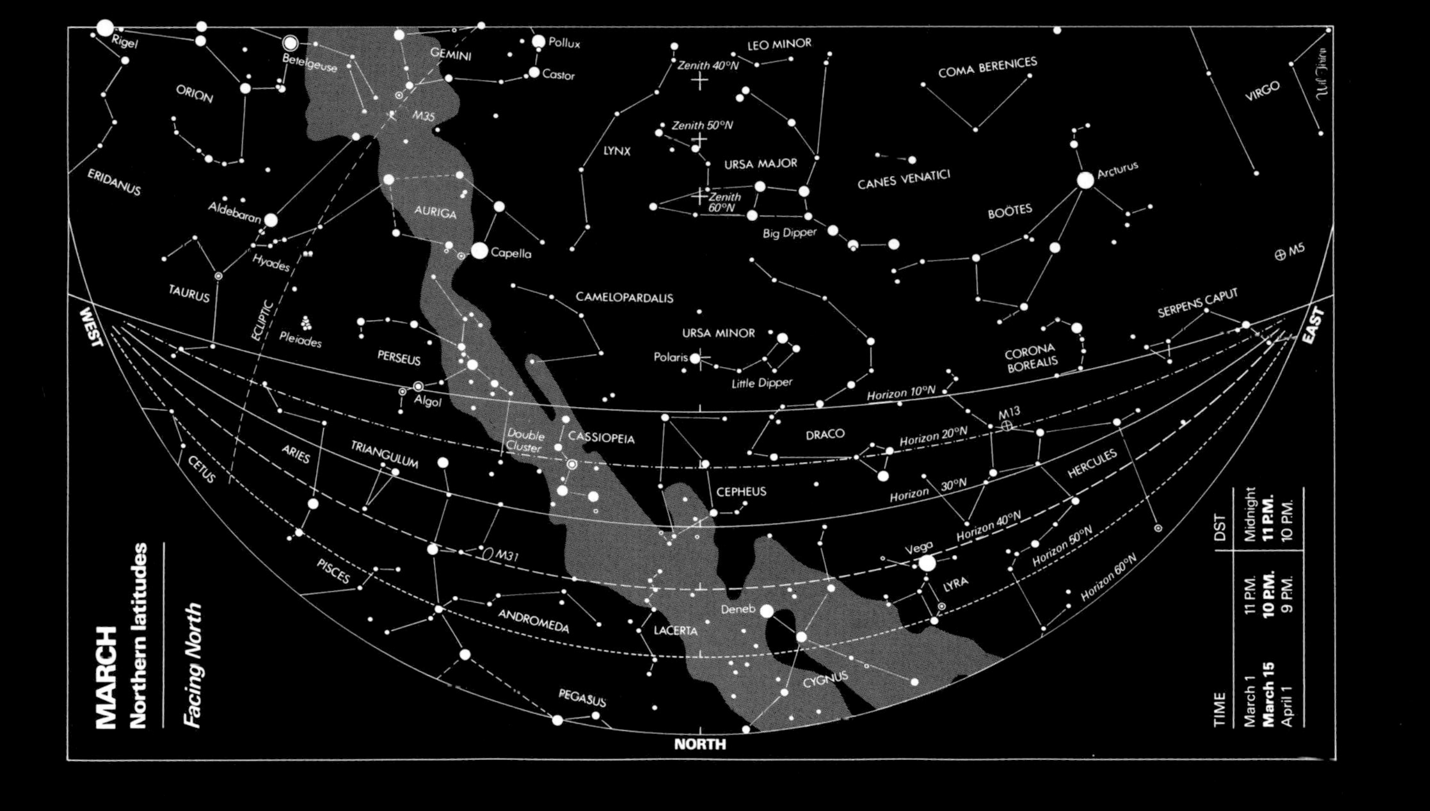
MARCH
Northern latitudes
Facing North
Rigel
Betelgeuse
GEMINI
Pollux
Castor
LEO MINOR
Zenith 40°N
COMA BERENICES
VIRGO
ORION
M35
Zenith 50°N
LYNX
URSA MAJOR
CANES VENATICI
ERIDANUS
Arcturus
AURIGA
Zenith 60°N
Aldebaran
Big Dipper
BOÖTES
Capella
Hyades
M5
TAURUS
CAMELOPARDALIS
SERPENS CAPUT
WEST
ECLIPTIC
Pleiades
URSA MINOR
EAST
PERSEUS
Polaris
Little Dipper
CORONA BOREALIS
Algol
Horizon 10°N
M13
Double Cluster
CASSIOPEIA
DRACO
Horizon 20°N
ARIES
TRIANGULUM
HERCULES
CETUS
CEPHEUS
Horizon 30°N
Horizon 40°N
M31
Vega
Horizon 50°N
PISCES
LYRA
Horizon 60°N
Deneb
ANDROMEDA
LACERTA
CYGNUS
PEGASUS
NORTH
TIME
March 1
March 15
April 1
11 P.M.
10 P.M.
9 P.M.
DST
Midnight
11 P.M.
10 P.M.
Wil Tirion

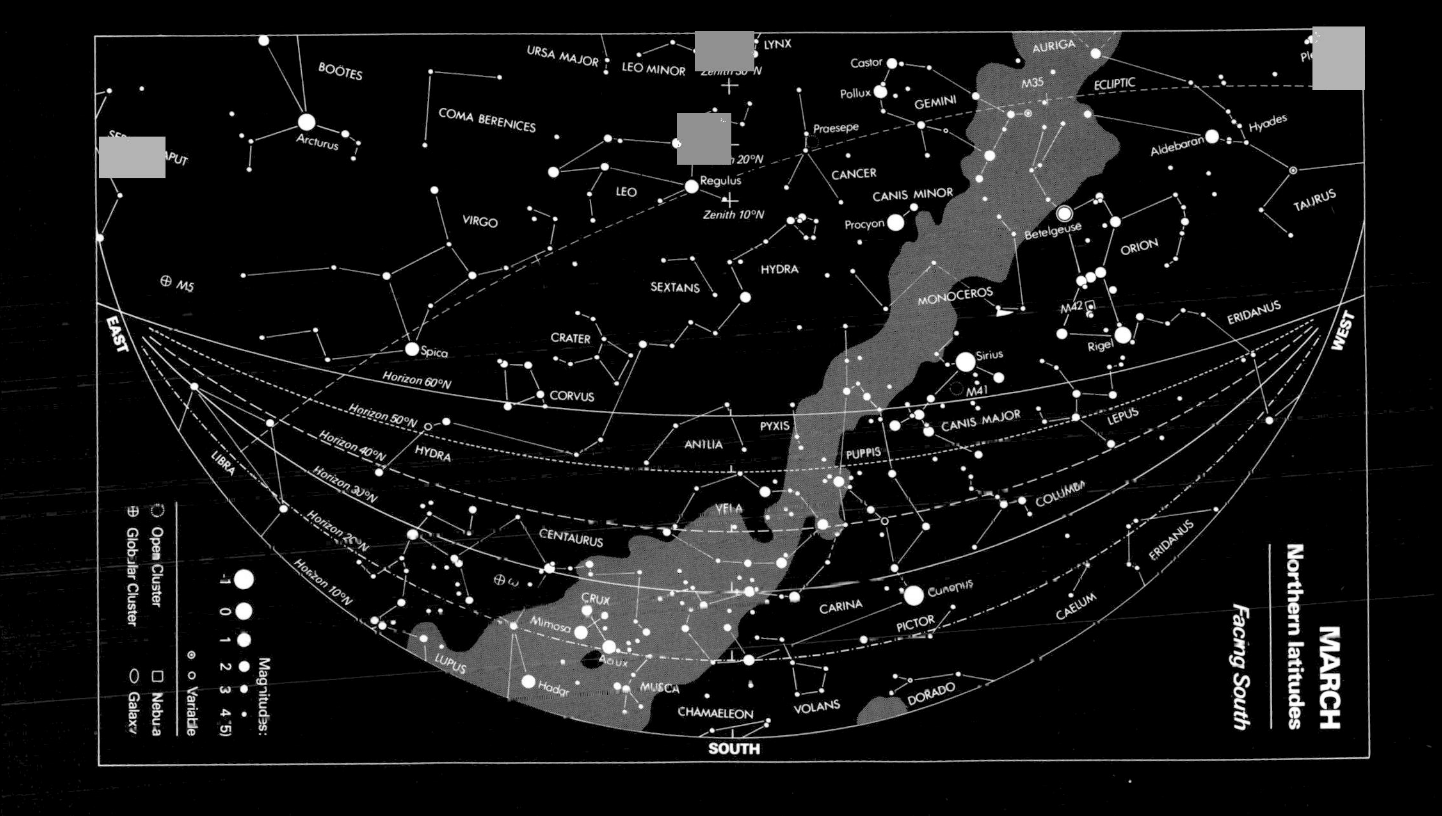
MARCH
Northern latitudes
Facing South
WEST
EAST
SOUTH
BOÖTES
Arcturus
URSA MAJOR
LEO MINOR
LYNX
COMA BERENICES
Castor
Pollux
GEMINI
AURIGA
M35
ECLIPTIC
Praesepe
Zenith 10°N
Regulus
LEO
CANCER
CANIS MINOR
Procyon
Betelgeuse
Hyades
Aldebaran
TAURUS
VIRGO
ORION
HYDRA
SEXTANS
MONOCEROS
M42
ERIDANUS
M5
Spica
CRATER
Sirius
Rigel
M41
Horizon 60°N
CORVUS
Horizon 50°N
PYXIS
CANIS MAJOR
LEPUS
Horizon 40°N
ANTLIA
HYDRA
PUPPIS
LIBRA
Horizon 30°N
COLUMBA
VELA
Horizon 20°N
CENTAURUS
ERIDANUS
Horizon 10°N
CRUX
Canopus
CARINA
CAELUM
Mimosa
PICTOR
Acrux
LUPUS
Hadar
MUSCA
DORADO
CHAMAELEON
VOLANS
Magnitudes:
-1
0
1
2
3
4
Variable
Open Cluster
Globular Cluster
Nebula
Galaxy

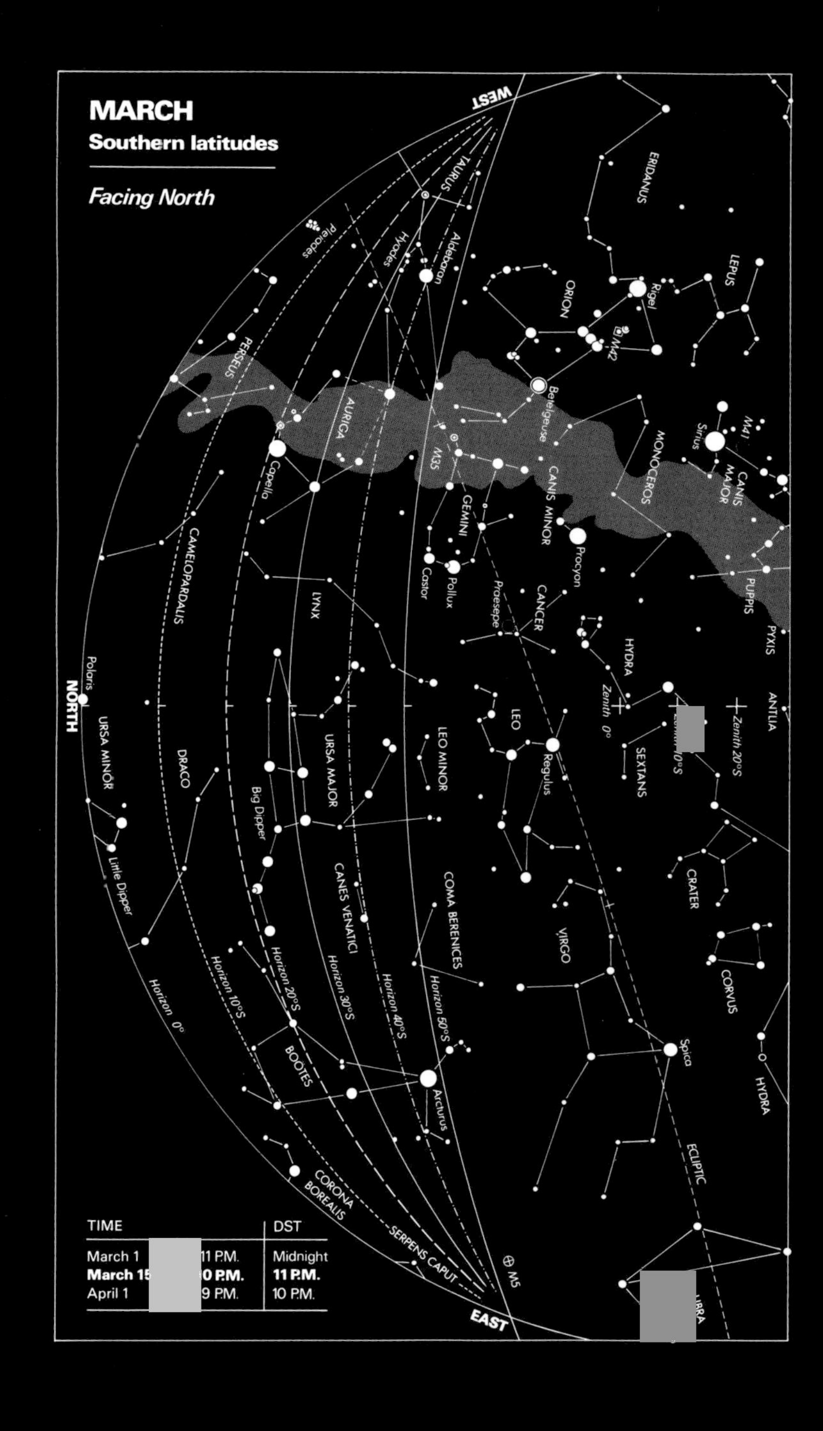
MARCH
Southern latitudes
Facing North
WEST
EAST
NORTH
TAURUS
Pleiades
Hyades
Aldebaran
ERIDANUS
LEPUS
ORION
Rigel
M42
Betelgeuse
PERSEUS
AURIGA
Capella
M35
GEMINI
Castor
Pollux
CANIS MINOR
Procyon
MONOCEROS
Sirius
M41
CANIS MAJOR
PUPPIS
PYXIS
CAMELOPARDALIS
LYNX
Praesepe
CANCER
HYDRA
ANTLIA
Polaris
URSA MINOR
DRACO
Big Dipper
URSA MAJOR
LEO MINOR
LEO
Regulus
SEXTANS
Zenith 0°
Zenith 20°S
Little Dipper
CANES VENATICI
COMA BERENICES
VIRGO
CRATER
CORVUS
Spica
HYDRA
Horizon 0°
Horizon 10°S
Horizon 20°S
Horizon 30°S
Horizon 40°S
Horizon 50°S
BOÖTES
Arcturus
CORONA BOREALIS
SERPENS CAPUT
M5
ECLIPTIC
TIME
DST
March 1
11 P.M.
Midnight
March 15
0 P.M.
11 P.M.
April 1
9 P.M.
10 P.M.

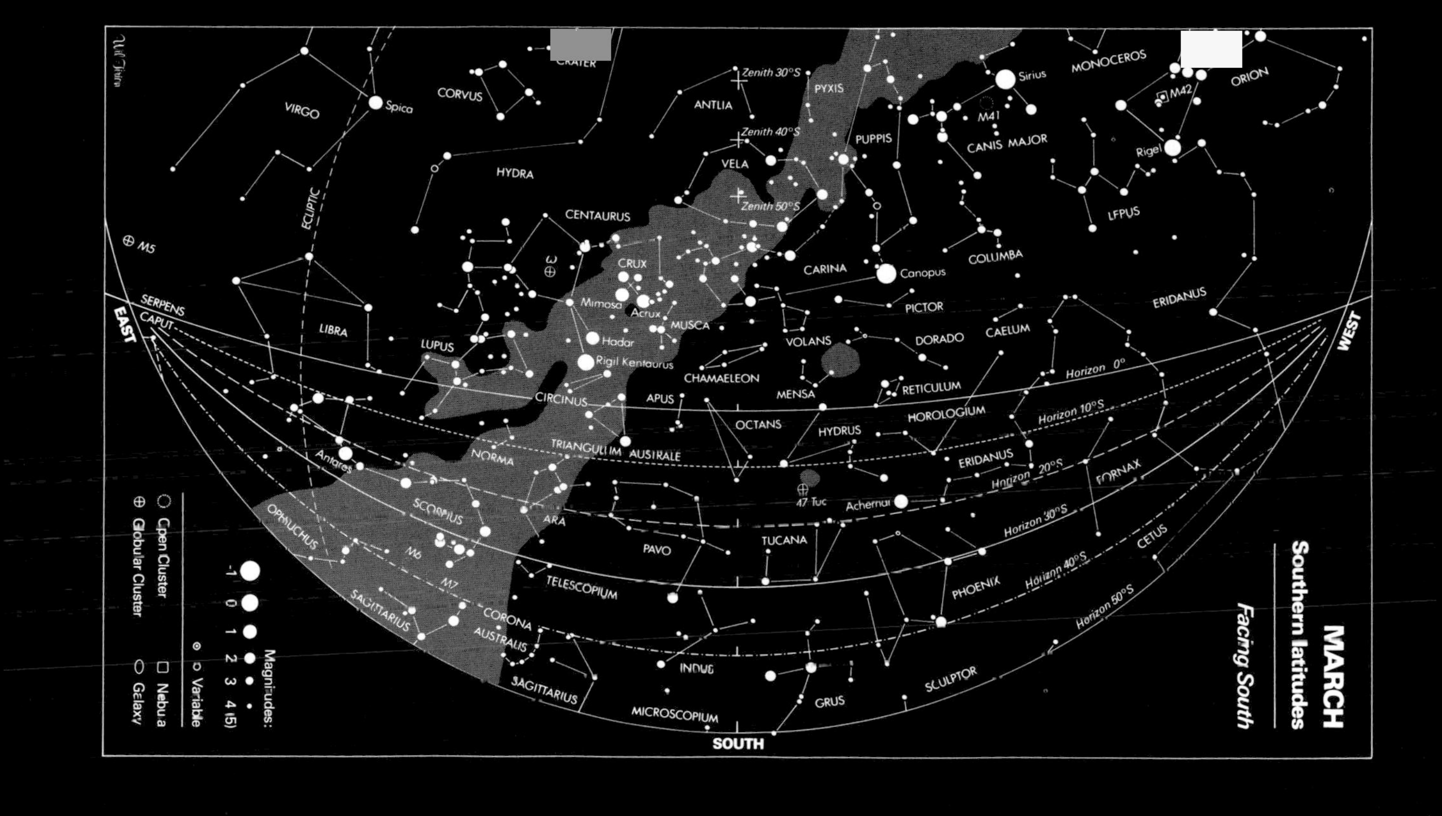

MARCH
Southern latitudes
Facing South
WEST
EAST
SOUTH
ORION
M42
Rigel
MONOCEROS
LEPUS
ERIDANUS
Sirius
M41
CANIS MAJOR
COLUMBA
CAELUM
PICTOR
DORADO
Canopus
PUPPIS
PYXIS
CARINA
VOLANS
RETICULUM
HOROLOGIUM
ERIDANUS
FORNAX
CETUS
PHOENIX
SCULPTOR
Achernar
HYDRUS
MENSA
47 Tuc
TUCANA
GRUS
CHAMAELEON
OCTANS
ANTLIA
VELA
Zenith 30°S
Zenith 40°S
Zenith 50°S
MUSCA
CRUX
Acrux
Mimosa
Hadar
Rigil Kentaurus
CENTAURUS
APUS
TRIANGULUM AUSTRALE
PAVO
INDUS
MICROSCOPIUM
CIRCINUS
ARA
TELESCOPIUM
CORONA AUSTRALIS
SAGITTARIUS
HYDRA
CORVUS
VIRGO
Spica
LIBRA
LUPUS
NORMA
SCORPIUS
M6
M7
Antares
OPHIUCHUS
ECLIPTIC
SERPENS CAPUT
M5
Horizon 0°
Horizon 10°S
Horizon 20°S
Horizon 30°S
Horizon 40°S
Horizon 50°S
Magnitudes:
-1 0 1 2 3 4 (5)
Variable
Open Cluster
Nebula
Globular Cluster
Galaxy
Wil Tirion

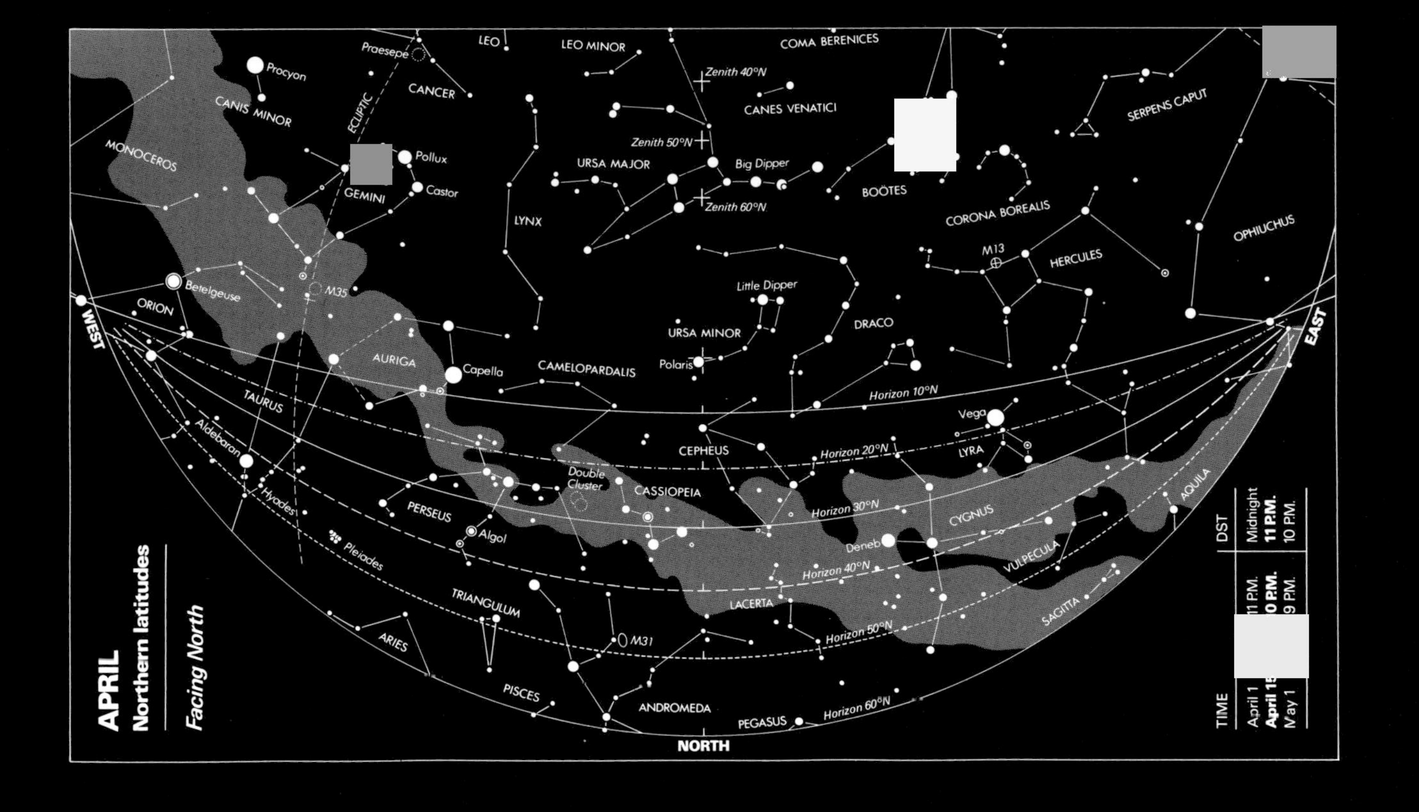

APRIL
Northern latitudes
Facing North
LEO
LEO MINOR
COMA BERENICES
Praesepe
Procyon
CANIS MINOR
CANCER
ECLIPTIC
Zenith 40°N
CANES VENATICI
SERPENS CAPUT
MONOCEROS
Pollux
GEMINI
Castor
Zenith 50°N
URSA MAJOR
Big Dipper
BOÖTES
Zenith 60°N
LYNX
CORONA BOREALIS
OPHIUCHUS
M13
HERCULES
Betelgeuse
M35
ORION
WEST
Little Dipper
URSA MINOR
DRACO
AURIGA
Capella
CAMELOPARDALIS
Polaris
EAST
Horizon 10°N
TAURUS
Vega
Aldebaran
CEPHEUS
Horizon 20°N
LYRA
Double Cluster
CASSIOPEIA
Hyades
PERSEUS
Horizon 30°N
CYGNUS
AQUILA
Algol
Deneb
Pleiades
Horizon 40°N
VULPECULA
TRIANGULUM
LACERTA
SAGITTA
ARIES
M31
Horizon 50°N
PISCES
ANDROMEDA
PEGASUS
Horizon 60°N
NORTH
TIME
DST
April 1
11 P.M.
Midnight
April 15
0 P.M.
11 P.M.
May 1
9 P.M.
10 P.M.

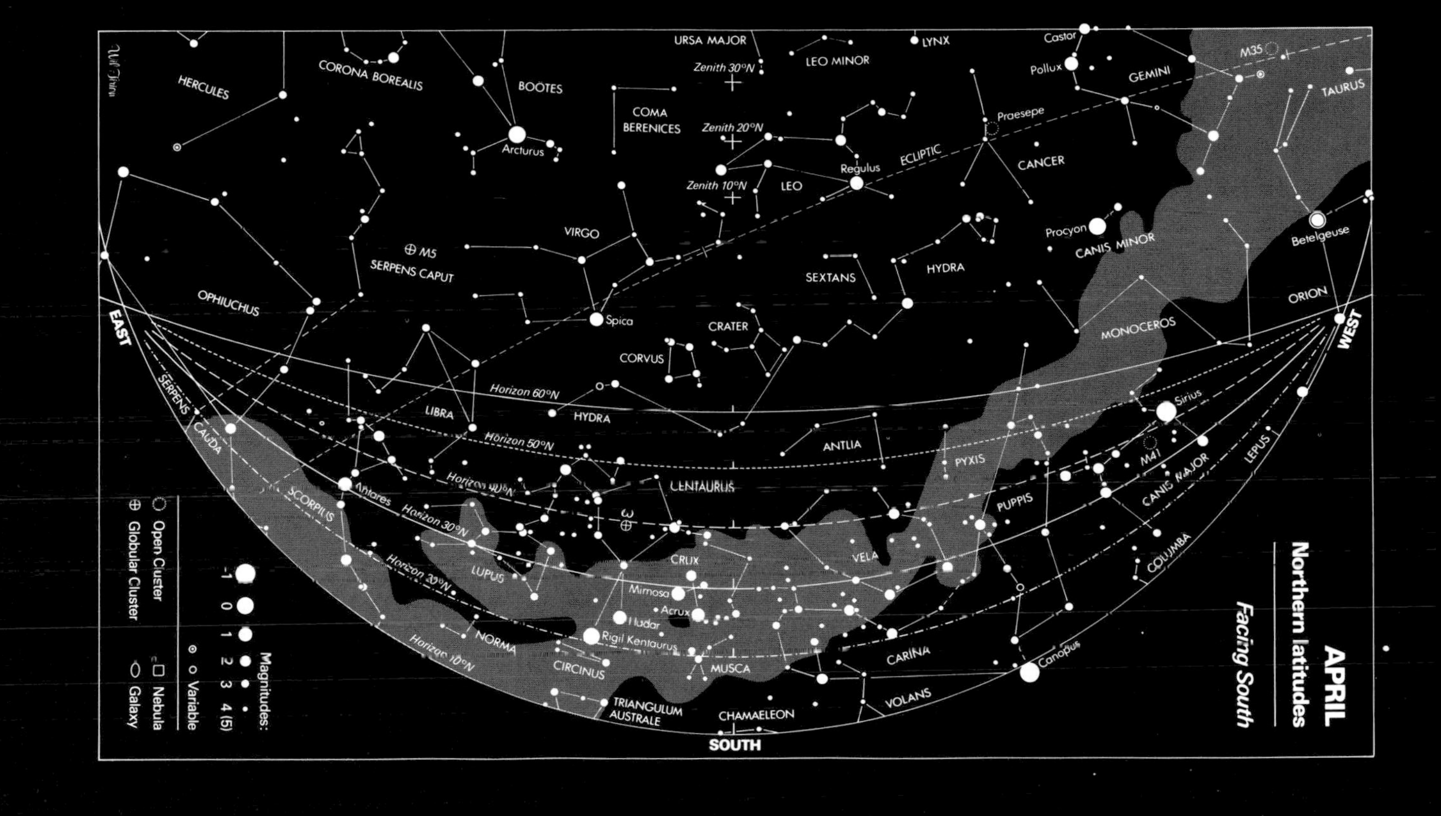
APRIL
Northern latitudes
Facing South
HERCULES
CORONA BOREALIS
BOOTES
Arcturus
COMA BERENICES
URSA MAJOR
Zenith 30°N
Zenith 20°N
Zenith 10°N
LEO MINOR
LYNX
Castor
Pollux
GEMINI
M35
TAURUS
Praesepe
Regulus
ECLIPTIC
CANCER
LEO
VIRGO
M5
SERPENS CAPUT
OPHIUCHUS
Spica
CRATER
CORVUS
SEXTANS
HYDRA
Procyon
CANIS MINOR
Betelgeuse
ORION
MONOCEROS
EAST
WEST
SERPENS CAUDA
Horizon 60°N
LIBRA
HYDRA
Horizon 50°N
ANTLIA
PYXIS
Sirius
M41
CANIS MAJOR
LEPUS
Antares
SCORPIUS
CENTAURUS
PUPPIS
Horizon 30°N
Horizon 20°N
Horizon 10°N
LUPUS
CRUX
VELA
COLUMBA
Mimosa
Acrux
Hadar
Rigil Kentaurus
NORMA
CIRCINUS
MUSCA
CARINA
Canopus
TRIANGULUM AUSTRALE
CHAMAELEON
VOLANS
SOUTH
Magnitudes:
-1 0 1 2 3 4 (5)
Variable
Open Cluster
Nebula
Globular Cluster
Galaxy
Wil Tirion

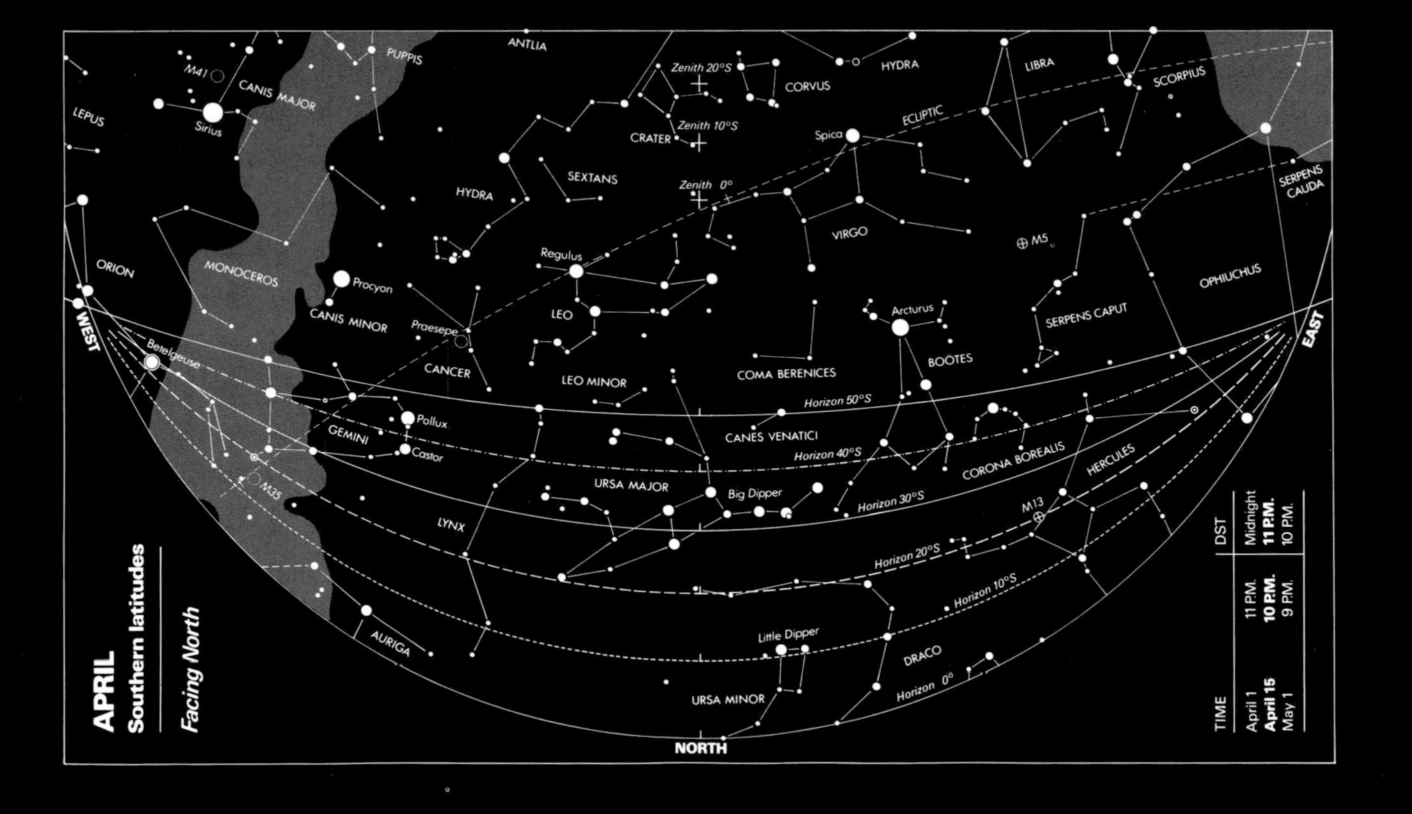
APRIL
Southern latitudes
Facing North
TIME
DST
April 1
11 P.M.
Midnight
April 15
10 P.M.
11 P.M.
May 1
9 P.M.
10 P.M.
LEPUS
M41
CANIS MAJOR
Sirius
PUPPIS
ANTLIA
Zenith 20°S
CORVUS
HYDRA
LIBRA
SCORPIUS
Zenith 10°S
CRATER
Spica
ECLIPTIC
SEXTANS
HYDRA
Zenith 0°
SERPENS CAUDA
ORION
MONOCEROS
Regulus
VIRGO
M5
OPHIUCHUS
Procyon
CANIS MINOR
Praesepe
LEO
Arcturus
SERPENS CAPUT
WEST
EAST
Betelgeuse
CANCER
LEO MINOR
COMA BERENICES
BOÖTES
Horizon 50°S
Pollux
GEMINI
Castor
CANES VENATICI
Horizon 40°S
CORONA BOREALIS
HERCULES
M35
URSA MAJOR
Big Dipper
Horizon 30°S
M13
LYNX
Horizon 20°S
Horizon 10°S
AURIGA
Little Dipper
DRACO
Horizon 0°
URSA MINOR
NORTH

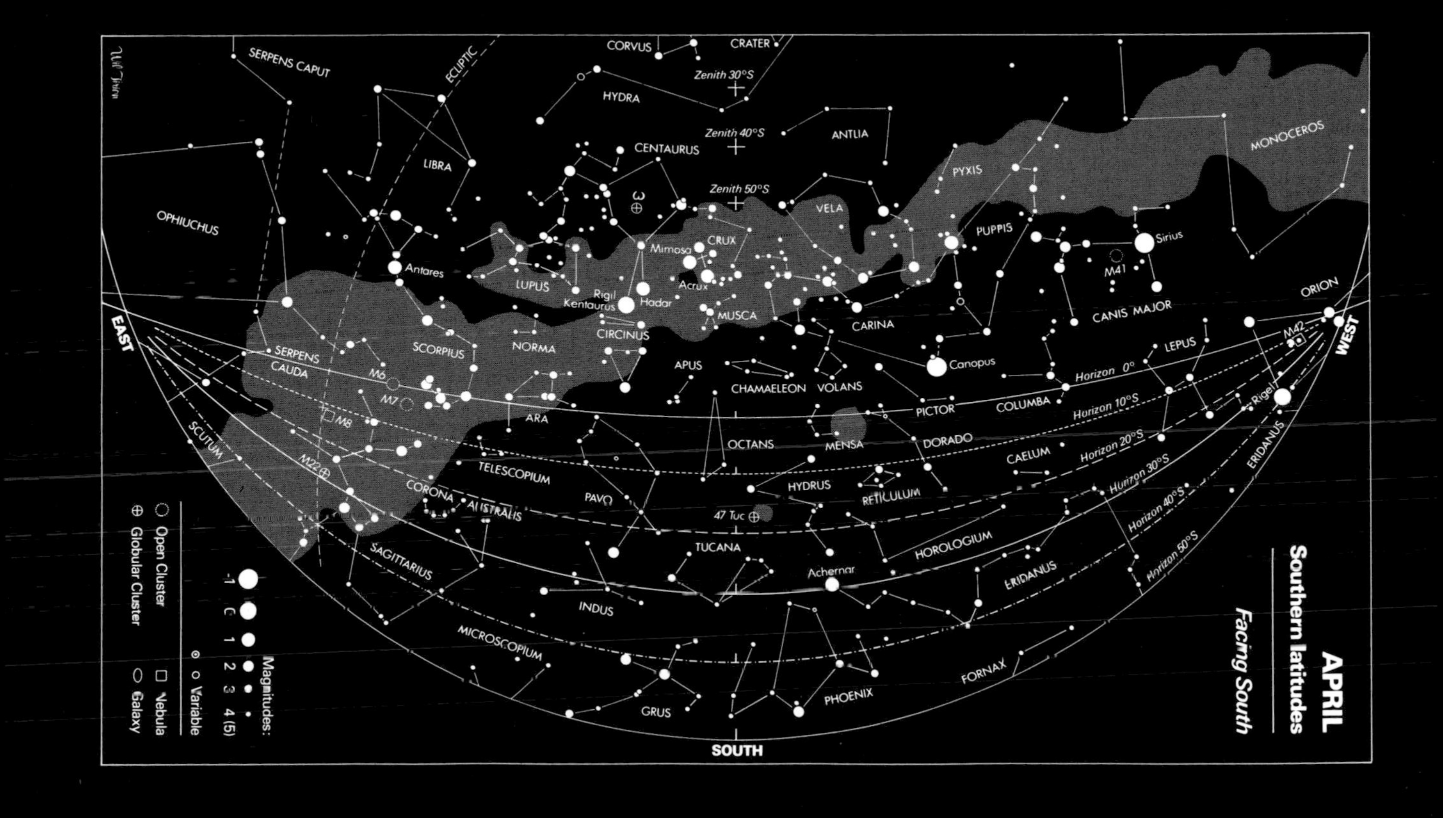

APRIL
Southern latitudes
Facing South
WEST
SOUTH
EAST
SERPENS CAPUT
ECLIPTIC
CORVUS
CRATER
Zenith 30°S
HYDRA
Zenith 40°S
ANTLIA
CENTAURUS
LIBRA
Zenith 50°S
MONOCEROS
PYXIS
OPHIUCHUS
ω
VELA
CRUX
Mimosa
PUPPIS
Sirius
M41
Antares
LUPUS
Acrux
Rigil Kentaurus
Hadar
MUSCA
CARINA
CANIS MAJOR
ORION
SERPENS CAUDA
SCORPIUS
NORMA
CIRCINUS
APUS
Canopus
LEPUS
M42
M6
M7
CHAMAELEON
VOLANS
Horizon 0°
M8
ARA
PICTOR
COLUMBA
Horizon 10°S
Rigel
SCUTUM
OCTANS
MENSA
DORADO
Horizon 20°S
ERIDANUS
M22
TELESCOPIUM
CAELUM
Horizon 30°S
HYDRUS
CORONA AUSTRALIS
PAVO
RETICULUM
47 Tuc
Horizon 40°S
Horizon 50°S
TUCANA
HOROLOGIUM
SAGITTARIUS
Achernar
ERIDANUS
INDUS
MICROSCOPIUM
FORNAX
GRUS
PHOENIX
Magnitudes:
-1
0
1
2
3
4 (5)
Variable
Open Cluster
Nebula
Globular Cluster
Galaxy
Wil Tirion

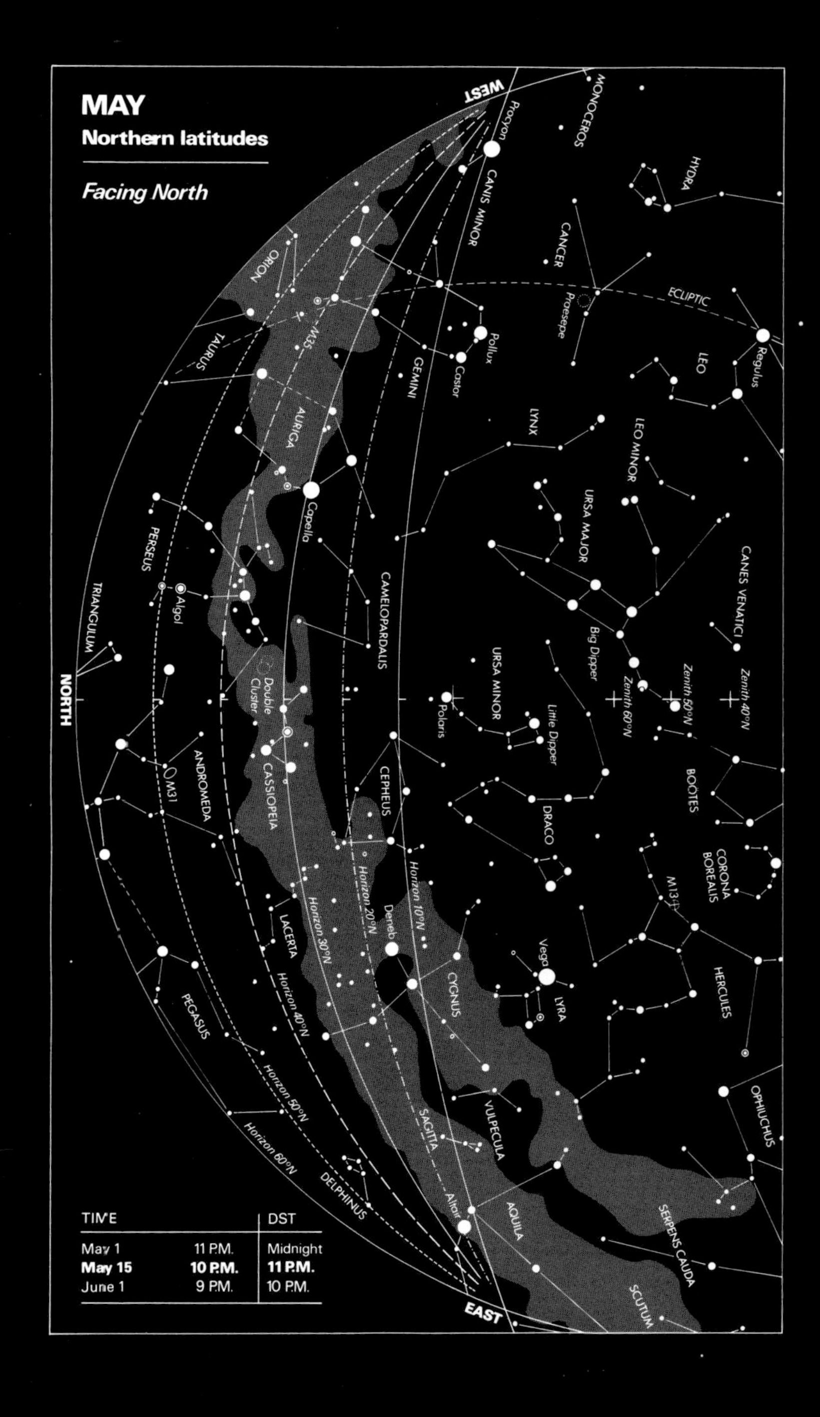
MAY
Northern latitudes
Facing North
WEST
NORTH
EAST
MONOCEROS
Procyon
CANIS MINOR
HYDRA
CANCER
ECLIPTIC
Praesepe
ORION
TAURUS
M35
GEMINI
Pollux
Castor
LEO
Regulus
LYNX
LEO MINOR
AURIGA
Capella
URSA MAJOR
PERSEUS
CANES VENATICI
Algol
CAMELOPARDALIS
Big Dipper
TRIANGULUM
URSA MINOR
Double Cluster
Polaris
Little Dipper
Zenith 60°N
Zenith 50°N
Zenith 40°N
ANDROMEDA
M31
CASSIOPEIA
CEPHEUS
BOÖTES
DRACO
CORONA BOREALIS
M13
Horizon 20°N
Horizon 10°N
Horizon 30°N
Deneb
LACERTA
Vega
CYGNUS
LYRA
HERCULES
Horizon 40°N
PEGASUS
Horizon 50°N
Horizon 60°N
OPHIUCHUS
SAGITTA
VULPECULA
DELPHINUS
Altair
AQUILA
SERPENS CAUDA
SCUTUM
TIME | DST
May 1 11 P.M. | Midnight
May 15 10 P.M. | 11 P.M.
June 1 9 P.M. | 10 P.M.

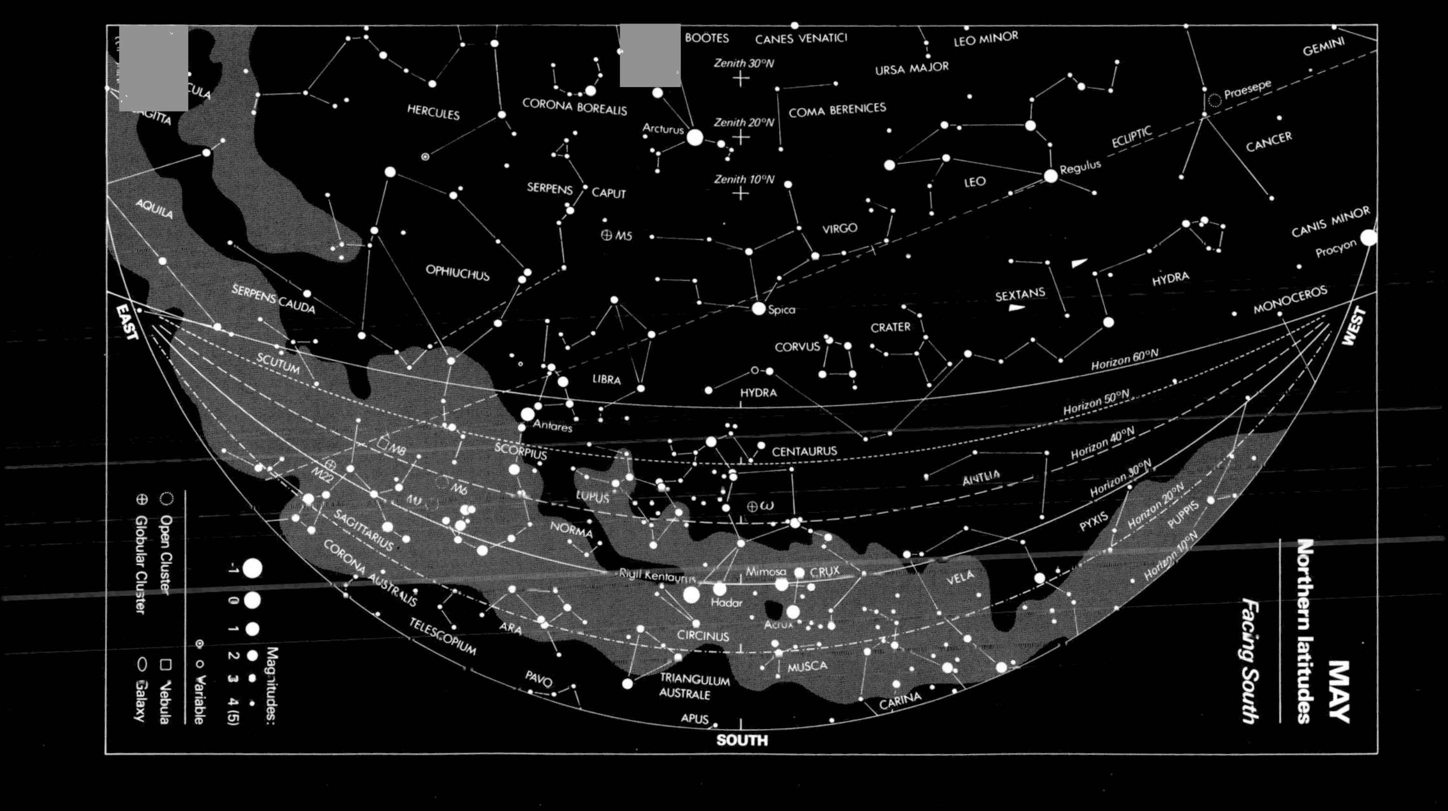

MAY
Northern latitudes
Facing South
BOÖTES
CANES VENATICI
LEO MINOR
GEMINI
URSA MAJOR
Praesepe
Zenith 30°N
Zenith 20°N
Zenith 10°N
HERCULES
CORONA BOREALIS
Arcturus
COMA BERENICES
ECLIPTIC
CANCER
Regulus
LEO
SERPENS CAPUT
AQUILA
M5
VIRGO
CANIS MINOR
Procyon
OPHIUCHUS
HYDRA
SERPENS CAUDA
Spica
SEXTANS
MONOCEROS
EAST
WEST
CRATER
CORVUS
SCUTUM
LIBRA
HYDRA
Horizon 60°N
Horizon 50°N
Horizon 40°N
Horizon 30°N
Horizon 20°N
Horizon 10°N
Antares
M8
SCORPIUS
CENTAURUS
M22
M6
LUPUS
ANTLIA
SAGITTARIUS
NORMA
PYXIS
PUPPIS
CORONA AUSTRALIS
Rigil Kentaurus
Mimosa
CRUX
VELA
Hadar
Acrux
TELESCOPIUM
ARA
CIRCINUS
MUSCA
PAVO
TRIANGULUM AUSTRALE
CARINA
APUS
SOUTH
Magnitudes:
-1 0 1 2 3 4 (5)
Variable
Open Cluster
Nebula
Globular Cluster
Galaxy

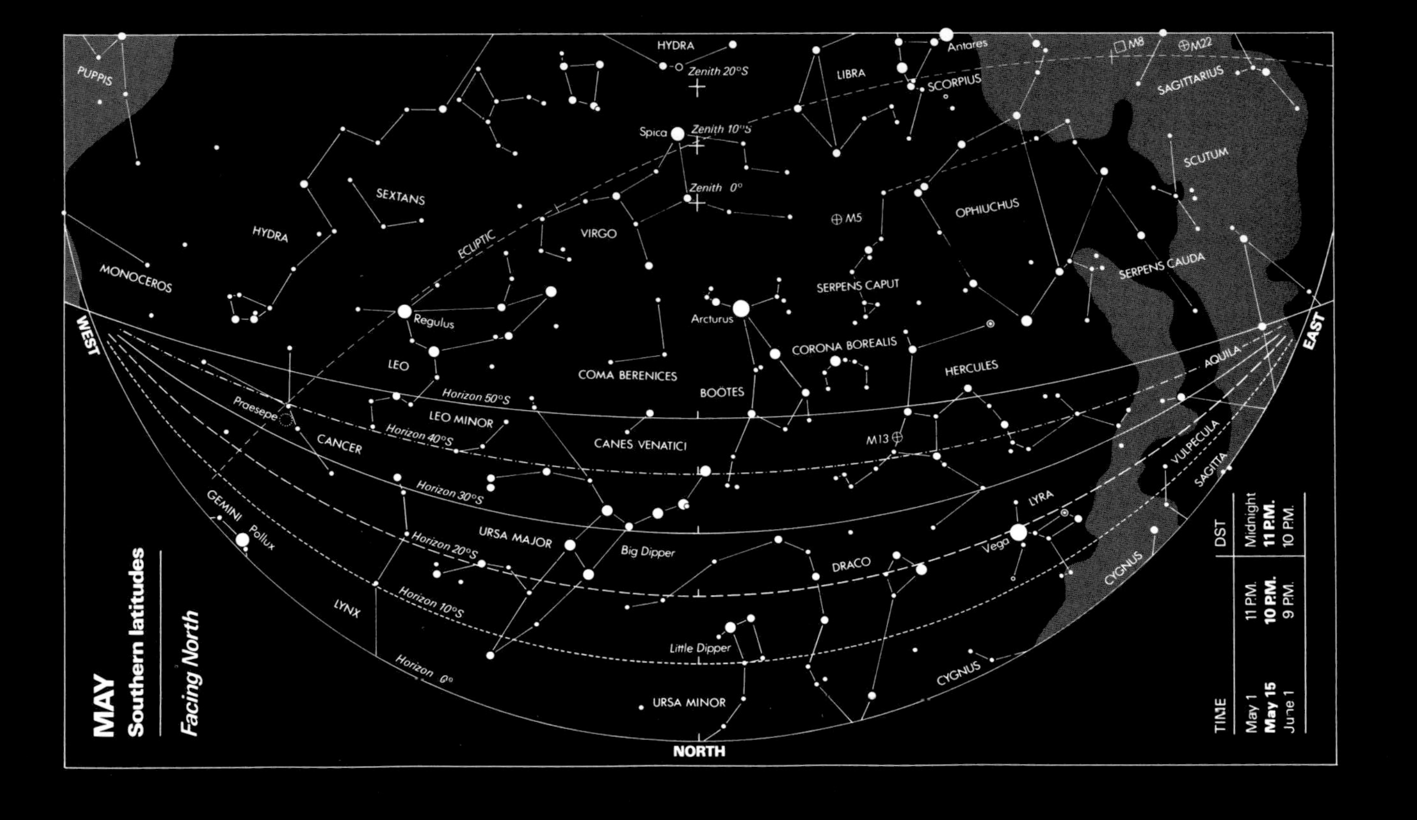
MAY
Southern latitudes
Facing North
PUPPIS
HYDRA
Zenith 20°S
LIBRA
Antares
SCORPIUS
M8
M22
SAGITTARIUS
Spica
Zenith 10°S
SCUTUM
SEXTANS
Zenith 0°
M5
OPHIUCHUS
HYDRA
ECLIPTIC
VIRGO
MONOCEROS
SERPENS CAPUT
SERPENS CAUDA
WEST
Regulus
Arcturus
EAST
CORONA BOREALIS
LEO
COMA BERENICES
HERCULES
AQUILA
BOÖTES
Horizon 50°S
Praesepe
LEO MINOR
CANCER
Horizon 40°S
CANES VENATICI
M13
VULPECULA
SAGITTA
Horizon 30°S
LYRA
GEMINI
Pollux
URSA MAJOR
Vega
Horizon 20°S
Big Dipper
DRACO
CYGNUS
LYNX
Horizon 10°S
Little Dipper
Horizon 0°
CYGNUS
URSA MINOR
NORTH
TIME
DST
May 1 11 P.M. Midnight
May 15 10 P.M. 11 P.M.
June 1 9 P.M. 10 P.M.

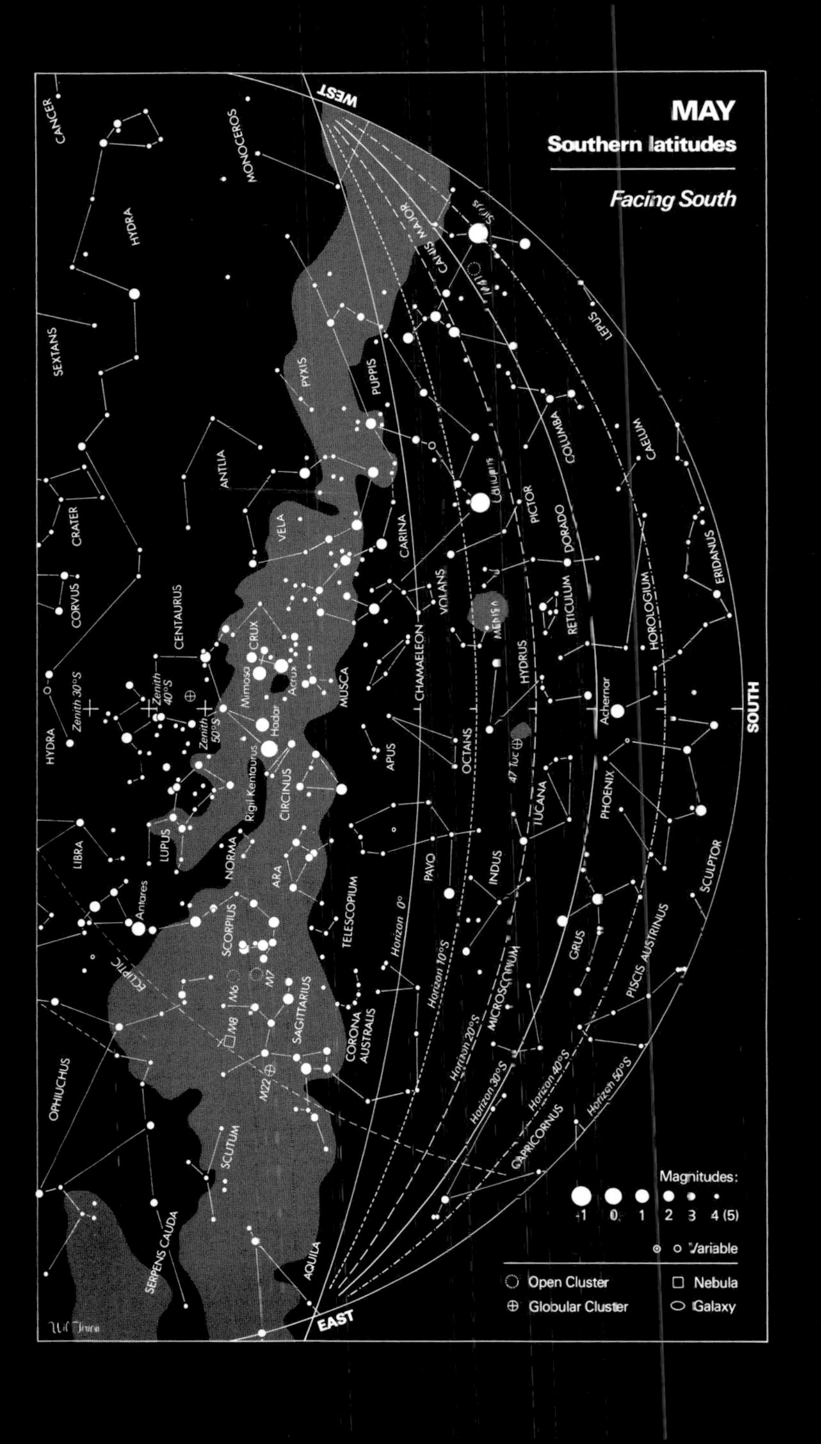
MAY
Southern latitudes
Facing South
WEST
EAST
SOUTH
Magnitudes:
-1 0 1 2 3 4 (5)
Variable
Open Cluster
Globular Cluster
Nebula
Galaxy
CANCER
MONOCEROS
HYDRA
CANIS MAJOR
LEPUS
SEXTANS
PYXIS
PUPPIS
COLUMBA
CAELUM
ANTLIA
CRATER
VELA
CARINA
Canopus
PICTOR
DORADO
ERIDANUS
CORVUS
CENTAURUS
VOLANS
RETICULUM
HOROLOGIUM
CRUX
CHAMAELEON
HYDRUS
Mimosa
Acrux
MUSCA
Achernar
Zenith 30°S
Zenith 40°S
Zenith 50°S
Hadar
Rigil Kentaurus
APUS
OCTANS
47 Tuc
TUCANA
PHOENIX
CIRCINUS
LUPUS
LIBRA
NORMA
PAVO
INDUS
SCULPTOR
ARA
Antares
SCORPIUS
TELESCOPIUM
Horizon 0°
Horizon 10°S
Horizon 20°S
Horizon 30°S
Horizon 40°S
Horizon 50°S
GRUS
PISCIS AUSTRINUS
ECLIPTIC
M6
M7
M8
M22
SAGITTARIUS
CORONA AUSTRALIS
MICROSCOPIUM
OPHIUCHUS
SCUTUM
CAPRICORNUS
SERPENS CAUDA
AQUILA

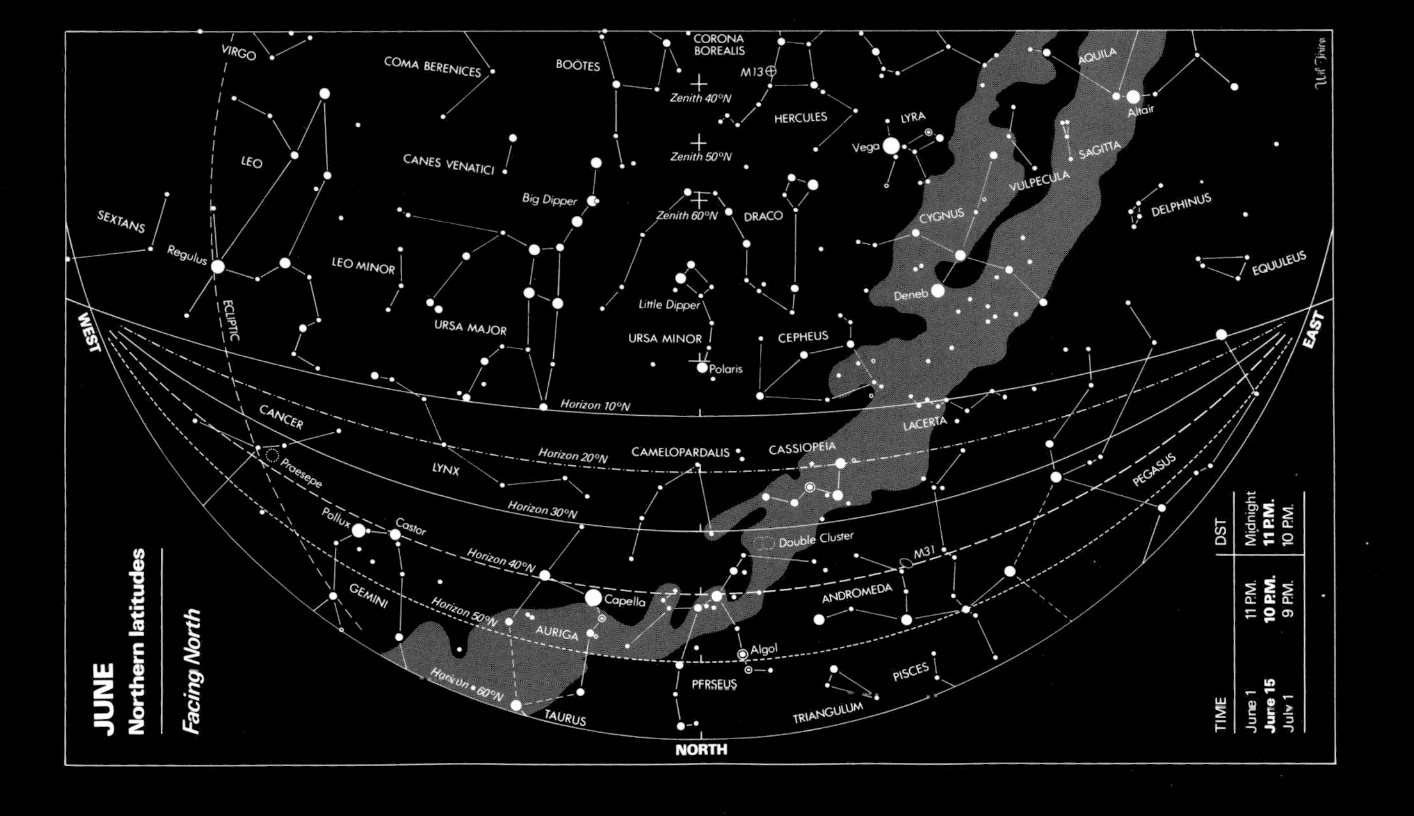
JUNE
Northern latitudes
Facing North
VIRGO
COMA BERENICES
BOÖTES
CORONA BOREALIS
M13
Zenith 40°N
HERCULES
LYRA
AQUILA
Altair
LEO
CANES VENATICI
Zenith 50°N
Vega
SAGITTA
VULPECULA
Big Dipper
Zenith 60°N
DRACO
CYGNUS
DELPHINUS
SEXTANS
Regulus
LEO MINOR
EQUULEUS
Deneb
ECLIPTIC
Little Dipper
URSA MAJOR
URSA MINOR
CEPHEUS
WEST
EAST
Polaris
Horizon 10°N
CANCER
LACERTA
Praesepe
Horizon 20°N
CAMELOPARDALIS
CASSIOPEIA
LYNX
PEGASUS
Horizon 30°N
Pollux
Castor
Double Cluster
M31
Horizon 40°N
GEMINI
ANDROMEDA
Capella
Horizon 50°N
AURIGA
Algol
PISCES
Horizon 60°N
PERSEUS
TAURUS
TRIANGULUM
NORTH
TIME
DST
June 1
11 P.M.
Midnight
June 15
10 P.M.
11 P.M.
July 1
9 P.M.
10 P.M.
Wil Tirion

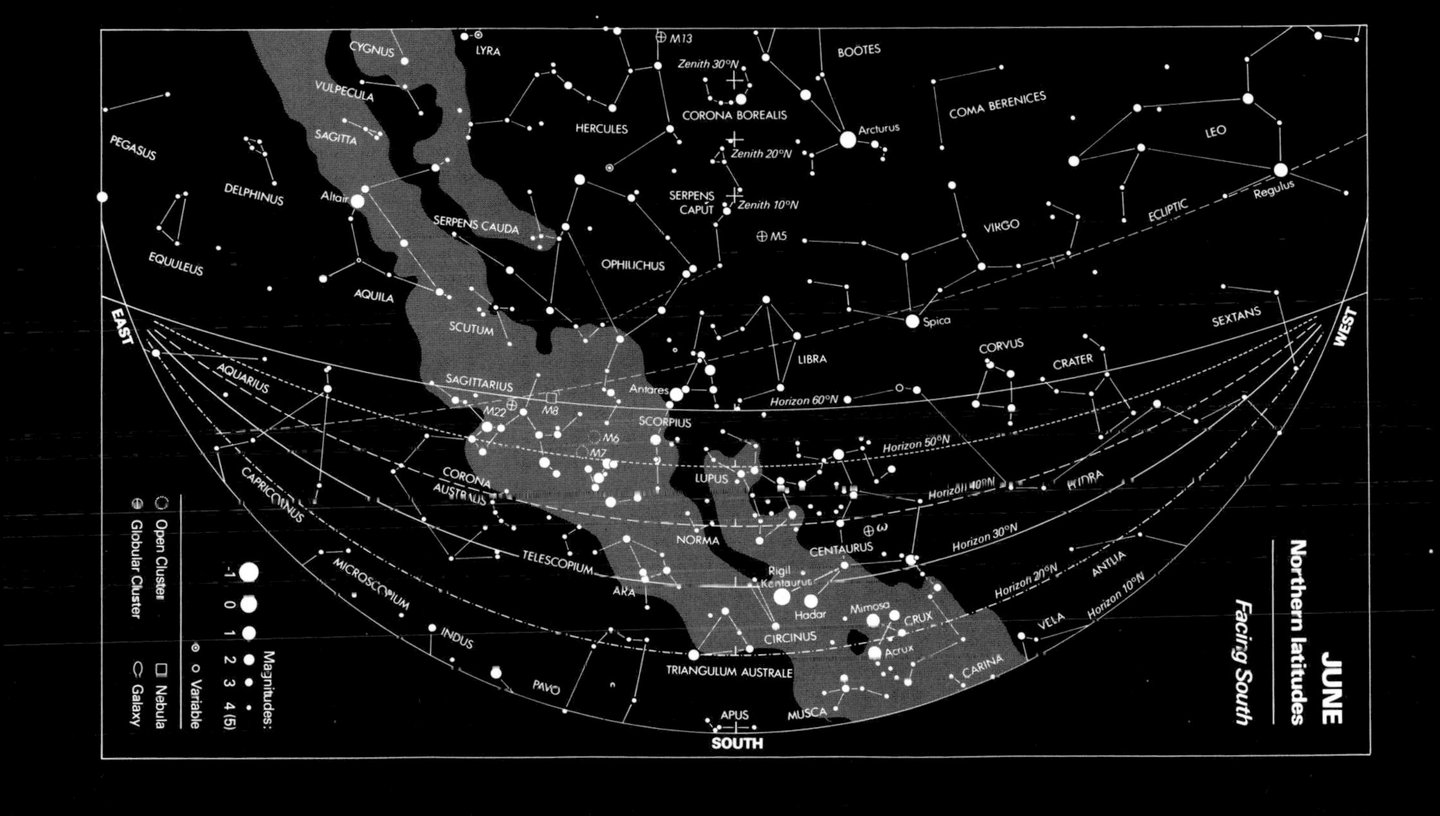
JUNE
Northern latitudes
Facing South
CYGNUS
LYRA
M13
Zenith 30°N
BOÖTES
VULPECULA
CORONA BOREALIS
HERCULES
Arcturus
COMA BERENICES
LEO
SAGITTA
Zenith 20°N
PEGASUS
DELPHINUS
Altair
SERPENS CAPUT
Zenith 10°N
Regulus
ECLIPTIC
VIRGO
SERPENS CAUDA
M5
EQUULEUS
OPHIUCHUS
AQUILA
SCUTUM
Spica
SEXTANS
EAST
WEST
LIBRA
CORVUS
CRATER
AQUARIUS
SAGITTARIUS
Antares
Horizon 60°N
M22
M8
SCORPIUS
M6
M7
Horizon 50°N
CORONA AUSTRALIS
LUPUS
CAPRICORNUS
HYDRA
NORMA
CENTAURUS
Horizon 30°N
TELESCOPIUM
ANTLIA
Horizon 20°N
MICROSCOPIUM
ARA
Rigil Kentaurus
Hadar
Mimosa
CRUX
Horizon 10°N
VELA
INDUS
CIRCINUS
Acrux
CARINA
TRIANGULUM AUSTRALE
PAVO
APUS
MUSCA
SOUTH
Magnitudes:
-1 0 1 2 3 4 (5)
Variable
Open Cluster
Nebula
Globular Cluster
Galaxy

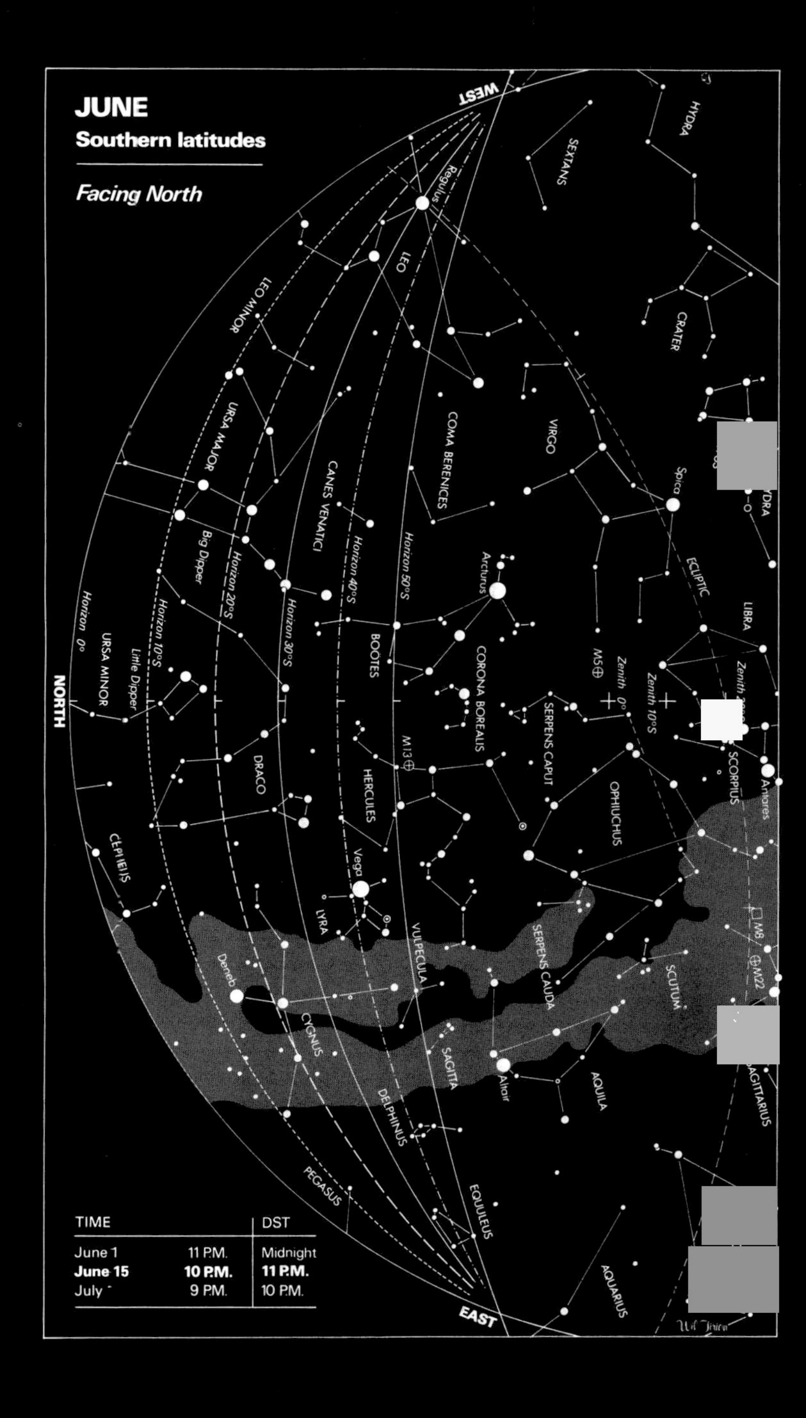

JUNE
Southern latitudes
Facing North
WEST
NORTH
EAST
HYDRA
SEXTANS
Regulus
LEO
LEO MINOR
CRATER
URSA MAJOR
COMA BERENICES
VIRGO
Spica
CANES VENATICI
Big Dipper
Horizon 20°S
Horizon 30°S
Horizon 40°S
Horizon 50°S
Horizon 10°S
Horizon 0°
Arcturus
ECLIPTIC
LIBRA
URSA MINOR
Little Dipper
BOÖTES
M5
Zenith 0°
Zenith 10°S
CORONA BOREALIS
SERPENS CAPUT
SCORPIUS
Antares
DRACO
M13
HERCULES
OPHIUCHUS
CEPHEUS
Vega
LYRA
VULPECULA
SERPENS CAUDA
M8
M22
SCUTUM
Deneb
CYGNUS
SAGITTA
Altair
AQUILA
SAGITTARIUS
DELPHINUS
PEGASUS
EQUULEUS
AQUARIUS
TIME
DST
June 1 11 P.M. Midnight
June 15 10 P.M. 11 P.M.
July 9 P.M. 10 P.M.
Wil Tirion

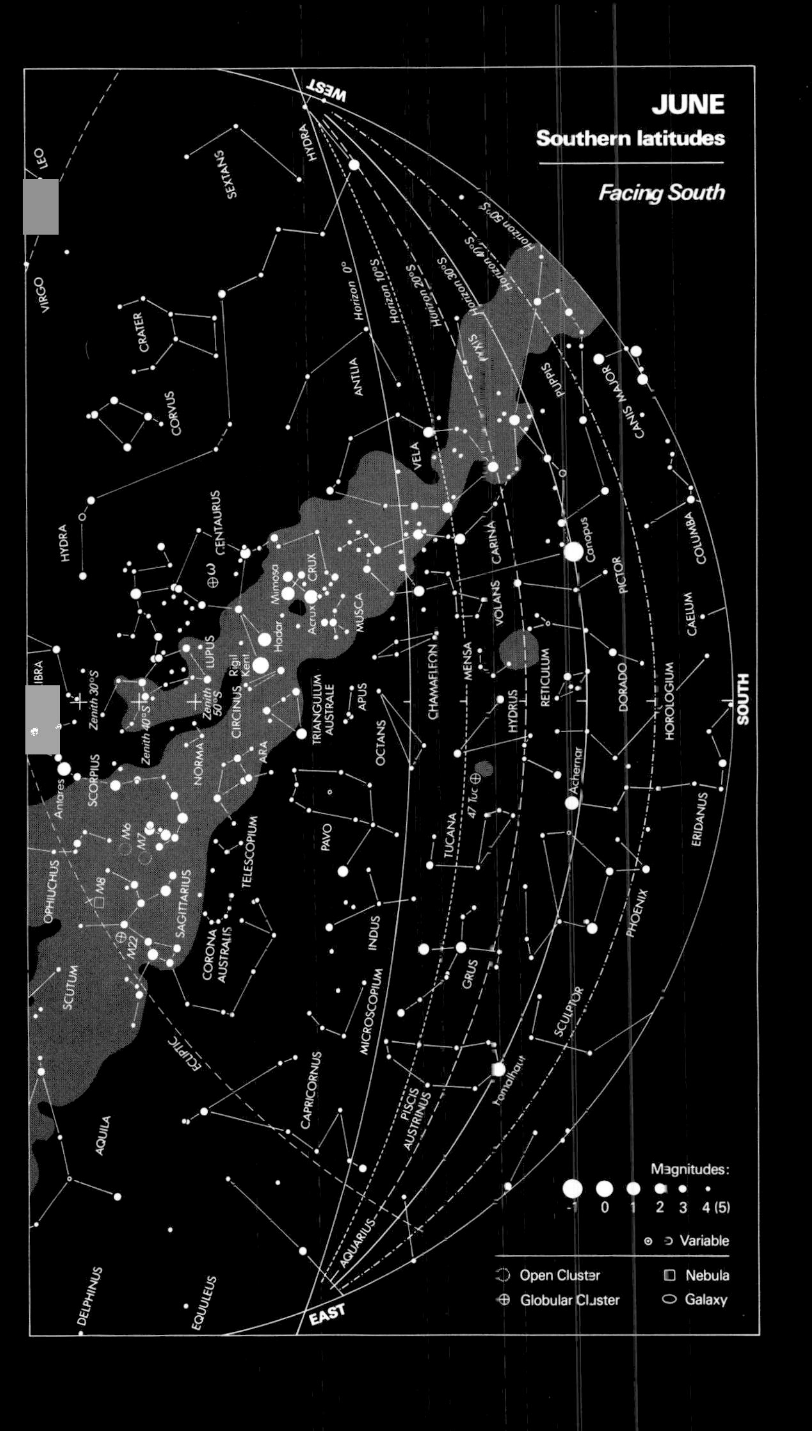
JUNE
Southern latitudes
Facing South
WEST
EAST
SOUTH
Horizon 0°
Horizon 10°S
Horizon 20°S
Horizon 30°S
Horizon 40°S
Horizon 50°S
Zenith 30°S
Zenith 40°S
Zenith 50°S
ECLIPTIC
LEO
VIRGO
SEXTANS
HYDRA
CRATER
CORVUS
ANTLIA
VELA
PUPPIS
CANIS MAJOR
CENTAURUS
CARINA
Canopus
COLUMBA
PICTOR
CRUX
Mimosa
Acrux
Hadar
MUSCA
VOLANS
CAELUM
LIBRA
LUPUS
Rigil Kent
CIRCINUS
TRIANGULUM AUSTRALE
APUS
CHAMAELEON
MENSA
RETICULUM
DORADO
HOROLOGIUM
HYDRUS
OCTANS
NORMA
ARA
SCORPIUS
Antares
M6
M7
M8
M22
OPHIUCHUS
SAGITTARIUS
TELESCOPIUM
PAVO
TUCANA
47 Tuc
Achernar
ERIDANUS
PHOENIX
INDUS
GRUS
CORONA AUSTRALIS
SCUTUM
MICROSCOPIUM
SCULPTOR
CAPRICORNUS
PISCIS AUSTRINUS
Fomalhaut
AQUILA
AQUARIUS
DELPHINUS
EQUULEUS
Magnitudes:
-1 0 1 2 3 4 (5)
Variable
Open Cluster
Globular Cluster
Nebula
Galaxy

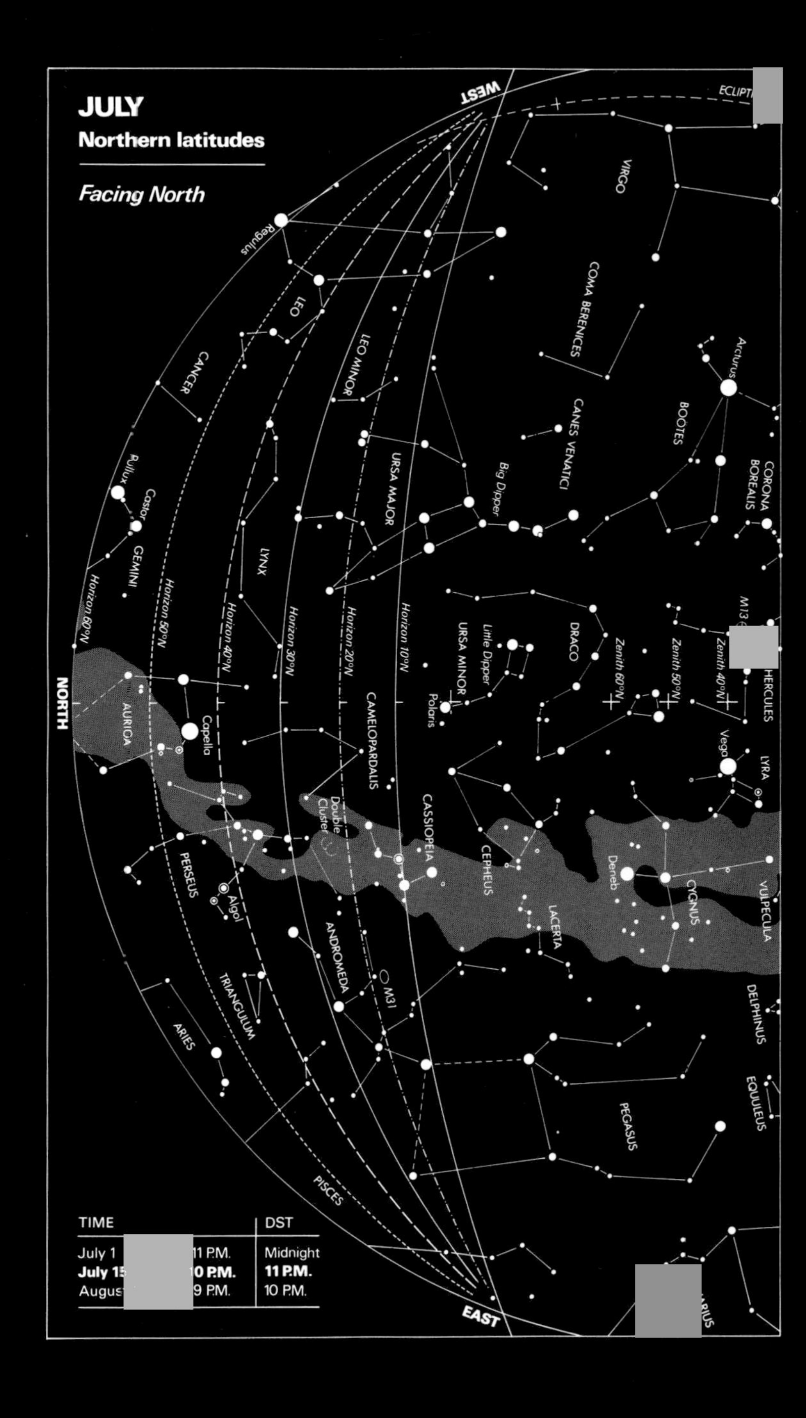
JULY
Northern latitudes
Facing North
WEST
ECLIPT
VIRGO
Regulus
COMA BERENICES
LEO
Arcturus
LEO MINOR
CANCER
CANES VENATICI
BOÖTES
URSA MAJOR
Big Dipper
CORONA BOREALIS
Pollux
Castor
GEMINI
LYNX
Horizon 60°N
Horizon 50°N
Horizon 40°N
Horizon 30°N
Horizon 20°N
Horizon 10°N
URSA MINOR
Little Dipper
DRACO
Zenith 60°N
Zenith 50°N
Zenith 40°N
M13
HERCULES
NORTH
AURIGA
Capella
CAMELOPARDALIS
Polaris
Vega
LYRA
Double Cluster
CASSIOPEIA
CEPHEUS
Deneb
CYGNUS
VULPECULA
PERSEUS
Algol
ANDROMEDA
LACERTA
M31
TRIANGULUM
DELPHINUS
ARIES
PEGASUS
EQUULEUS
PISCES
TIME
DST
July 1
11 P.M.
Midnight
July 15
10 P.M.
11 P.M.
August
9 P.M.
10 P.M.
EAST
RIUS

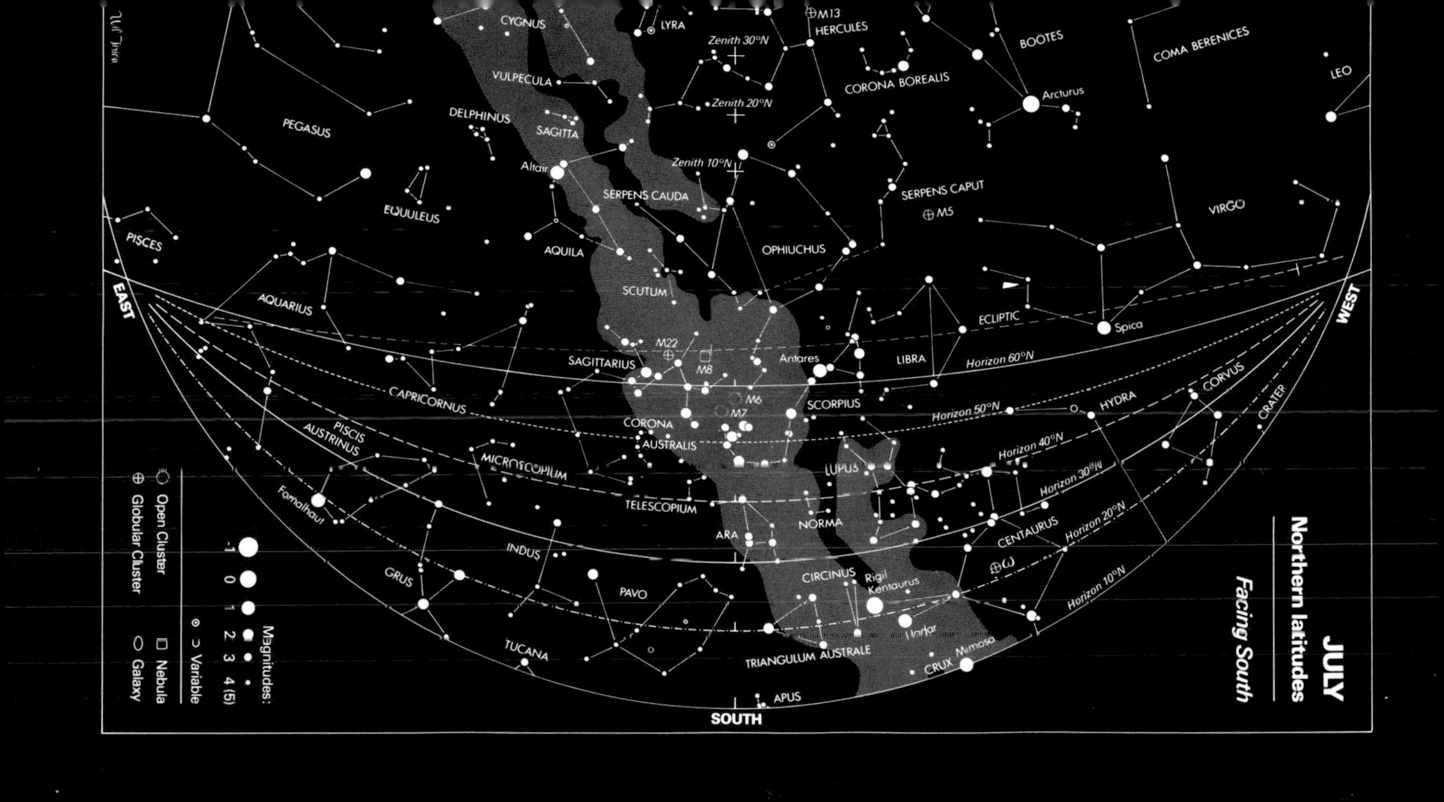

JULY
Northern latitudes
Facing South
WEST
EAST
SOUTH
CYGNUS
LYRA
M13
HERCULES
Zenith 30°N
Zenith 20°N
Zenith 10°N
BOÖTES
COMA BERENICES
LEO
CORONA BOREALIS
Arcturus
VULPECULA
DELPHINUS
SAGITTA
PEGASUS
Altair
SERPENS CAUDA
SERPENS CAPUT
M5
VIRGO
EQUULEUS
PISCES
AQUILA
OPHIUCHUS
SCUTUM
AQUARIUS
ECLIPTIC
Spica
M22
M8
Antares
LIBRA
Horizon 60°N
SAGITTARIUS
CAPRICORNUS
M6
M7
SCORPIUS
Horizon 50°N
HYDRA
CORVUS
CRATER
CORONA
AUSTRALIS
PISCIS
AUSTRINUS
Horizon 40°N
MICROSCOPIUM
LUPUS
Horizon 30°N
Fomalhaut
TELESCOPIUM
NORMA
CENTAURUS
Horizon 20°N
ARA
INDUS
ω
GRUS
CIRCINUS
Rigil Kentaurus
Horizon 10°N
PAVO
Hadar
TUCANA
TRIANGULUM AUSTRALE
CRUX
Mimosa
APUS
Magnitudes:
-1 0 1 2 3 4 (5)
Variable
Open Cluster
Nebula
Globular Cluster
Galaxy

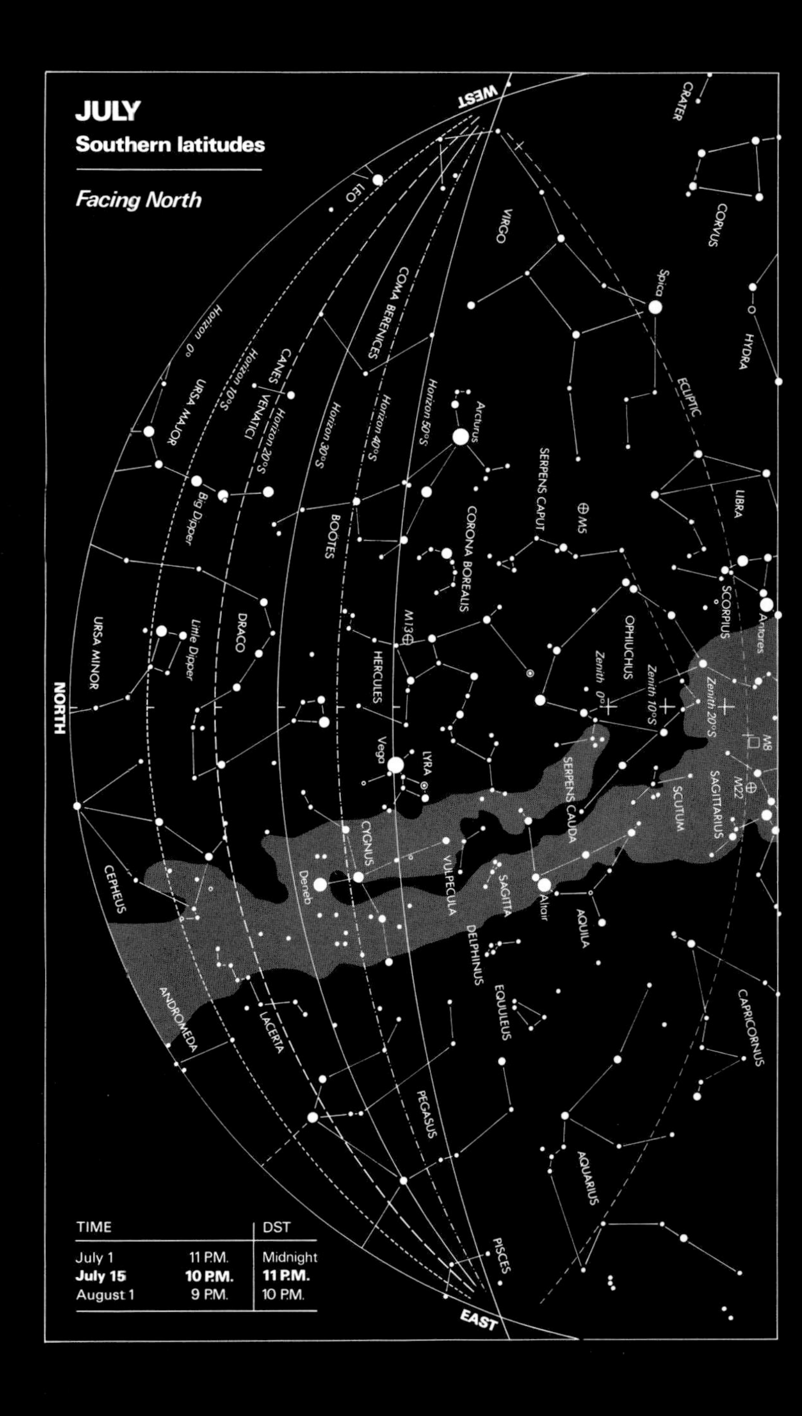

JULY
Southern latitudes
Facing North
WEST
EAST
NORTH
CRATER
CORVUS
HYDRA
LEO
VIRGO
Spica
COMA BERENICES
ECLIPTIC
CANES VENATICI
URSA MAJOR
Big Dipper
Arcturus
SERPENS CAPUT
M5
LIBRA
BOOTES
CORONA BOREALIS
SCORPIUS
Antares
URSA MINOR
Little Dipper
DRACO
M13
HERCULES
OPHIUCHUS
Zenith 0°
Zenith 10°S
Zenith 20°S
M8
M22
Vega
LYRA
SERPENS CAUDA
SCUTUM
SAGITTARIUS
CYGNUS
Deneb
VULPECULA
SAGITTA
Altair
AQUILA
CEPHEUS
DELPHINUS
EQUULEUS
CAPRICORNUS
ANDROMEDA
LACERTA
PEGASUS
AQUARIUS
PISCES
Horizon 0°
Horizon 10°S
Horizon 20°S
Horizon 30°S
Horizon 40°S
Horizon 50°S
TIME | DST
July 1 11 P.M. | Midnight
July 15 10 P.M. | 11 P.M.
August 1 9 P.M. | 10 P.M.

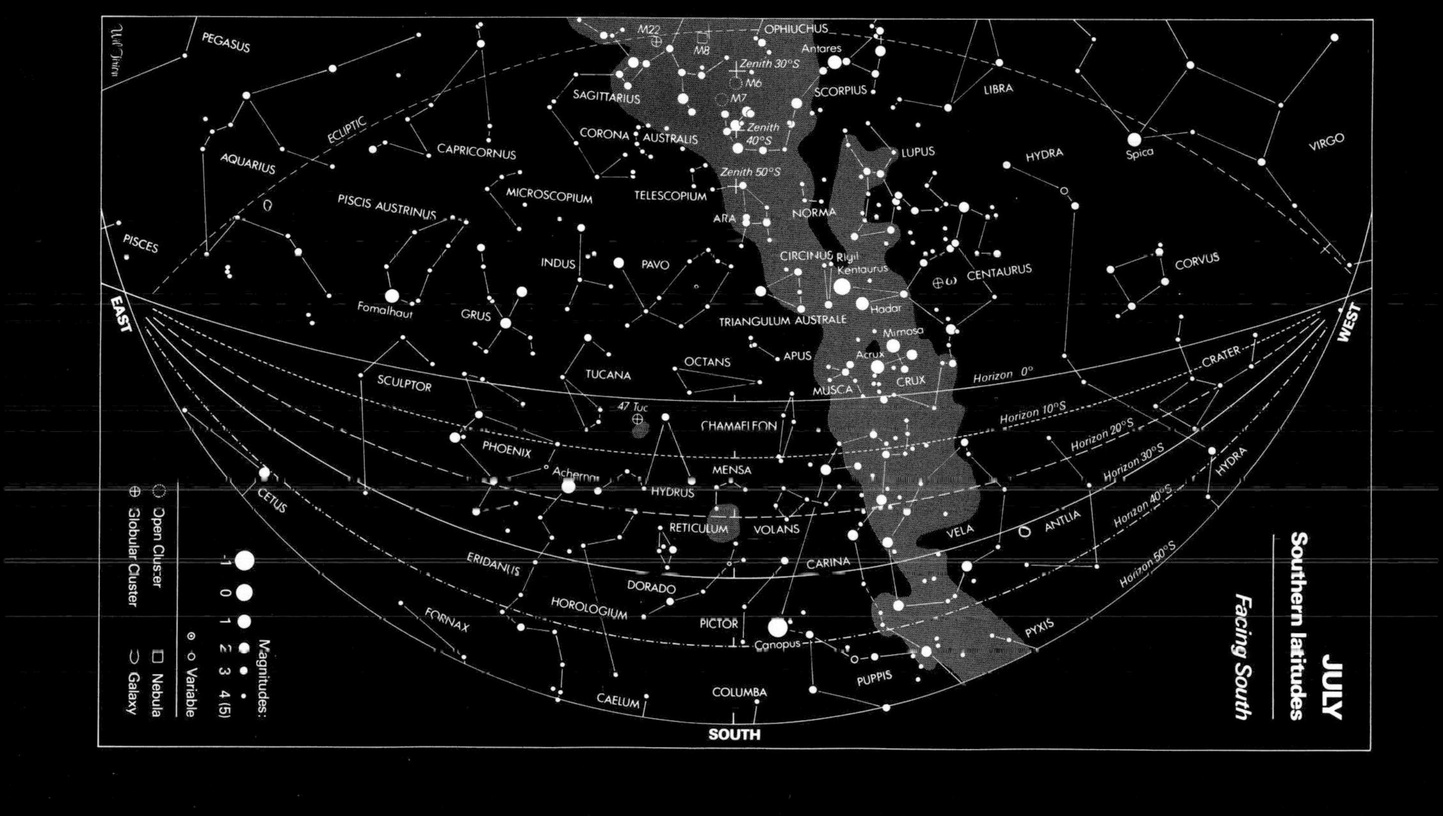

JULY
Southern latitudes
Facing South
WEST
EAST
SOUTH
VIRGO
Spica
CORVUS
CRATER
HYDRA
LIBRA
CENTAURUS
LUPUS
SCORPIUS
Antares
OPHIUCHUS
NORMA
CIRCINUS
Rigil Kentaurus
Hadar
Mimosa
Acrux
CRUX
MUSCA
TRIANGULUM AUSTRALE
APUS
ARA
SAGITTARIUS
CORONA AUSTRALIS
TELESCOPIUM
PAVO
OCTANS
CHAMAELEON
MENSA
VOLANS
CARINA
VELA
ANTLIA
PYXIS
PUPPIS
Canopus
PICTOR
COLUMBA
RETICULUM
DORADO
HYDRUS
TUCANA
47 Tuc
INDUS
MICROSCOPIUM
CAPRICORNUS
GRUS
PHOENIX
Achernar
HOROLOGIUM
CAELUM
ERIDANUS
FORNAX
SCULPTOR
PISCIS AUSTRINUS
Fomalhaut
AQUARIUS
PEGASUS
PISCES
CETUS
ECLIPTIC
M6
M7
M8
M22
Zenith 30°S
Zenith 40°S
Zenith 50°S
Horizon 0°
Horizon 10°S
Horizon 20°S
Horizon 30°S
Horizon 40°S
Horizon 50°S
Magnitudes:
-1 0 1 2 3 4 (5)
Variable
Open Cluster
Globular Cluster
Nebula
Galaxy
Wil Tirion

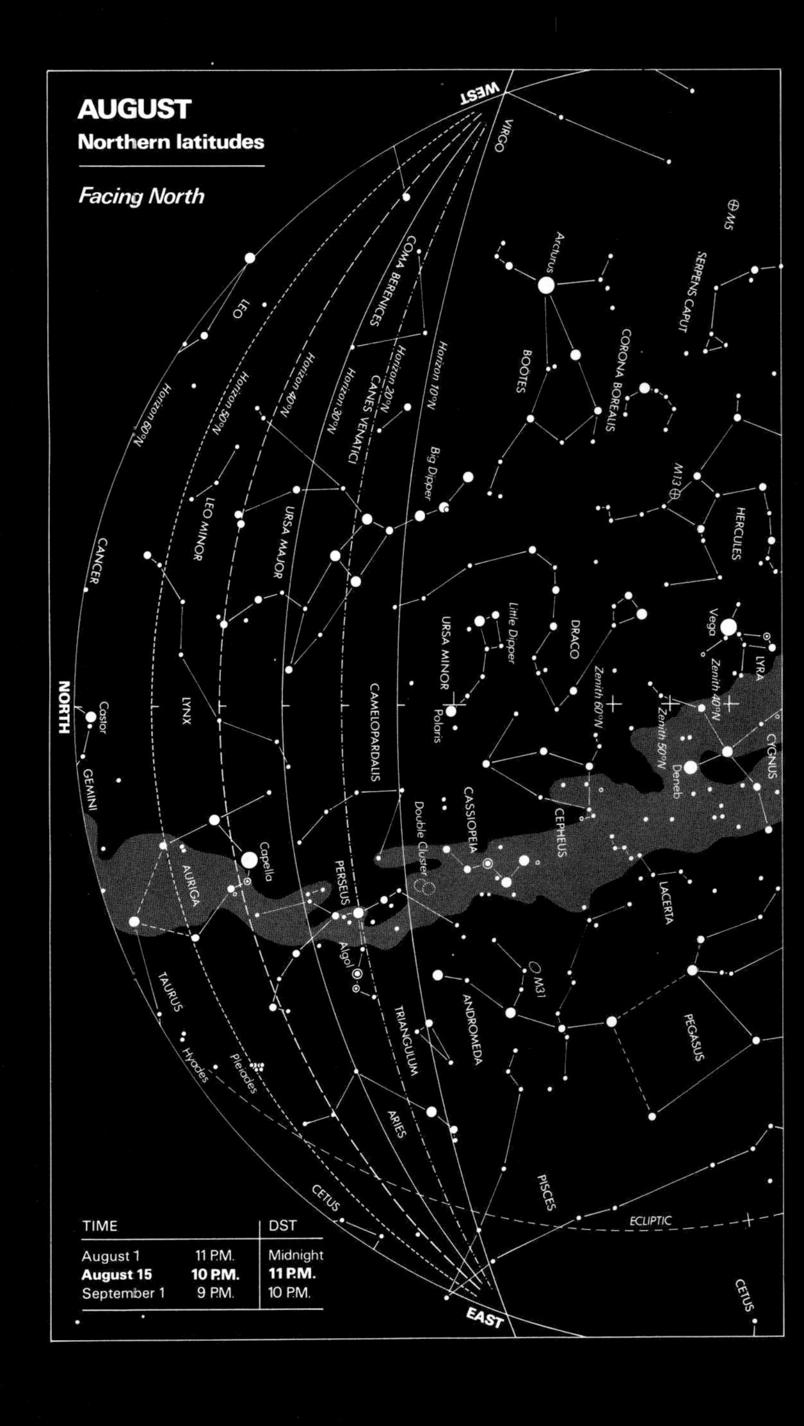

TIME		DST
August 1	11 P.M.	Midnight
August 15	**10 P.M.**	**11 P.M.**
September 1	9 P.M.	10 P.M.

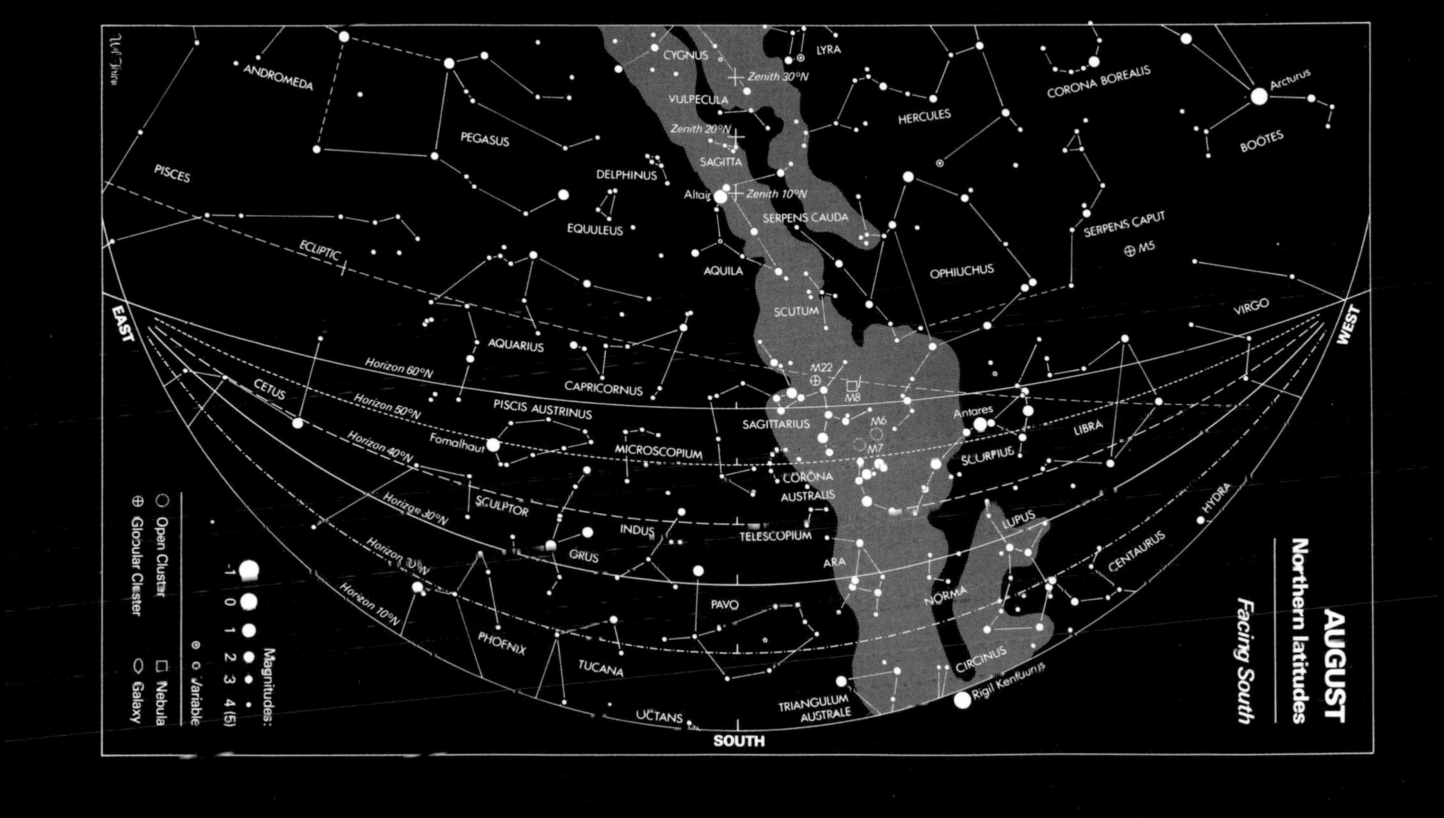
AUGUST
Northern latitudes
Facing South
WEST
EAST
SOUTH
ANDROMEDA
PEGASUS
PISCES
CYGNUS
LYRA
Zenith 30°N
VULPECULA
Zenith 20°N
SAGITTA
HERCULES
CORONA BOREALIS
Arcturus
BOÖTES
DELPHINUS
Altair
Zenith 10°N
SERPENS CAUDA
EQUULEUS
ECLIPTIC
AQUILA
OPHIUCHUS
SERPENS CAPUT
M5
SCUTUM
VIRGO
AQUARIUS
Horizon 60°N
CAPRICORNUS
M22
M8
CETUS
Horizon 50°N
PISCIS AUSTRINUS
SAGITTARIUS
M6
M7
Antares
LIBRA
Horizon 40°N
Fomalhaut
MICROSCOPIUM
CORONA AUSTRALIS
SCORPIUS
SCULPTOR
INDUS
TELESCOPIUM
LUPUS
HYDRA
CENTAURUS
GRUS
ARA
NORMA
Horizon 10°N
PAVO
PHOENIX
TUCANA
CIRCINUS
Rigil Kentaurus
UCTANS
TRIANGULUM AUSTRALE
Magnitudes:
-1 0 1 2 3 4 (5)
Variable
Open Cluster
Nebula
Globular Cluster
Galaxy

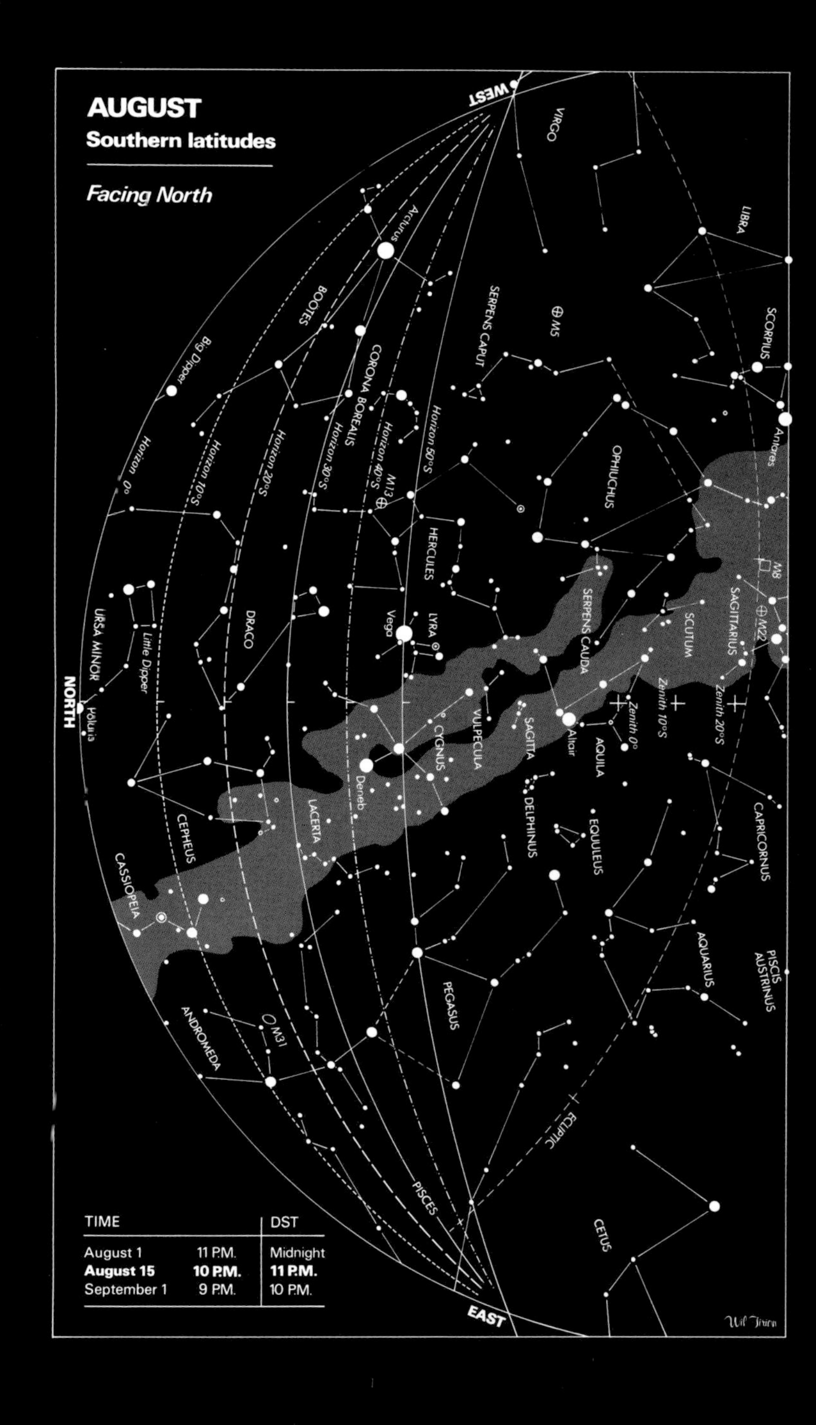

TIME		DST
August 1	11 P.M.	Midnight
August 15	**10 P.M.**	**11 P.M.**
September 1	9 P.M.	10 P.M.

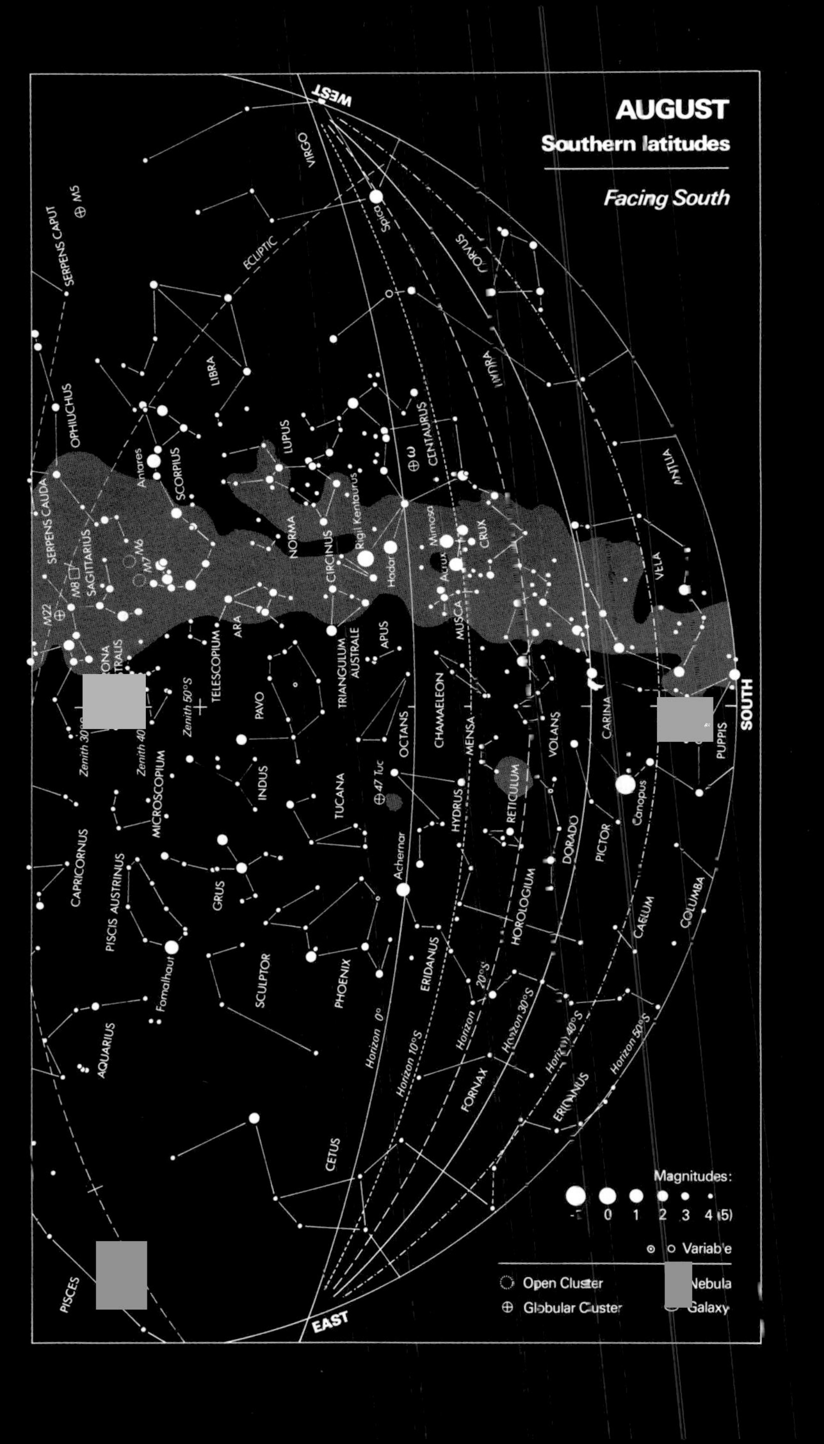

AUGUST
Southern latitudes
Facing South
WEST
EAST
SOUTH
ECLIPTIC
VIRGO
Spica
CORVUS
HYDRA
ANTLIA
VELA
CRUX
Mimosa
Acrux
MUSCA
CENTAURUS
Rigil Kentaurus
Hadar
LUPUS
LIBRA
SCORPIUS
Antares
OPHIUCHUS
SERPENS CAPUT
SERPENS CAUDA
M5
SAGITTARIUS
M22
M8
M7
M6
NORMA
CIRCINUS
ARA
TELESCOPIUM
TRIANGULUM AUSTRALE
APUS
PAVO
OCTANS
CHAMAELEON
MENSA
VOLANS
CARINA
Canopus
PUPPIS
PICTOR
DORADO
RETICULUM
HYDRUS
TUCANA
47 Tuc
INDUS
MICROSCOPIUM
GRUS
CAPRICORNUS
PISCIS AUSTRINUS
Fomalhaut
SCULPTOR
PHOENIX
Achernar
ERIDANUS
HOROLOGIUM
CAELUM
COLUMBA
FORNAX
CETUS
AQUARIUS
PISCES
Zenith 30°S
Zenith 40°S
Zenith 50°S
Horizon 0°
Horizon 10°S
Horizon 20°S
Horizon 30°S
Horizon 40°S
Horizon 50°S
Magnitudes:
-1 0 1 2 3 4 (5)
Variable
Open Cluster
Globular Cluster
Nebula
Galaxy

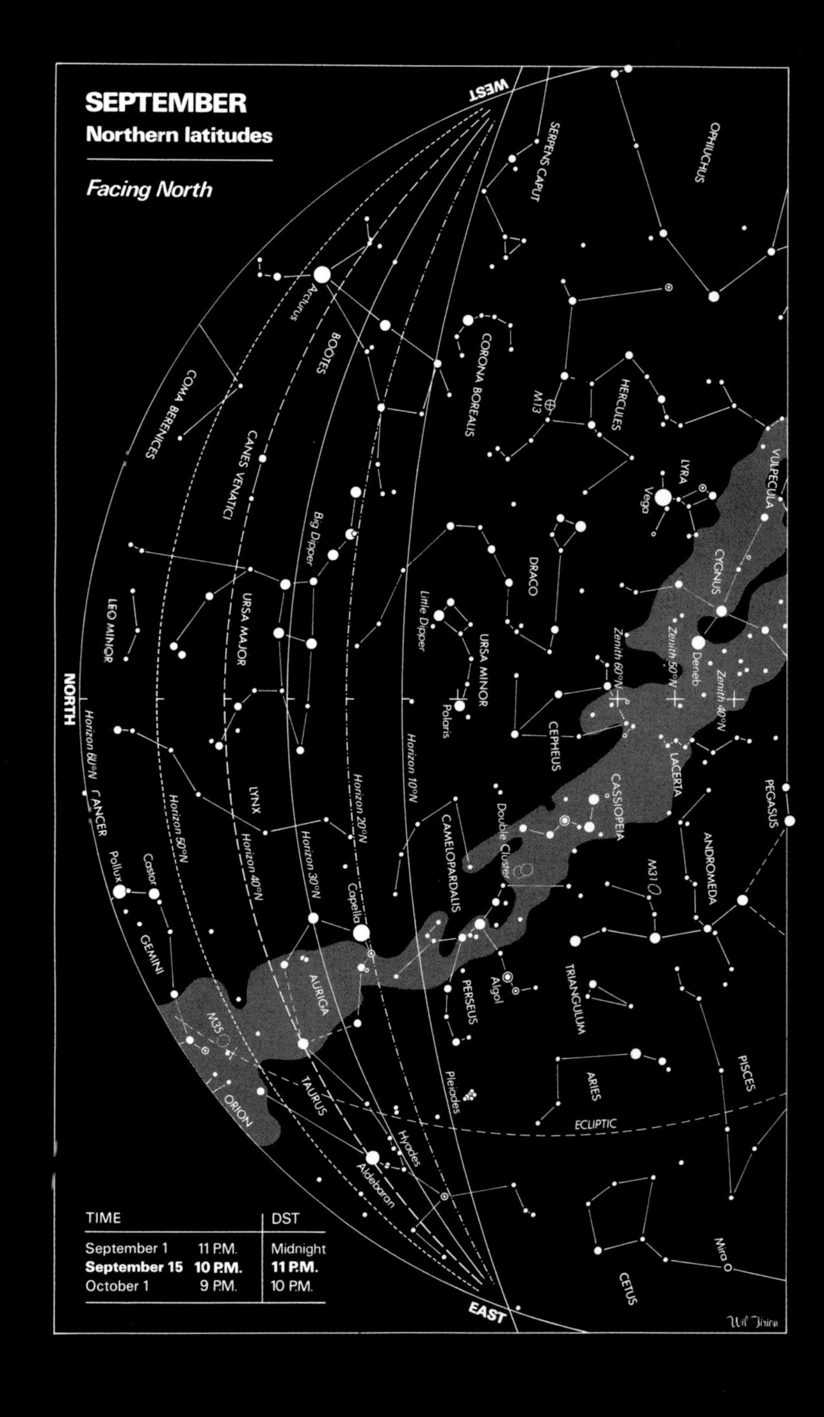
SEPTEMBER
Northern latitudes
Facing North
TIME | DST
September 1 11 P.M. Midnight
September 15 10 P.M. 11 P.M.
October 1 9 P.M. 10 P.M.
WEST
EAST
NORTH
OPHIUCHUS
SERPENS CAPUT
Arcturus
BOÖTES
CORONA BOREALIS
COMA BERENICES
CANES VENATICI
M13
HERCULES
LYRA
Vega
VULPECULA
CYGNUS
Deneb
Big Dipper
DRACO
Little Dipper
URSA MINOR
URSA MAJOR
LEO MINOR
Polaris
Zenith 60°N
Zenith 50°N
Zenith 40°N
CEPHEUS
LACERTA
PEGASUS
Horizon 60°N
Horizon 50°N
Horizon 40°N
Horizon 30°N
Horizon 20°N
Horizon 10°N
LYNX
CANCER
CAMELOPARDALIS
Double Cluster
CASSIOPEIA
ANDROMEDA
M31
Pollux
Castor
Capella
GEMINI
AURIGA
PERSEUS
Algol
TRIANGULUM
M35
ORION
TAURUS
Pleiades
ARIES
PISCES
ECLIPTIC
Hyades
Aldebaran
Mira
CETUS
Wil Tirion

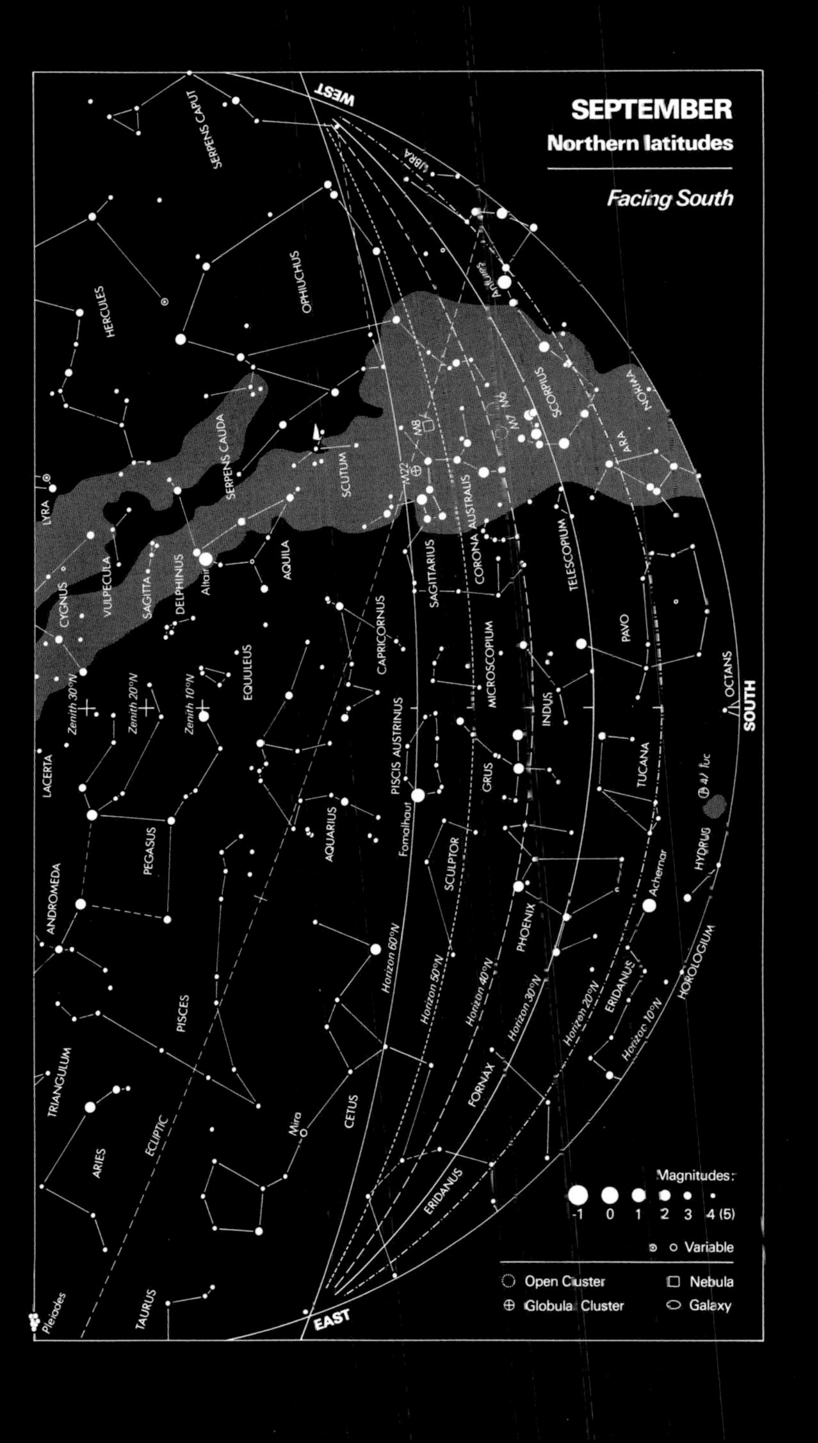
SEPTEMBER
Northern latitudes
Facing South
WEST
EAST
SOUTH
SERPENS CAPUT
LIBRA
HERCULES
OPHIUCHUS
Antares
SCORPIUS
NORMA
M6
M7
M8
M22
SERPENS CAUDA
SCUTUM
ARA
LYRA
CYGNUS
VULPECULA
SAGITTA
DELPHINUS
Altair
AQUILA
SAGITTARIUS
CORONA AUSTRALIS
TELESCOPIUM
PAVO
CAPRICORNUS
MICROSCOPIUM
INDUS
OCTANS
EQUULEUS
Zenith 30°N
Zenith 20°N
Zenith 10°N
PISCIS AUSTRINUS
GRUS
TUCANA
47 Tuc
LACERTA
PEGASUS
AQUARIUS
Fomalhaut
SCULPTOR
HYDRUS
ANDROMEDA
PHOENIX
Achernar
HOROLOGIUM
Horizon 60°N
Horizon 50°N
Horizon 40°N
Horizon 30°N
Horizon 20°N
Horizon 10°N
ERIDANUS
PISCES
TRIANGULUM
FORNAX
ARIES
ECLIPTIC
Mira
CETUS
Pleiades
TAURUS
Magnitudes:
-1 0 1 2 3 4 (5)
Variable
Open Cluster
Nebula
Globular Cluster
Galaxy

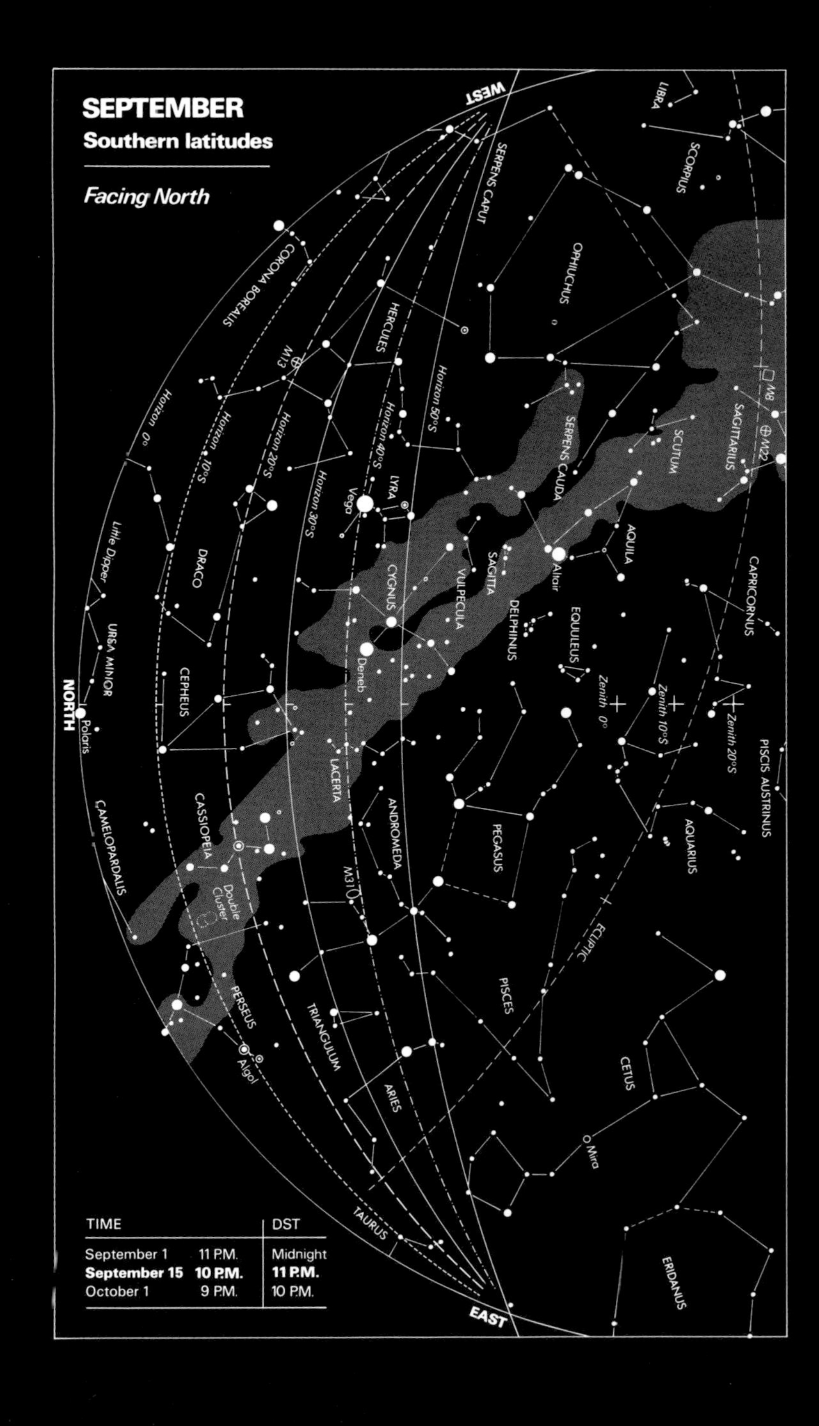
SEPTEMBER
Southern latitudes
Facing North
WEST
EAST
NORTH
TIME | DST
September 1 11 P.M. | Midnight
September 15 10 P.M. | 11 P.M.
October 1 9 P.M. | 10 P.M.
LIBRA
SCORPIUS
SERPENS CAPUT
OPHIUCHUS
CORONA BOREALIS
HERCULES
M13
Horizon 0°
Horizon 10°S
Horizon 20°S
Horizon 30°S
Horizon 40°S
Horizon 50°S
M8
M22
SAGITTARIUS
SCUTUM
SERPENS CAUDA
LYRA
Vega
Little Dipper
DRACO
AQUILA
Altair
SAGITTA
VULPECULA
CYGNUS
Deneb
CAPRICORNUS
DELPHINUS
EQUULEUS
URSA MINOR
Polaris
CEPHEUS
Zenith 0°
Zenith 10°S
Zenith 20°S
PISCIS AUSTRINUS
LACERTA
CAMELOPARDALIS
CASSIOPEIA
ANDROMEDA
PEGASUS
AQUARIUS
M31
Double Cluster
ECLIPTIC
PISCES
PERSEUS
TRIANGULUM
Algol
CETUS
ARIES
Mira
TAURUS
ERIDANUS

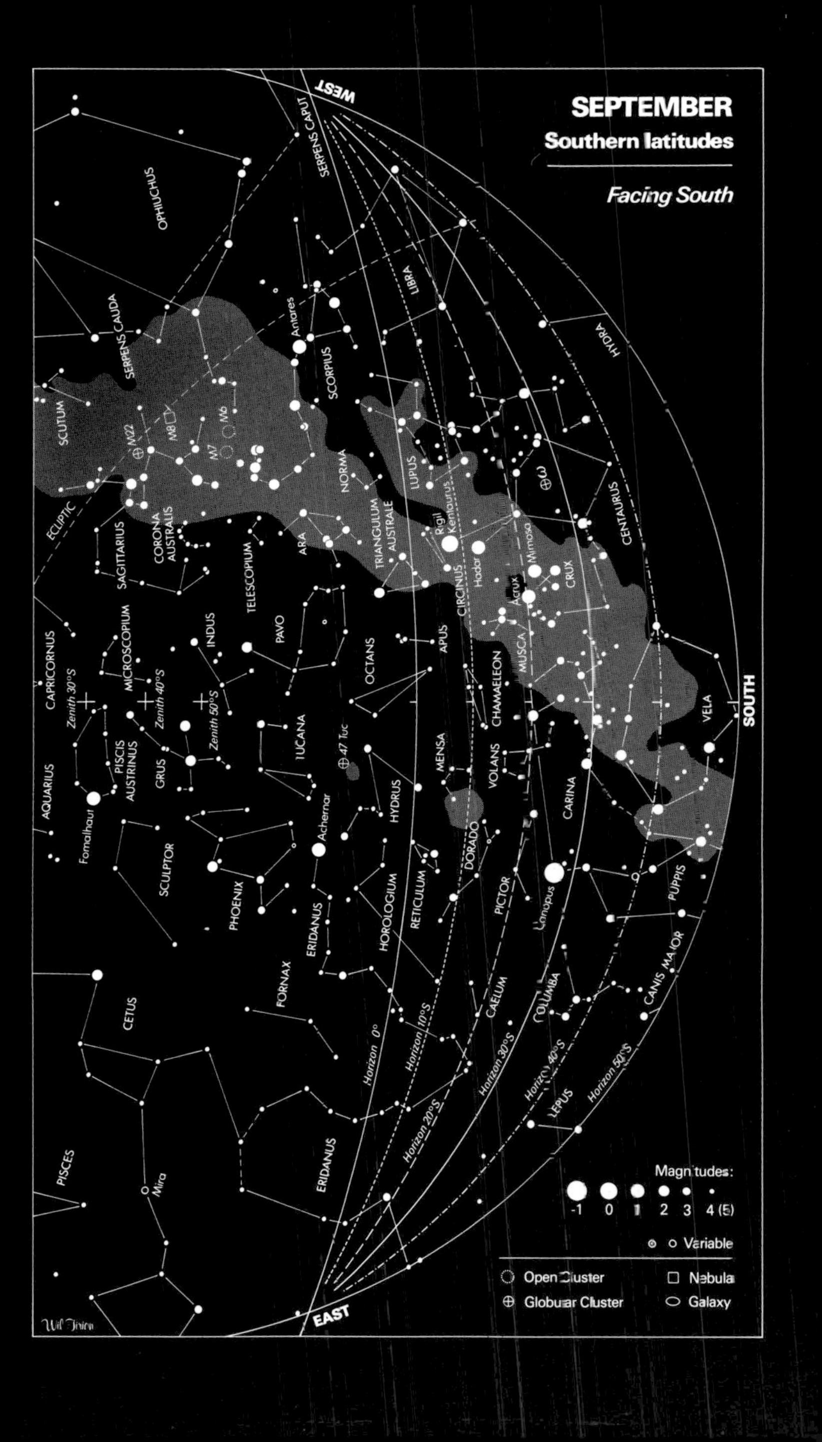

SEPTEMBER
Southern latitudes
Facing South
WEST
EAST
SOUTH
SERPENS CAPUT
OPHIUCHUS
LIBRA
HYDRA
SERPENS CAUDA
Antares
SCORPIUS
SCUTUM
M22
M8
M6
M7
NORMA
LUPUS
ω
CENTAURUS
ECLIPTIC
SAGITTARIUS
CORONA AUSTRALIS
TELESCOPIUM
ARA
TRIANGULUM AUSTRALE
Rigil Kentaurus
Hadar
Mimosa
CRUX
Acrux
CIRCINUS
PAVO
INDUS
MICROSCOPIUM
CAPRICORNUS
APUS
MUSCA
CHAMAELEON
OCTANS
VELA
Zenith 30°S
Zenith 40°S
Zenith 50°S
TUCANA
47 Tuc
MENSA
VOLANS
CARINA
AQUARIUS
PISCIS AUSTRINUS
GRUS
HYDRUS
Fomalhaut
Achernar
DORADO
SCULPTOR
PHOENIX
ERIDANUS
HOROLOGIUM
RETICULUM
PICTOR
Canopus
PUPPIS
CANIS MAJOR
FORNAX
CETUS
CAELUM
COLUMBA
Horizon 0°
Horizon 10°S
Horizon 20°S
Horizon 30°S
Horizon 40°S
Horizon 50°S
LEPUS
PISCES
Mira
Magnitudes:
-1 0 1 2 3 4 (5)
Variable
Open Cluster
Nebula
Globular Cluster
Galaxy
Wil Tirion

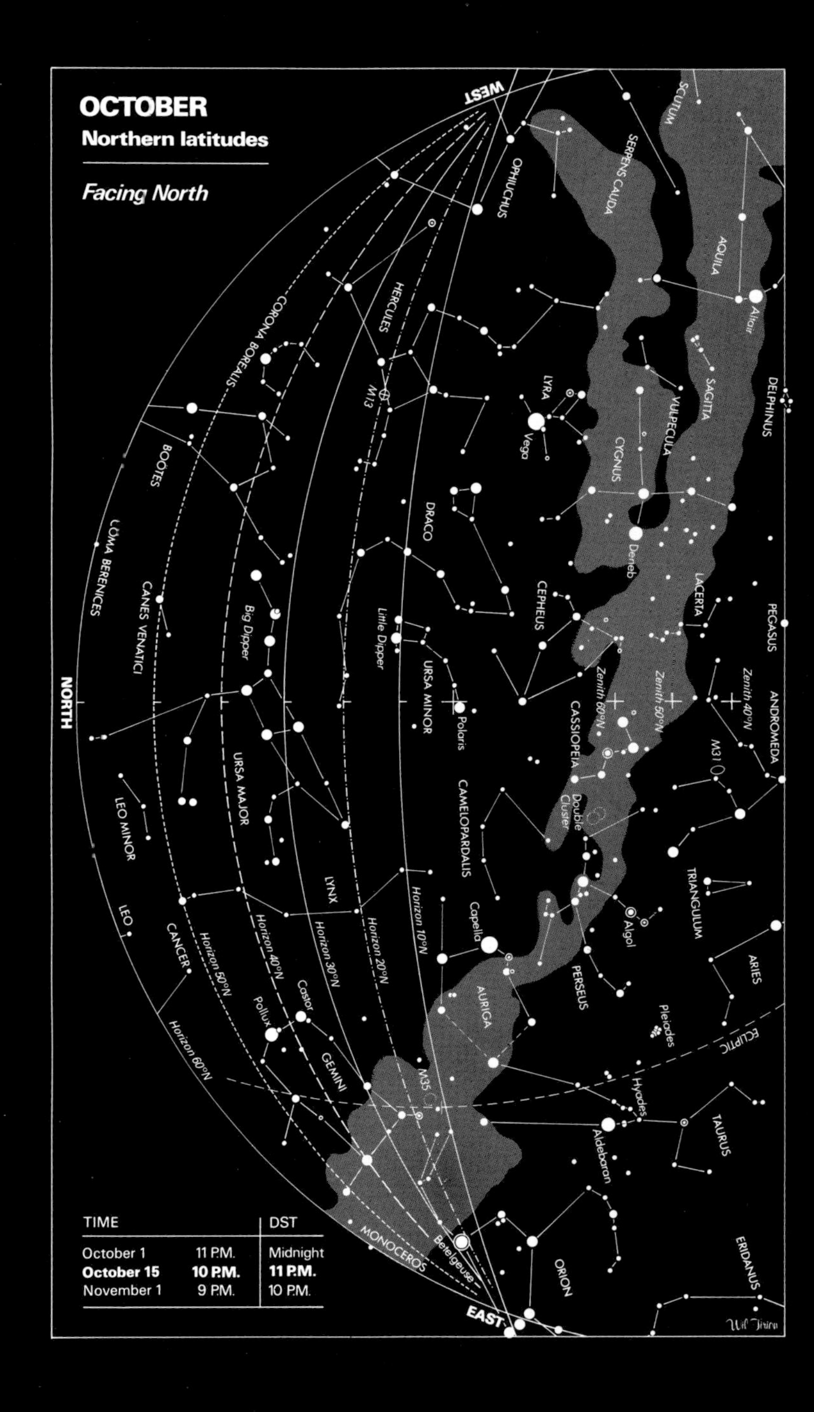
OCTOBER
Northern latitudes
Facing North
TIME
DST
October 1
11 P.M.
Midnight
October 15
10 P.M.
11 P.M.
November 1
9 P.M.
10 P.M.
WEST
NORTH
EAST
OPHIUCHUS
SERPENS CAUDA
SCUTUM
AQUILA
Altair
DELPHINUS
SAGITTA
VULPECULA
CYGNUS
Deneb
LYRA
Vega
HERCULES
M13
CORONA BOREALIS
BOÖTES
COMA BERENICES
CANES VENATICI
Big Dipper
URSA MAJOR
DRACO
Little Dipper
URSA MINOR
Polaris
CEPHEUS
LACERTA
PEGASUS
Zenith 60°N
Zenith 50°N
Zenith 40°N
CASSIOPEIA
ANDROMEDA
M31
Double Cluster
CAMELOPARDALIS
LEO MINOR
LEO
CANCER
LYNX
TRIANGULUM
Algol
ARIES
PERSEUS
Capella
AURIGA
Pleiades
ECLIPTIC
Hyades
TAURUS
Aldebaran
Castor
Pollux
GEMINI
M35
MONOCEROS
Betelgeuse
ORION
ERIDANUS
Horizon 10°N
Horizon 20°N
Horizon 30°N
Horizon 40°N
Horizon 50°N
Horizon 60°N
Wil Tirion

OCTOBER
Northern latitudes
Facing South
WEST
EAST
SOUTH
OPHIUCHUS
HERCULES
SERPENS CAUDA
SCUTUM
CYGNUS
VULPECULA
SAGITTA
Altair
AQUILA
DELPHINUS
EQUULEUS
CAPRICORNUS
SAGITTARIUS
CORONA AUSTRALIS
TELESCOPIUM
MICROSCOPIUM
INDUS
PAVO
LACERTA
PEGASUS
PISCIS AUSTRINUS
GRUS
OCTANS
ANDROMEDA
Zenith 30°N
Zenith 20°N
Zenith 10°N
AQUARIUS
Fomalhaut
SCULPTOR
PHOENIX
TUCANA
47 Tuc
PISCES
Achernar
HYDRUS
Horizon 60°N
Horizon 50°N
Horizon 40°N
Horizon 30°N
Horizon 20°N
Horizon 10°N
TRIANGULUM
ARIES
ECLIPTIC
Mira
CETUS
FORNAX
HOROLOGIUM
RETICULUM
DORADO
ERIDANUS
Pleiades
Hyades
TAURUS
CAELUM
Aldebaran
LEPUS
Rigel
ORION
Magnitudes:
-1
0
1
2
3
4 (5)
Variable
Open Cluster
Nebula
Globular Cluster
Galaxy

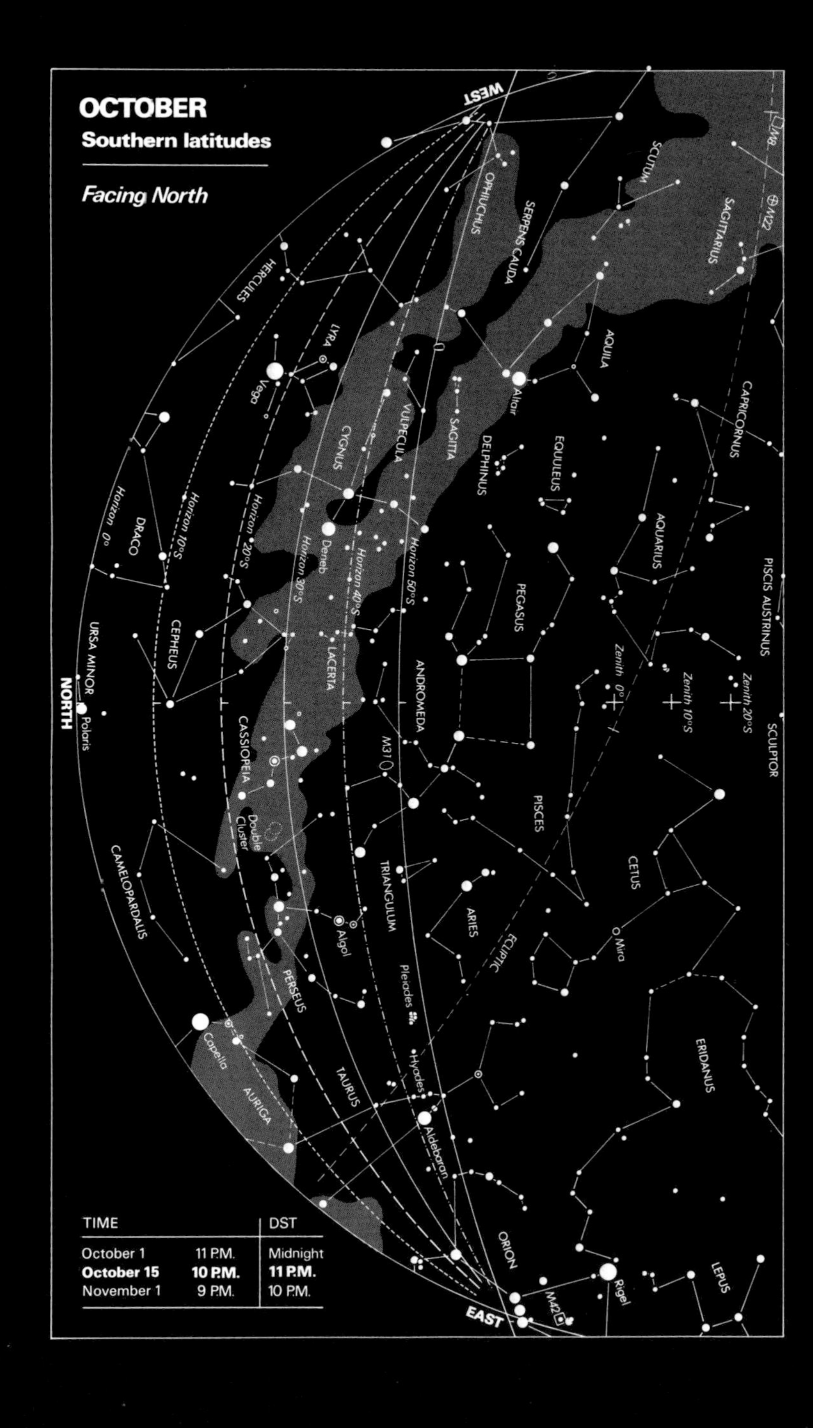
OCTOBER
Southern latitudes
Facing North
WEST
EAST
NORTH
TIME | DST
October 1 11 P.M. | Midnight
October 15 10 P.M. | 11 P.M.
November 1 9 P.M. | 10 P.M.
Horizon 0°
Horizon 10°S
Horizon 20°S
Horizon 30°S
Horizon 40°S
Horizon 50°S
Zenith 0°
Zenith 10°S
Zenith 20°S
ECLIPTIC
HERCULES
OPHIUCHUS
SERPENS CAUDA
SCUTUM
SAGITTARIUS
M8
M22
LYRA
Vega
CYGNUS
Deneb
VULPECULA
SAGITTA
AQUILA
Altair
DELPHINUS
EQUULEUS
CAPRICORNUS
AQUARIUS
PISCIS AUSTRINUS
DRACO
CEPHEUS
URSA MINOR
Polaris
LACERTA
PEGASUS
ANDROMEDA
M31
CASSIOPEIA
Double Cluster
PISCES
CETUS
SCULPTOR
CAMELOPARDALIS
TRIANGULUM
ARIES
Algol
Mira
PERSEUS
Pleiades
Capella
AURIGA
TAURUS
Hyades
Aldebaran
ERIDANUS
ORION
M42
Rigel
LEPUS

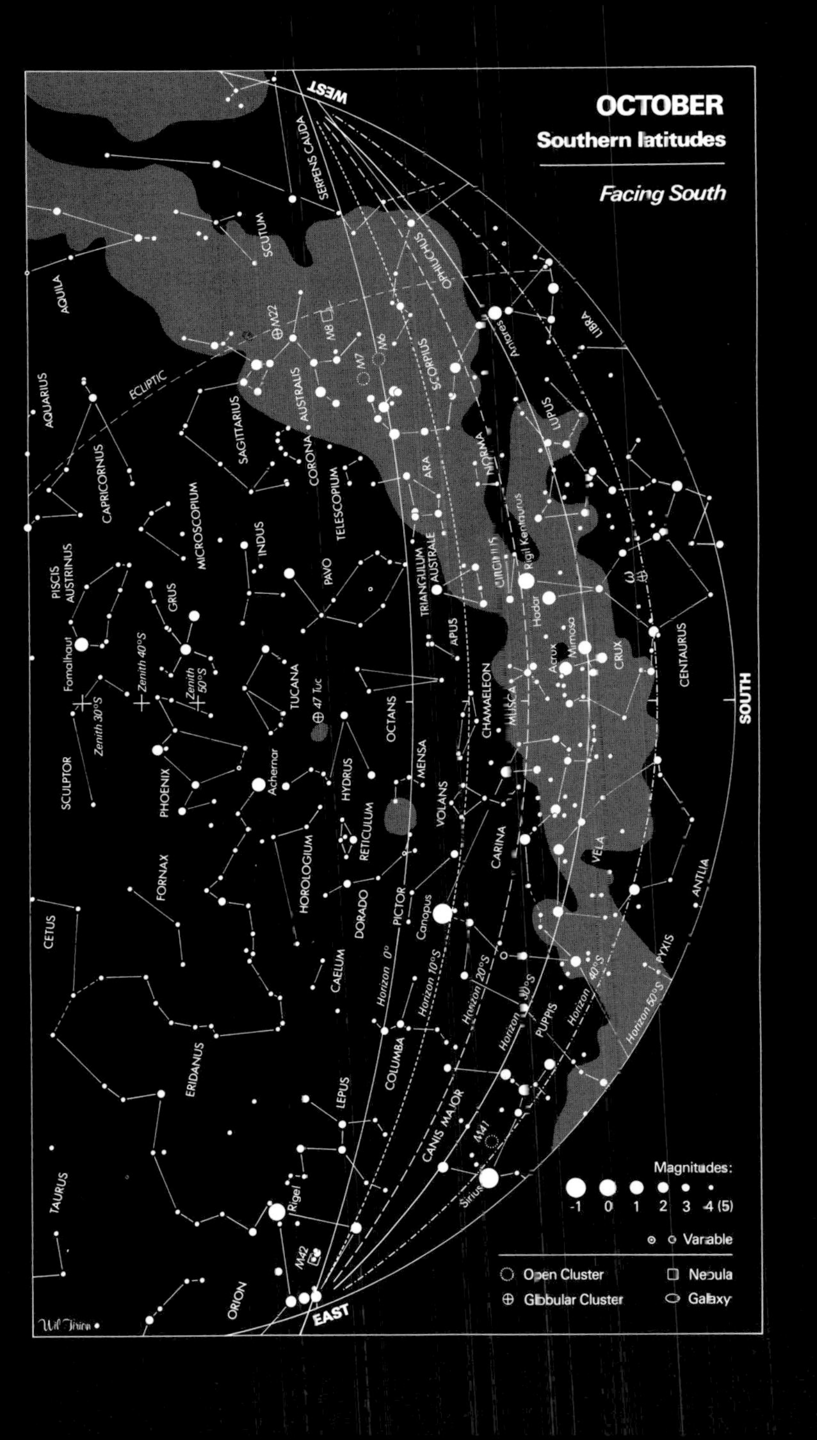

OCTOBER
Southern latitudes
Facing South
WEST
EAST
SOUTH
SERPENS CAUDA
SCUTUM
AQUILA
OPHIUCHUS
LIBRA
Antares
M22
M8
M6
M7
SCORPIUS
CORONA AUSTRALIS
SAGITTARIUS
ECLIPTIC
AQUARIUS
CAPRICORNUS
MICROSCOPIUM
TELESCOPIUM
ARA
NORMA
LUPUS
INDUS
PAVO
TRIANGULUM AUSTRALE
CIRCINUS
Rigil Kentaurus
Hadar
Mimosa
Acrux
CRUX
CENTAURUS
PISCIS AUSTRINUS
GRUS
Fomalhaut
Zenith 40°S
Zenith 50°S
Zenith 30°S
TUCANA
47 Tuc
APUS
CHAMAELEON
MUSCA
OCTANS
MENSA
SCULPTOR
PHOENIX
Achernar
HYDRUS
VOLANS
CARINA
VELA
ANTLIA
FORNAX
HOROLOGIUM
RETICULUM
DORADO
PICTOR
Canopus
CETUS
CAELUM
Horizon 0°
Horizon 10°S
Horizon 20°S
Horizon 30°S
Horizon 40°S
Horizon 50°S
PUPPIS
PYXIS
ERIDANUS
COLUMBA
LEPUS
CANIS MAJOR
M41
Sirius
TAURUS
Rigel
M42
ORION
Magnitudes:
-1 0 1 2 3 4 (5)
Variable
Open Cluster
Nebula
Globular Cluster
Galaxy
Wil Tirion

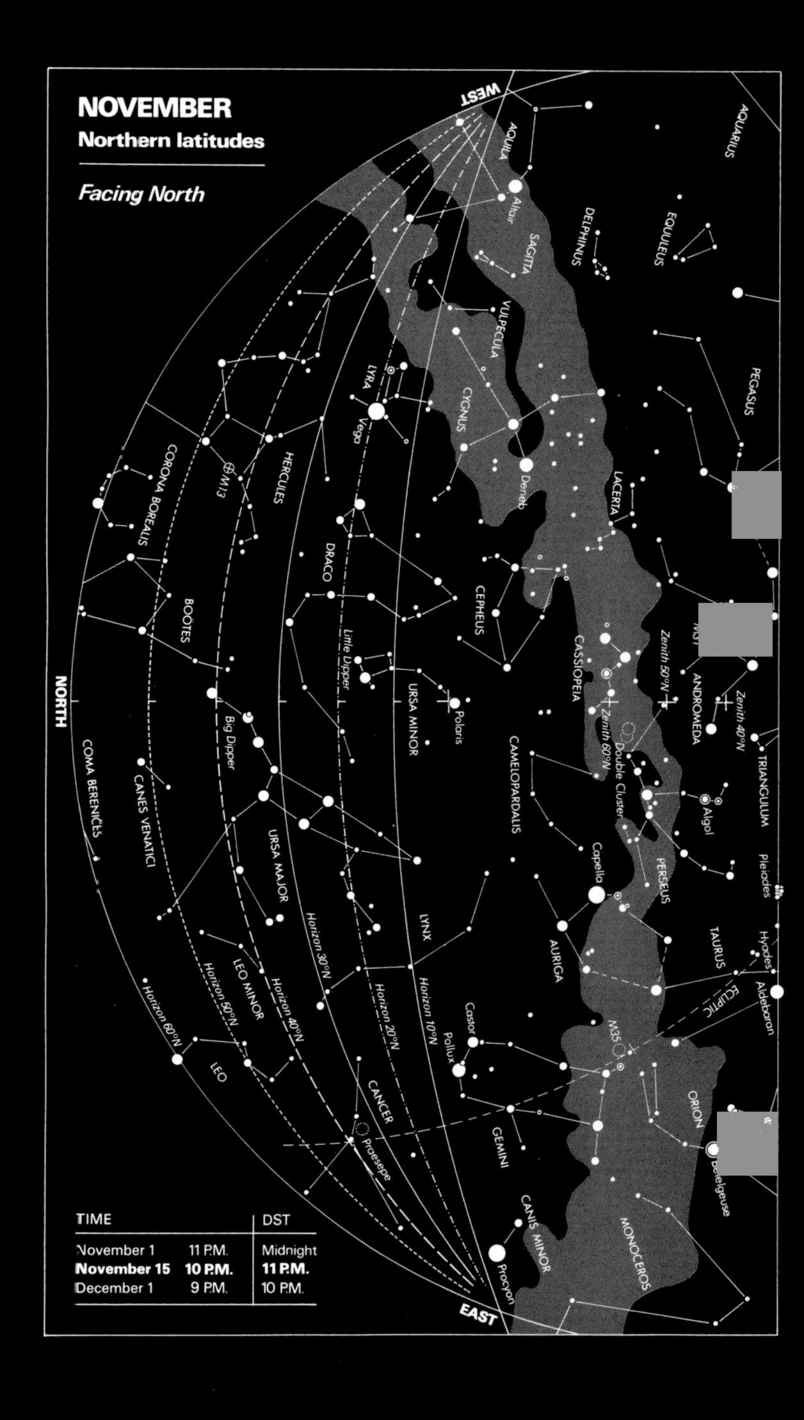
NOVEMBER
Northern latitudes
Facing North
TIME
DST
November 1 11 P.M. Midnight
November 15 10 P.M. 11 P.M.
December 1 9 P.M. 10 P.M.
WEST
NORTH
EAST
AQUILA
Altair
SAGITTA
DELPHINUS
EQUULEUS
AQUARIUS
VULPECULA
PEGASUS
LYRA
Vega
CYGNUS
Deneb
LACERTA
HERCULES
M13
CORONA BOREALIS
DRACO
CEPHEUS
BOÖTES
Little Dipper
URSA MINOR
Polaris
CASSIOPEIA
Zenith 50°N
ANDROMEDA
Zenith 40°N
Zenith 60°N
TRIANGULUM
Big Dipper
CAMELOPARDALIS
Double Cluster
Algol
COMA BERENICES
CANES VENATICI
URSA MAJOR
Capella
PERSEUS
Pleiades
Horizon 30°N
LYNX
AURIGA
TAURUS
Hyades
Horizon 50°N
LEO MINOR
Horizon 40°N
Horizon 20°N
Horizon 10°N
ECLIPTIC
Aldebaran
Horizon 60°N
Castor
Pollux
M35
LEO
CANCER
GEMINI
ORION
Praesepe
Betelgeuse
CANIS MINOR
MONOCEROS
Procyon

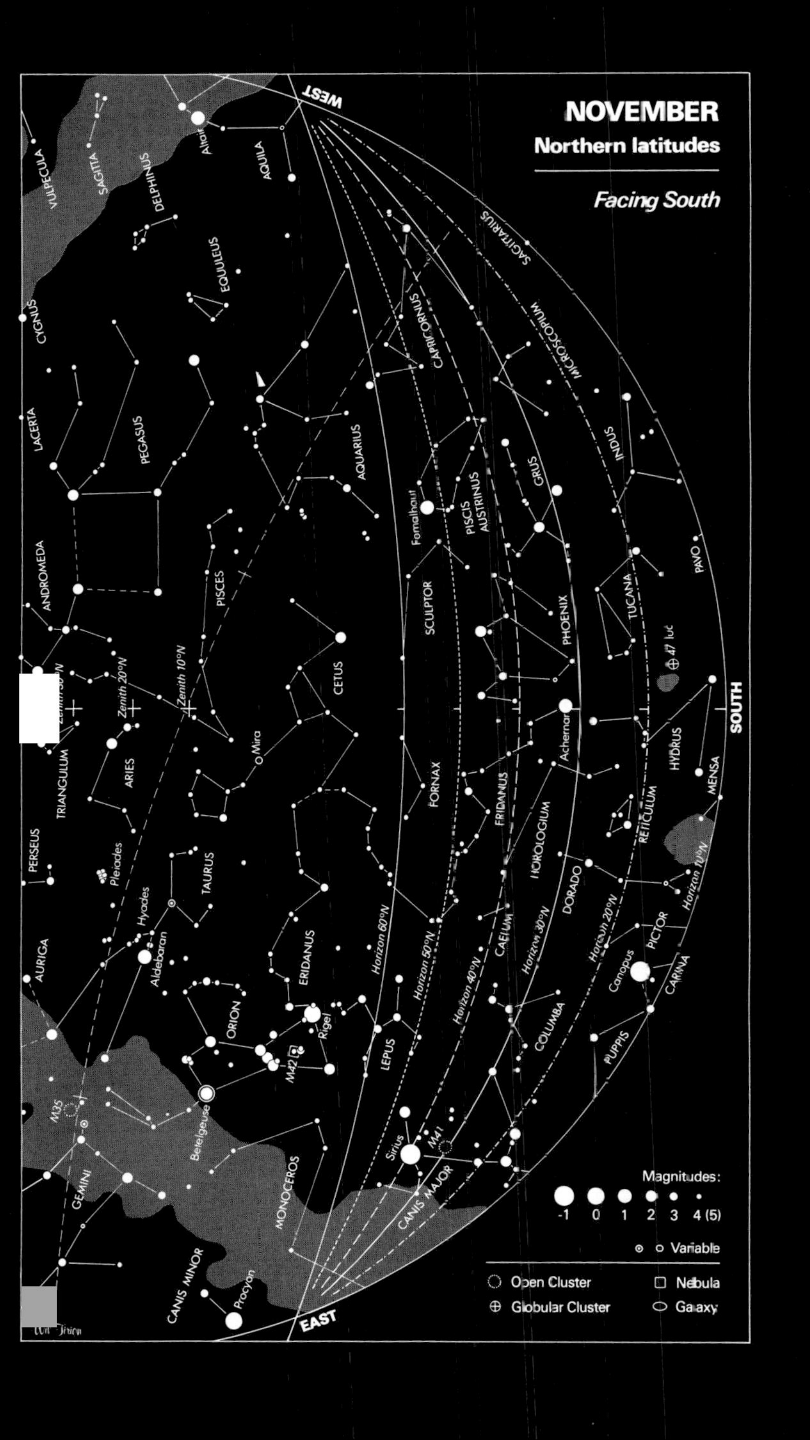
NOVEMBER
Northern latitudes
Facing South
WEST
EAST
SOUTH
VULPECULA
SAGITTA
DELPHINUS
Altair
AQUILA
EQUULEUS
CYGNUS
LACERTA
PEGASUS
ANDROMEDA
PISCES
AQUARIUS
CAPRICORNUS
SAGITTARIUS
MICROSCOPIUM
INDUS
GRUS
PISCIS AUSTRINUS
Fomalhaut
PAVO
TUCANA
47 Tuc
PHOENIX
SCULPTOR
CETUS
Mira
Zenith 20°N
Zenith 10°N
TRIANGULUM
ARIES
PERSEUS
Pleiades
Hyades
TAURUS
Aldebaran
AURIGA
ORION
Betelgeuse
Rigel
M42
M35
GEMINI
CANIS MINOR
Procyon
MONOCEROS
ERIDANUS
LEPUS
Sirius
M41
CANIS MAJOR
FORNAX
Achernar
HYDRUS
MENSA
RETICULUM
HOROLOGIUM
DORADO
CAELUM
COLUMBA
PICTOR
Canopus
CARINA
PUPPIS
Horizon 60°N
Horizon 50°N
Horizon 40°N
Horizon 30°N
Horizon 20°N
Horizon 10°N
Magnitudes:
-1 0 1 2 3 4 (5)
Variable
Open Cluster
Nebula
Globular Cluster
Galaxy

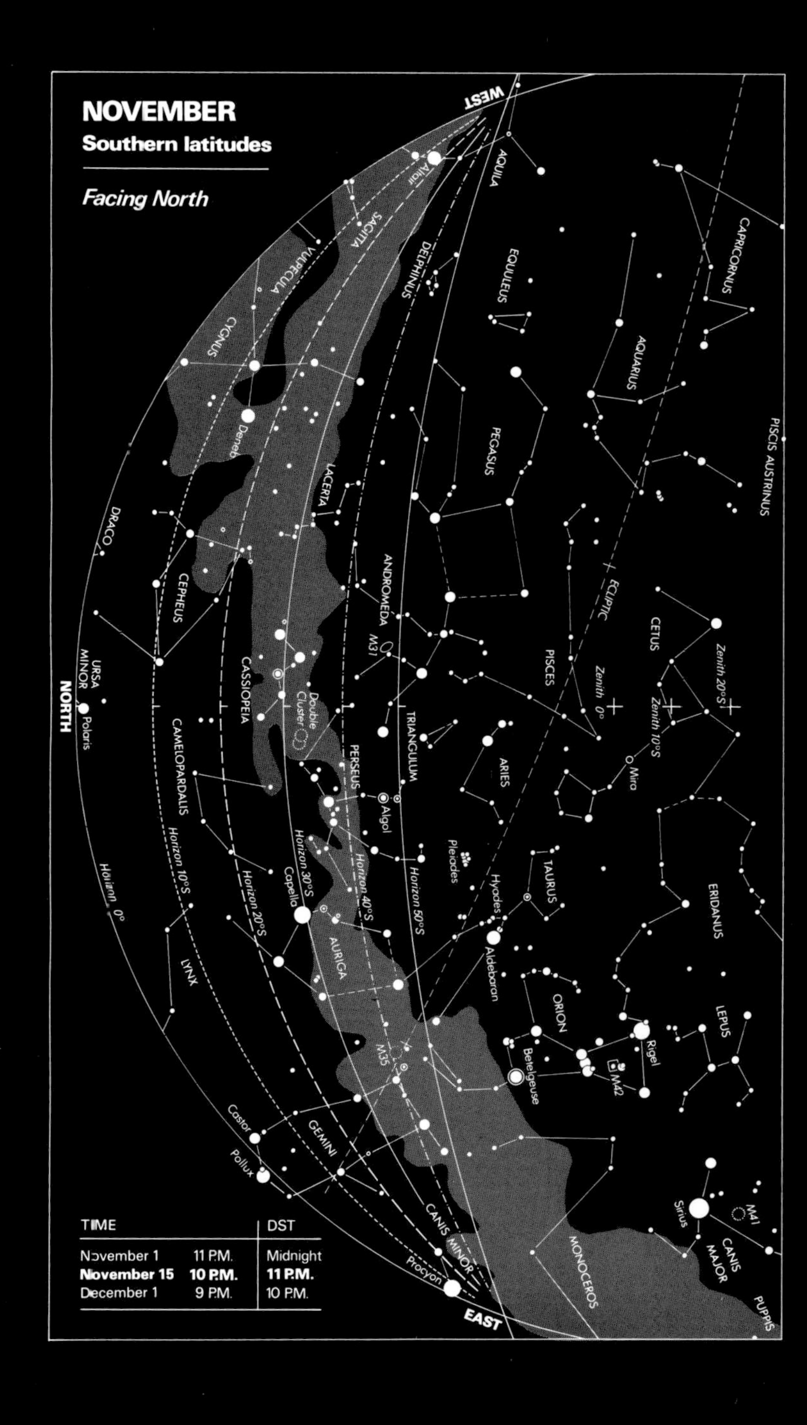
NOVEMBER
Southern latitudes
Facing North
WEST
EAST
NORTH
AQUILA
Altair
SAGITTA
VULPECULA
DELPHINUS
EQUULEUS
CAPRICORNUS
AQUARIUS
CYGNUS
Deneb
PEGASUS
PISCIS AUSTRINUS
LACERTA
DRACO
CEPHEUS
ANDROMEDA
M31
ECLIPTIC
CETUS
PISCES
Zenith 0°
Zenith 10°S
Zenith 20°S
URSA MINOR
Polaris
CASSIOPEIA
Double Cluster
TRIANGULUM
CAMELOPARDALIS
PERSEUS
Algol
ARIES
Mira
Pleiades
Hyades
TAURUS
ERIDANUS
Horizon 0°
Horizon 10°S
Horizon 20°S
Horizon 30°S
Horizon 40°S
Horizon 50°S
Capella
AURIGA
LYNX
Aldebaran
ORION
LEPUS
Rigel
M35
Betelgeuse
M42
Castor
GEMINI
Pollux
Sirius
M41
CANIS MAJOR
CANIS MINOR
Procyon
MONOCEROS
PUPPIS
TIME
DST
November 1 11 P.M. Midnight
November 15 10 P.M. 11 P.M.
December 1 9 P.M. 10 P.M.

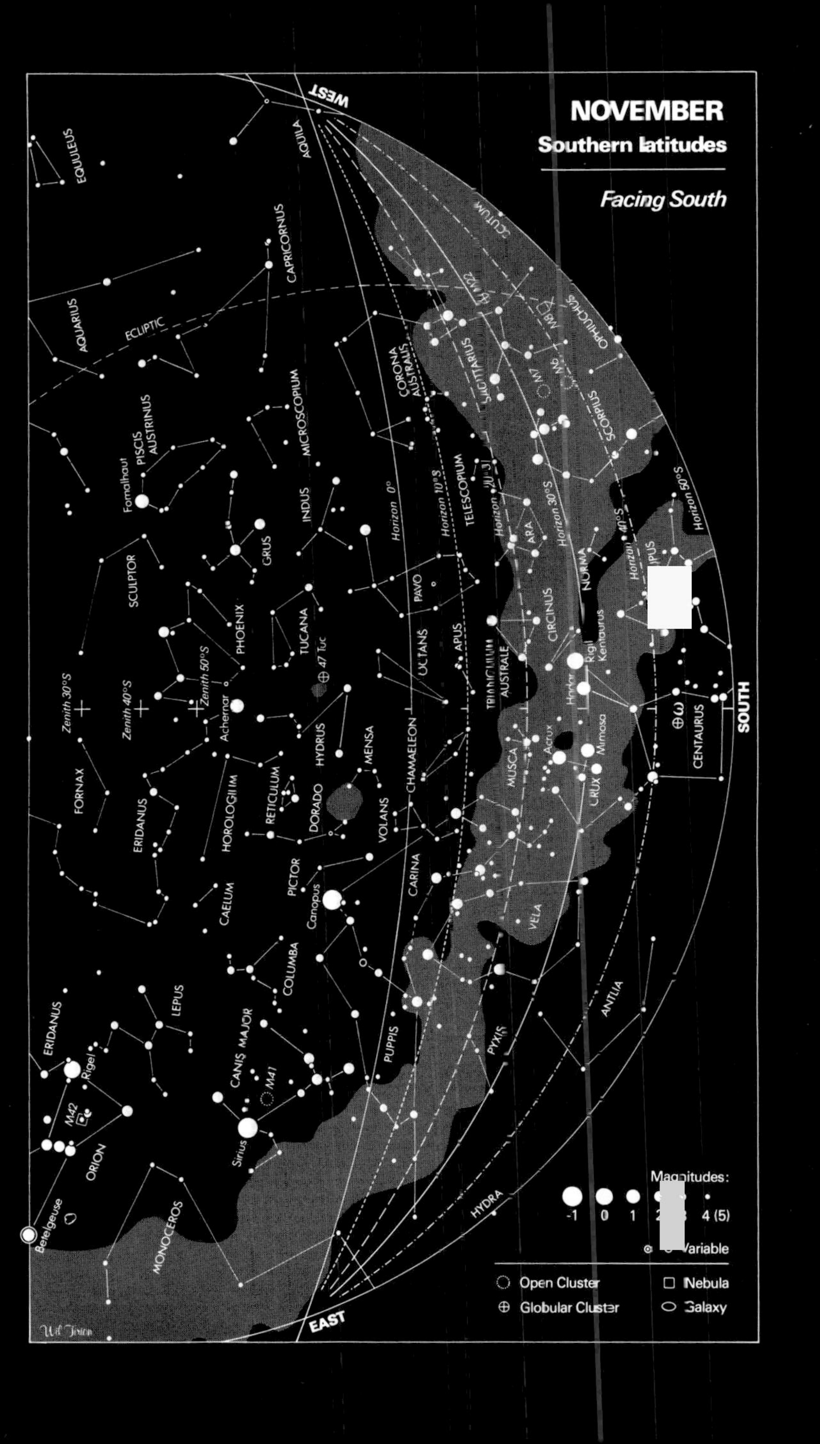
NOVEMBER
Southern latitudes
Facing South
WEST
EAST
SOUTH
EQUULEUS
AQUILA
CAPRICORNUS
AQUARIUS
ECLIPTIC
SCUTUM
SAGITTARIUS
OPHIUCHUS
SCORPIUS
CORONA AUSTRALIS
MICROSCOPIUM
PISCIS AUSTRINUS
Fomalhaut
INDUS
TELESCOPIUM
GRUS
SCULPTOR
PHOENIX
TUCANA
47 Tuc
PAVO
ARA
NORMA
CIRCINUS
APUS
OCTANS
TRIANGULUM AUSTRALE
Rigil Kentaurus
Hadar
CENTAURUS
FORNAX
ERIDANUS
Achernar
HOROLOGIUM
RETICULUM
HYDRUS
DORADO
MENSA
CHAMAELEON
VOLANS
MUSCA
Acrux
Mimosa
CRUX
CAELUM
PICTOR
Canopus
CARINA
VELA
COLUMBA
LEPUS
CANIS MAJOR
M41
Sirius
PUPPIS
PYXIS
ANTLIA
HYDRA
Rigel
M42
ORION
Betelgeuse
MONOCEROS
M7
M8
Zenith 30°S
Zenith 40°S
Zenith 50°S
Horizon 0°
Horizon 10°S
Horizon 30°S
Horizon 40°S
Horizon 50°S
Magnitudes:
-1
0
1
4 (5)
Variable
Open Cluster
Nebula
Globular Cluster
Galaxy
Wil Tirion

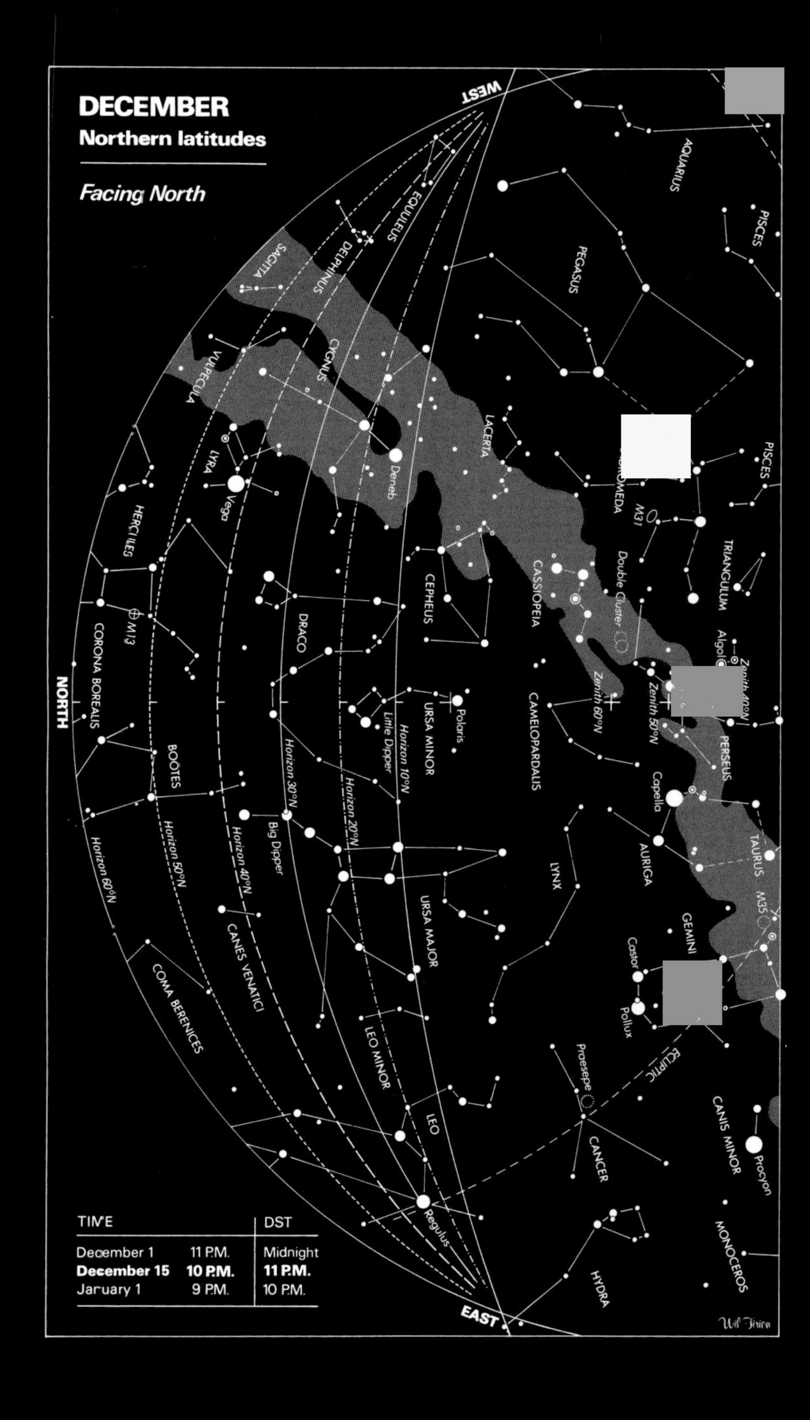
DECEMBER
Northern latitudes
Facing North
TIME
DST
December 1 11 P.M. Midnight
December 15 10 P.M. 11 P.M.
January 1 9 P.M. 10 P.M.
WEST
EAST
NORTH
AQUARIUS
PISCES
PEGASUS
EQUULEUS
DELPHINUS
SAGITTA
VULPECULA
CYGNUS
LACERTA
ANDROMEDA
M31
TRIANGULUM
LYRA
Vega
Deneb
HERCULES
M13
CORONA BOREALIS
CEPHEUS
CASSIOPEIA
Double Cluster
Algol
DRACO
URSA MINOR
Polaris
Little Dipper
CAMELOPARDALIS
Zenith 60°N
Zenith 50°N
PERSEUS
Capella
AURIGA
TAURUS
M35
BOÖTES
Horizon 60°N
Horizon 50°N
Horizon 40°N
Horizon 30°N
Horizon 20°N
Horizon 10°N
Big Dipper
URSA MAJOR
LYNX
GEMINI
Castor
Pollux
CANES VENATICI
COMA BERENICES
LEO MINOR
LEO
Regulus
Praesepe
CANCER
ECLIPTIC
CANIS MINOR
Procyon
MONOCEROS
HYDRA
Wil Tirion

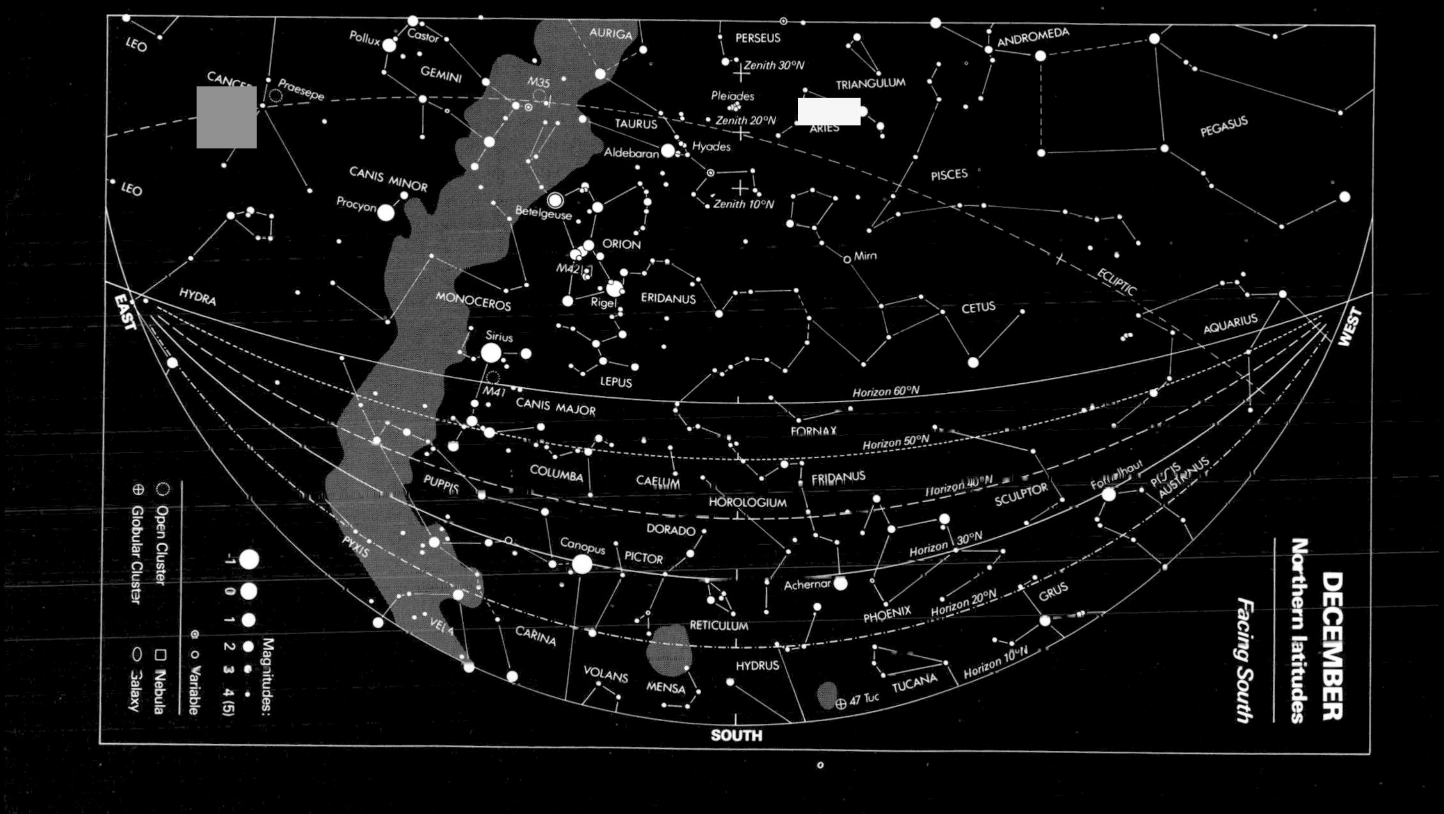
DECEMBER
Northern latitudes
Facing South
WEST
EAST
SOUTH
LEO
Pollux
Castor
GEMINI
AURIGA
PERSEUS
Zenith 30°N
ANDROMEDA
TRIANGULUM
PEGASUS
CANCER
Praesepe
M35
Pleiades
Zenith 20°N
ARIES
TAURUS
Hyades
Aldebaran
PISCES
CANIS MINOR
Procyon
Betelgeuse
Zenith 10°N
ORION
M42
Mira
ECLIPTIC
HYDRA
MONOCEROS
Rigel
ERIDANUS
CETUS
AQUARIUS
Sirius
LEPUS
M41
CANIS MAJOR
Horizon 60°N
FORNAX
Horizon 50°N
COLUMBA
CAELUM
ERIDANUS
PUPPIS
Horizon 40°N
SCULPTOR
Fomalhaut
PISCIS AUSTRINUS
HOROLOGIUM
DORADO
Horizon 30°N
PYXIS
Canopus
PICTOR
Achernar
GRUS
PHOENIX
Horizon 20°N
VELA
RETICULUM
CARINA
HYDRUS
Horizon 10°N
VOLANS
MENSA
TUCANA
47 Tuc
Magnitudes:
-1 0 1 2 3 4 (5)
Variable
Open Cluster
Globular Cluster
Nebula
Galaxy

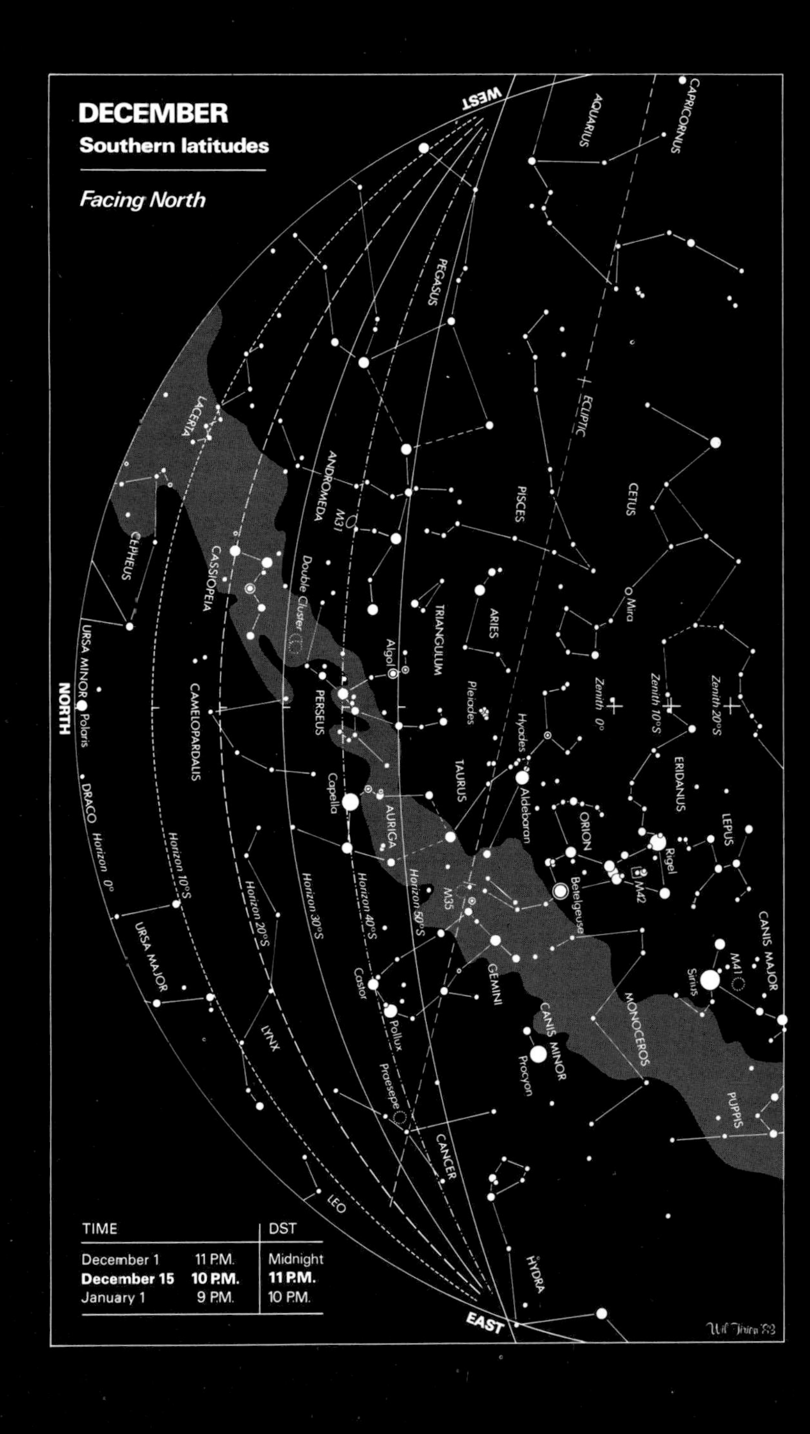

DECEMBER
Southern latitudes
Facing North
WEST
EAST
NORTH
CAPRICORNUS
AQUARIUS
PEGASUS
ECLIPTIC
CETUS
PISCES
ANDROMEDA
M31
LACERTA
CEPHEUS
CASSIOPEIA
Double Cluster
TRIANGULUM
ARIES
Mira
Algol
PERSEUS
Pleiades
Hyades
Zenith 0°
Zenith 10°S
Zenith 20°S
URSA MINOR
Polaris
CAMELOPARDALIS
TAURUS
Aldebaran
ERIDANUS
DRACO
Horizon 0°
Horizon 10°S
Horizon 20°S
Horizon 30°S
Horizon 40°S
Horizon 50°S
Capella
AURIGA
ORION
Rigel
LEPUS
M42
Betelgeuse
M35
URSA MAJOR
GEMINI
CANIS MAJOR
M41
Sirius
Castor
Pollux
CANIS MINOR
Procyon
MONOCEROS
LYNX
Praesepe
PUPPIS
CANCER
LEO
HYDRA
TIME
DST
December 1 11 P.M. Midnight
December 15 10 P.M. 11 P.M.
January 1 9 P.M. 10 P.M.
Wil Tirion '83

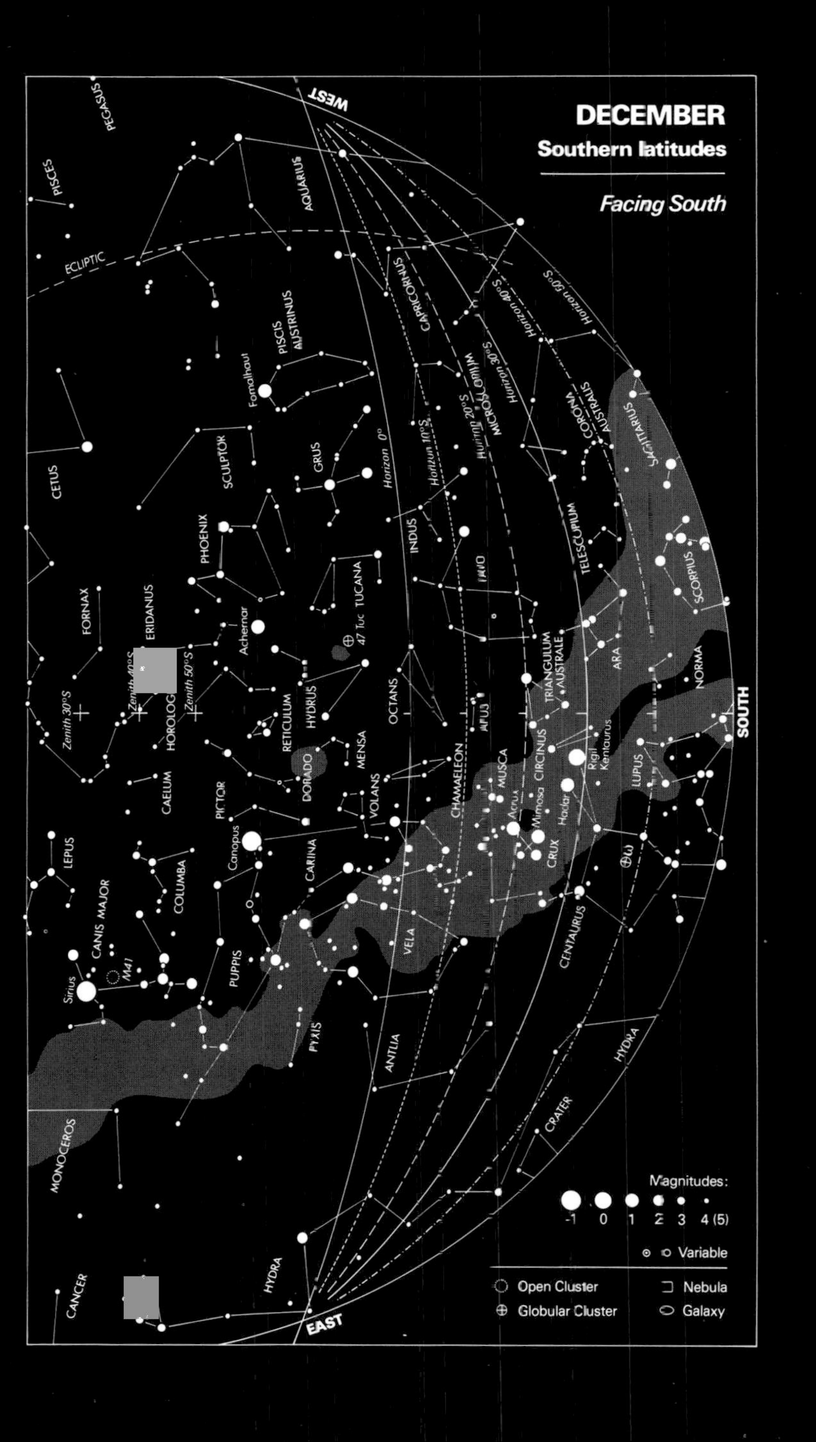
DECEMBER
Southern latitudes
Facing South
Magnitudes:
-1 0 1 2 3 4 (5)
Variable
Open Cluster
Globular Cluster
Nebula
Galaxy
WEST
EAST
SOUTH

ANDROMEDA

Andromeda represents the daughter of Queen Cassiopeia who was chained to a rock as a sacrifice to the sea monster Cetus until saved by Perseus, whom she subsequently married. The constellation originated in ancient times. Despite its fame, Andromeda is not particularly striking: its brightest star is of only 2nd magnitude. Its most prominent feature is a line of four stars extending from the Square of Pegasus; one of these stars marks a corner of the Square, although it is actually part of Andromeda. This star, known both as Sirrah and Alpheratz, marks the head of the chained Andromeda; another star in the line, Mirach, represents her waist and a third, Alamak, is her chained foot. The most celebrated object in the constellation is the Andromeda Galaxy, M31, a spiral galaxy like our own Milky Way; it is the most distant object visible to the naked eye. Two stars leading from Mirach, β (beta) Andromedae, act as a guide to it.

α (alpha) Andromedae, 0h 08m +29°, (Sirrah or Alpheratz), mag. 2.1, is a blue-white star 120 l.y. away.

β (beta) And, 1h 10m +36°, (Mirach), mag. 2.1, is a red giant 75 l.y. away.

γ (gamma) And, 2h 04m +42°, (Alamak or Almach), 330 l.y. away, is an outstanding triple star. Its two brightest components, of mags. 2.3 and 4.8, form one of the finest pairs for small telescopes: their colours are yellow and blue. The fainter, blue star also has a close 6th-mag. blue companion that orbits it every 61 years. They are closest around the year 2012 and are unresolvable by amateur telescopes for several years either side.

δ (delta) And, 0h 39m +31°, mag. 3.3, is an orange giant 175 l.y. away.

μ (mu) And, 0h 57m +38°, mag. 3.9, is a white star 75 l.y. away. ▶

The Andromeda Galaxy, M31, and its two smaller companion galaxies: M32 just below centre, and the larger but fainter NGC 205 to its upper right. Hale Observatories photograph.

1 ANDROMEDA
And · Andromedae

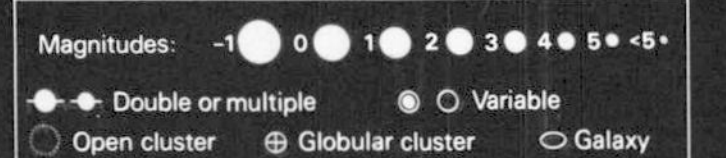

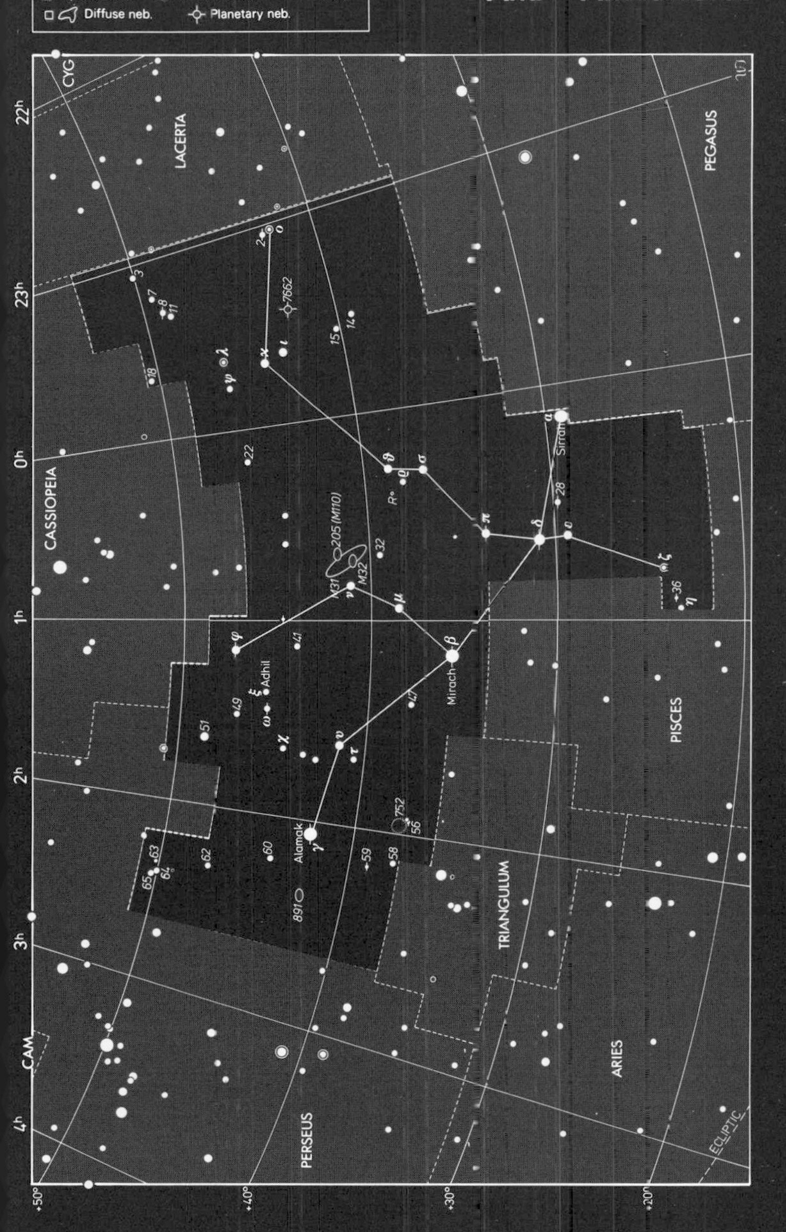

ANTLIA The Air Pump

An obscure southern constellation introduced on a map published in 1756 by the French astronomer Nicolas Louis de Lacaille to commemorate the air pump invented by the French physicist Denis Papin. Lacaille, the first person to map the southern skies completely (which he did from an observatory at the Cape of Good Hope), introduced 14 new constellations to fill gaps between existing ones. Most of these new figures are unremarkable, as is Antlia.

α (alpha) Antliae, 10h 27m –31°, mag. 4.3, the constellation's brightest star, is an orange giant 280 l.y. distant.

δ (delta) Ant, 10h 30m –31°, 420 l.y. away, is a blue-white star of mag. 5.6 with a mag. 9.6 companion visible in small telescopes.

ζ^1 ζ^2 (zeta^1 zeta^2) Ant, 9h 31m –32°, is a wide pair of stars of mags. 5.8 and 5.9 visible in binoculars. Small telescopes show that ζ^1 Ant is itself double, with components of mags. 6.2 and 7.0.

◀ π (pi) And, 0h 37m +34°, 460 l.y. away, is a blue-white star of mag. 4.4 with a mag. 8.6 companion visible in small telescopes.

56 And, 1h 56m +37°, 360 l.y. away, is an orange giant star of mag. 5.7, 360 l.y. away, with an unrelated red giant companion, mag. 5.9 and 590 l.y. away, easily split in binoculars. The two stars are found near the star cluster NGC 752 but are much closer to us.

M31 (NGC 224), 0h 43m +41°, the Andromeda Galaxy, is a spiral galaxy 2.4 million l.y. away. It is visible to the naked eye as an elliptical fuzzy patch, and becomes more prominent in binoculars or a telescope with low magnification (too high a power reduces the contrast and renders the fainter parts of the galaxy less visible). Dark lanes can be seen in the spiral arms surrounding the nucleus. But the full extent of the galaxy becomes apparent only on long-exposure photographs; visual observers see just its brightest, central portion. If the entire Andromeda Galaxy were bright enough to be seen by the naked eye, it would appear five or six times the diameter of the full Moon. M31 is accompanied by two small satellite galaxies, the equivalents of our Magellanic Clouds. The brighter of these, M32 (NGC 221), is visible in small telescopes as a fuzzy, 8th-mag. star-like glow ½° south of M31's core. The second companion, NGC 205 (also known as M110), is larger but more elusive, and over 1° northwest of M31.

NGC 752, 1h 58m +38°, is a widely spread cluster of about 60 stars of 9th mag. and fainter, 1300 l.y. away, best seen in a telescope.

NGC 7662, 23h 26m +43°, is one of the brightest and easiest planetary nebulae to see with a small telescope. At low powers it appears as a fuzzy, 9th-mag. blue-green star, but magnifications of ×100 or so reveal its slightly elliptical disk. Larger apertures show a central hole; the central star is a difficult object for amateur telescopes. NGC 7662 lies about 4000 l.y. away.

2
ANTLIA
Ant · Antliae

Magnitudes: -1 0 1 2 3 4 5 <5

Double or multiple · Variable · Open cluster · Globular cluster · Galaxy · Diffuse neb. · Planetary neb.

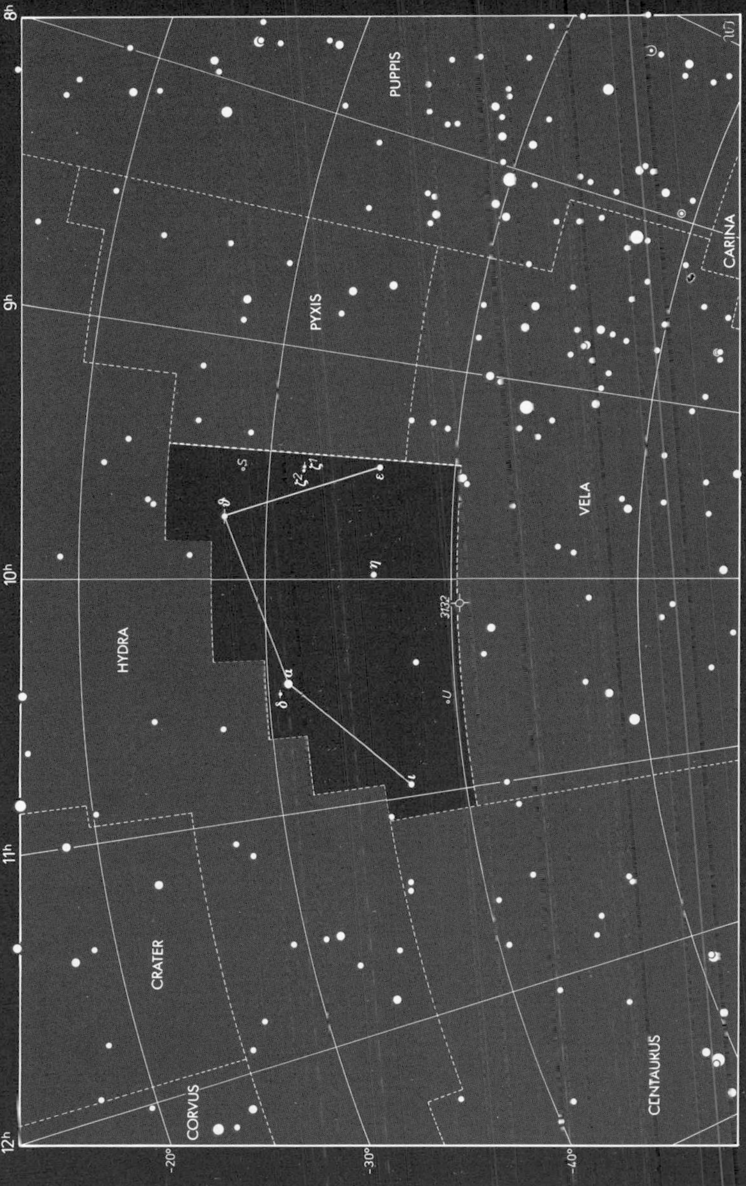

8h 9h 10h 11h 12h

-20° -30° -40°

PUPPIS
PYXIS
CARINA
VELA
HYDRA
CRATER
CORVUS
CENTAURUS

3132

APUS The Bird of Paradise

A faint constellation near the south celestial pole, introduced in the 1590s by the Dutch navigators Pieter Dirkszoon Keyser and Frederick de Houtman during voyages to the southern hemisphere. They introduced a total of 12 constellations in all. This one represents the bird of paradise, native to New Guinea.

α (alpha) Apodis, 14h 48m –79°, mag. 3.8, is an orange giant star 230 l.y. away.

β (beta) Aps, 16h 43m –78°, mag. 4.2, is an orange giant 190 l.y. away.

γ (gamma) Aps, 16h 33m –79°, mag. 3.9, is an orange star 68 l.y. away.

δ^1 δ^2 (delta^1 delta^2) Aps, 16h 20m –79°, is a naked-eye or binocular pair of orange giant stars of mags. 4.7 and 5.3, 390 and 310 l.y. away.

NGC 2997 is a handsome 11th-mag. spiral galaxy in Antlia. Anglo-Australian Telescope Board.

3
APUS
Aps · Apodis

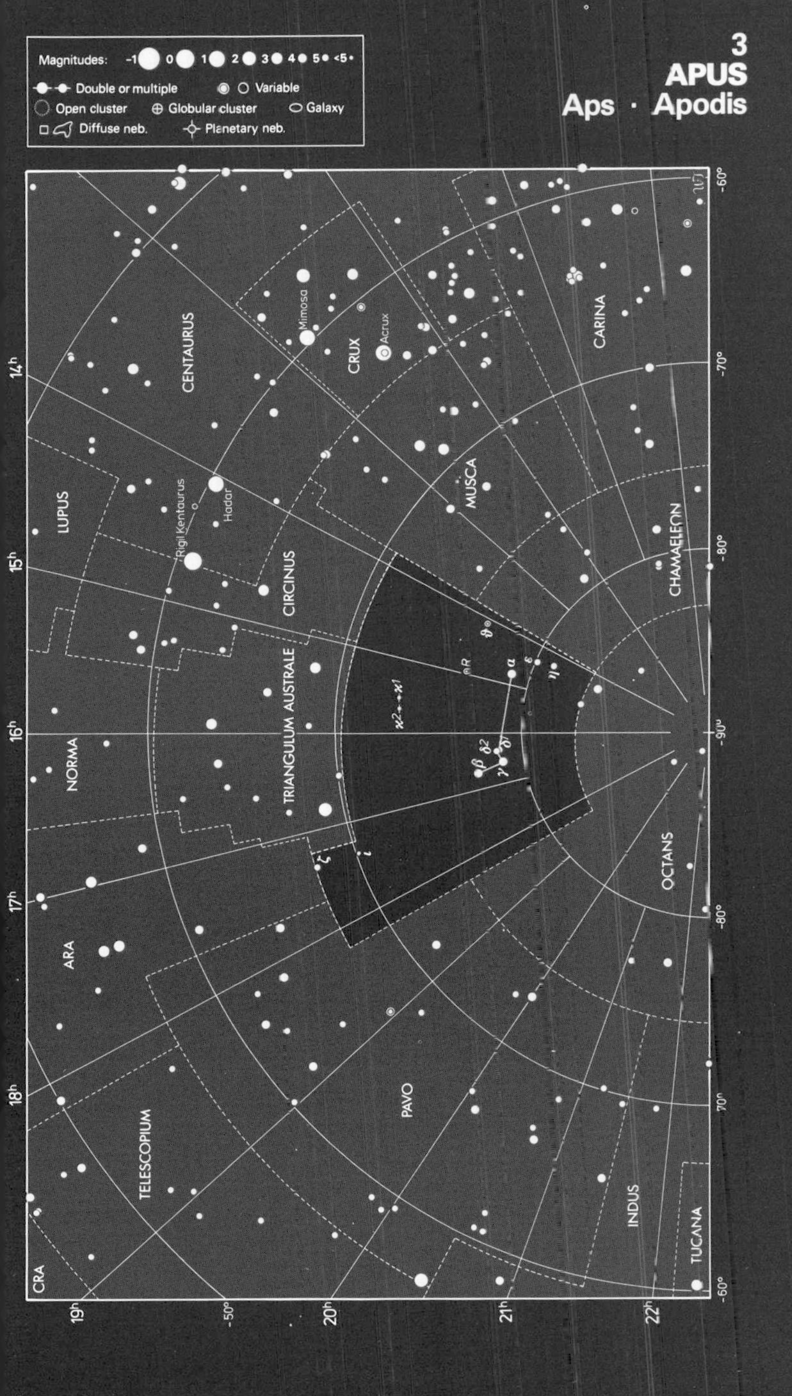

AQUARIUS The Water Carrier

Aquarius is one of the most ancient constellations. The Babylonians saw in this area of sky the figure of a man pouring water from a jar. In Greek mythology the figure represents Ganymede, a shepherd boy carried off by Zeus to Mount Olympus, where he became wine-waiter to the gods. The most prominent part of Aquarius is the Y-shaped figure of four stars representing the Water Jar itself, centred on the star ζ (zeta) Aquarii. Aquarius is in an area of 'watery' constellations that includes Pisces, Cetus and Capricornus. The Sun is in the constellation of Aquarius from late February to early March. Aquarius will one day contain the vernal equinox, the point at which the Sun crosses into the northern celestial hemisphere each year. This astronomically important point, from which the coordinate of right ascension is measured, will move into Aquarius from neighbouring Pisces because of the effect of precession in about 600 years' time. Hence the so-called Age of Aquarius is a long way off yet.

Three main meteor showers radiate from Aquarius each year. The first, the Eta Aquarids, is the richest, reaching a maximum of 35 meteors per hour around May 5–6. The Delta Aquarids, around July 29, produce about 20 meteors per hour. A weaker stream, the Iota Aquarids, produces a maximum of 8 meteors per hour on August 6. Each shower is named after the bright star closest to its radiant.

α (alpha) Aquarii, 22h 06m 0°, (Sadalmelik, from the Arabic for 'the lucky stars of the king'), mag. 2.9, is a yellow supergiant 550 l.y. away.

β (beta) Aqr, 21h 32m –6°, (Sadalsuud, from the Arabic for 'luckiest of the lucky stars'), mag. 2.9, is a yellow supergiant 680 l.y. away.

γ (gamma) Aqr, 22h 22m –1°, (Sadachbia), mag. 3.8, is a white star 180 l.y. away.

δ (delta) Aqr, 22h 55m –16°, (Skat), mag. 3.3, is a white star 68 l.y. away.

ε (epsilon) Aqr, 20h 48m –9°, (Albali), mag. 3.8, is a white star 110 l.y. away.

ζ (zeta) Aqr, 22h 29m 0°, 98 l.y. away, is a celebrated binary consisting of twin white stars of mags. 4.3 and 4.5 orbiting each other every 856 years. At present the two stars are close together as seen from Earth, and 75 mm aperture with high magnification is needed to split them, but they will become progressively easier after the year 2000.

M2 (NGC 7089), 21h 34m –1°, is a mag. 6.5 globular cluster easily visible in binoculars or small telescopes, but requiring 200 mm aperture to be resolved into individual stars. M2 is a rich and highly concentrated globular about 37,000 l.y. away.

M72 (NGC 6981), 20h 54m –13°, is a 9th-mag. globular cluster 56,000 l.y. away, much smaller and less impressive than M2. ▶

4
AQUARIUS
Aqr · Aquarii

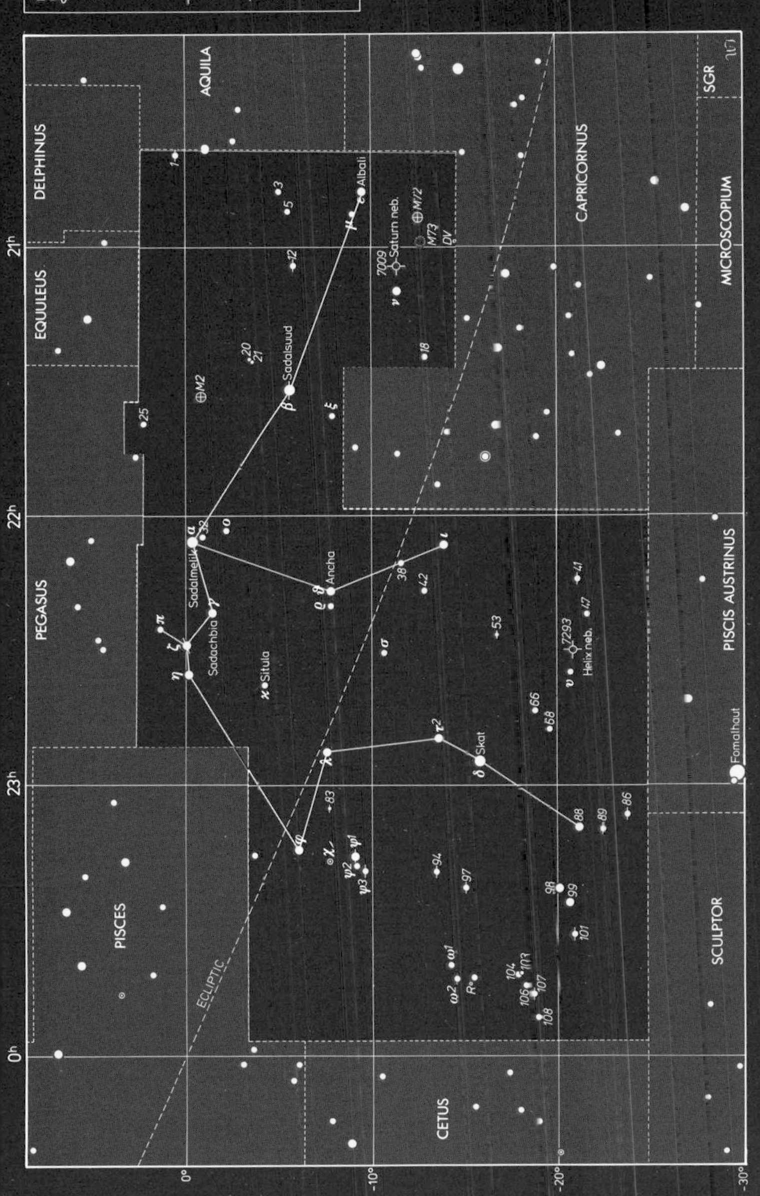

AQUILA The Eagle

A constellation dating from ancient times, representing the bird that in Greek mythology carried the thunderbolts of Zeus. According to myth, Zeus sent an eagle (or turned himself into an eagle) to abduct the shepherd boy Ganymede, represented by neighbouring Aquarius. Aquila's brightest star, Altair, forms one corner of the Summer Triangle that is completed by Deneb in Cygnus and Vega in Lyra. Altair's name comes from the Arabic *al-nasr al-tair*, 'the flying eagle'. Two fainter stars, β (beta) and γ (gamma) Aquilae, stand like sentinels either side of it; they are called Alshain and Tarazed, both from the Persian *shahin-i tarazu*, a translation of an Arabic name meaning 'the balance' which was jointly applied to these two stars and Altair. Aquila lies in the Milky Way and contains rich starfields, particularly towards Scutum in the south. It is an abundant area for novae.

α (alpha) Aquilae, 19h 51m +9°, (Altair, 'the flying eagle'), mag. 0.77, is a white star 17 l.y. away, among the Sun's closest neighbours.

β (beta) Aql, 19h 55m +6°, (Alshain), mag. 3.7, is a yellow star 49 l.y. away.

γ (gamma) Aql, 19h 46m +11°, (Tarazed), mag. 2.7, is a yellow giant star 270 l.y. away.

δ (delta) Aql, 19h 25m +3°, mag. 3.4, is a white star 49 l.y. away.

η (eta) Aql, 19h 52m +1°, 2100 l.y. away, is one of the brightest Cepheid variable stars. Its brightness ranges from mag. 3.5 to 4.4 with a period of 7.2 days.

15 Aql, 19h 05m –4°, 360 l.y. away, is a yellow giant star of mag. 5.4 with a purplish mag. 7.2 companion easily visible in small telescopes.

57 Aql, 19h 55m –8°, 650 l.y. away, is an easy double for small telescopes, consisting of a bluish star of mag. 5.7 with a mag. 6.5 companion.

NGC 6709, 18h 52m +10°, is a loosely scattered cluster of about 40 stars of mags. 9 to 11, about 3000 l.y. away.

◀ NGC 7009, 21h 04m –11°, is a famous planetary nebula, 3000 l.y. away, known as the Saturn Nebula because of its resemblance to that planet when seen in large telescopes. But in most amateur telescopes, of 75 mm aperture or more, it appears as an 8th-mag. blue-green ellipse. It has a central star of mag. 11.5.

NGC 7293, 22h 30m –21°, is the nearest planetary nebula to the Sun, only about 300 l.y. away, and is commonly known as the Helix Nebula. It is the largest planetary nebula in apparent size, covering ¼° of sky, half the apparent size of the Moon. Despite its size the Helix Nebula appears quite faint and is best found with binoculars or very low power on a telescope, when it appears as a circular misty patch, not as impressive as its large size would suggest.

5
AQUILA
Aql · Aquilae

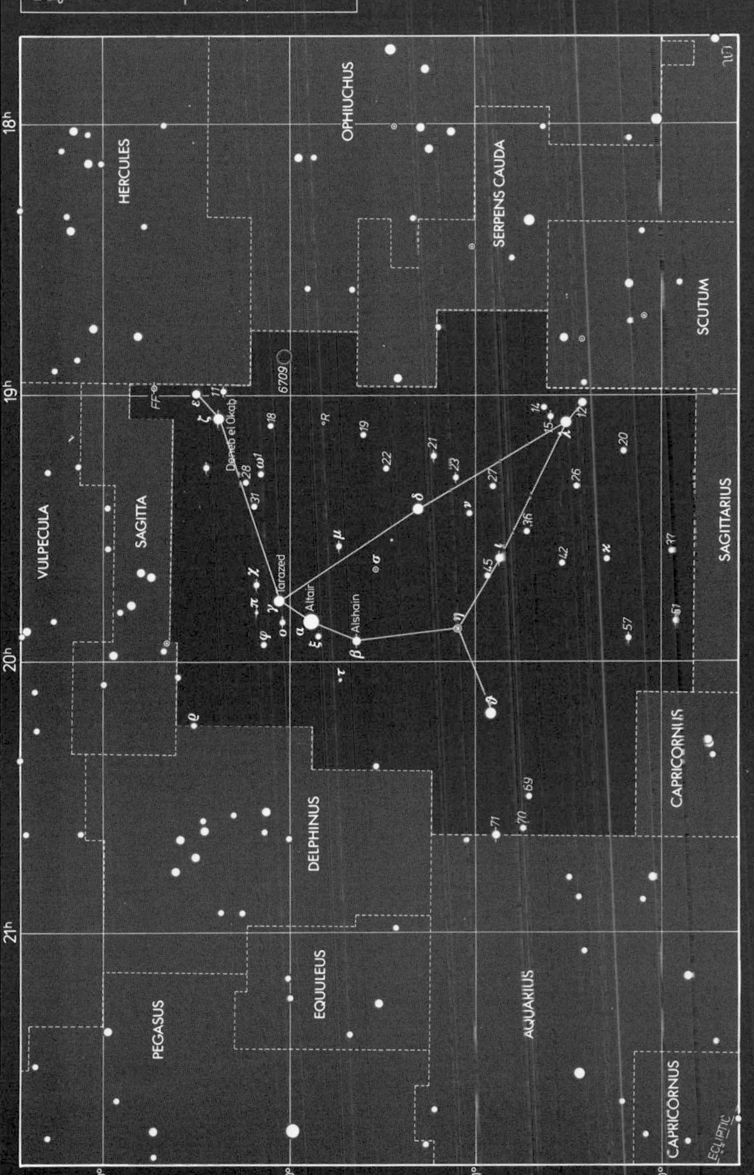

ARA The Altar

This constellation, although faint and relatively little-known, originated with the Greeks, who visualized it as the altar on which the gods of Olympus swore an oath of allegiance before their defeat of the Titans. Ara lies in a rich part of the Milky Way, south of Scorpius.

α (alpha) Arae, 17h 32m –50°, mag. 3.0, is a blue-white star 460 l.y. away.

β (beta) Ara, 17h 25m –56°, mag. 2.9, is an orange supergiant 460 l.y. away.

γ (gamma) Ara, 17h 25m –56°, mag. 3.3, is a blue supergiant 1800 l.y. away.

δ (delta) Ara, 17h 31m –61°, mag. 3.6, is a blue-white star 190 l.y. away.

ζ (zeta) Ara, 16h 59m –56°, mag. 3.1, is an orange giant star 140 l.y. away.

NGC 6193, 16h 41m –49°, is a 5th-mag. cluster of about 30 stars 4400 l.y. away. The brightest member is a blue-white star of mag. 5.7 which small telescopes show has a companion of mag. 6.9. Associated with the cluster is an irregular patch of faint nebulosity, NGC 6188.

NGC 6397, 17h 41m –54°, is a 6th-mag. globular cluster that appears almost as large as the Moon, easily visible with binoculars or a small telescope. It is one of the closest globulars to us, 7200 l.y. away.

The stars of the globular cluster NGC 6397 in Ara appear more widely scattered than do the stars of many other globulars. Anglo-Australian Telescope Board.

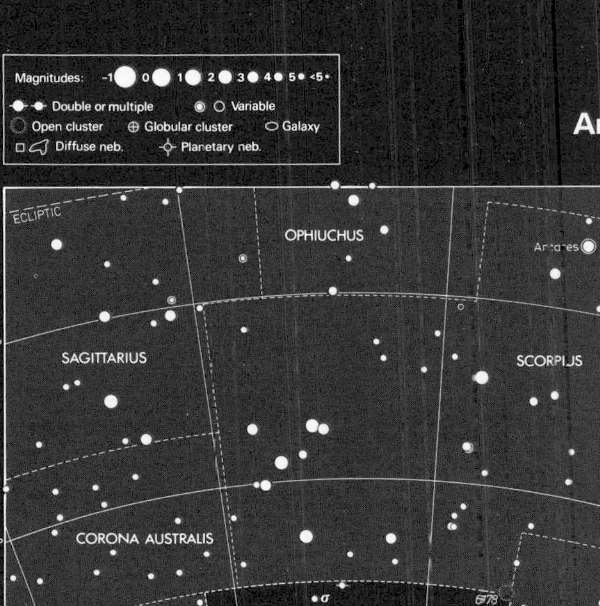

6
ARA
Ara · Arae

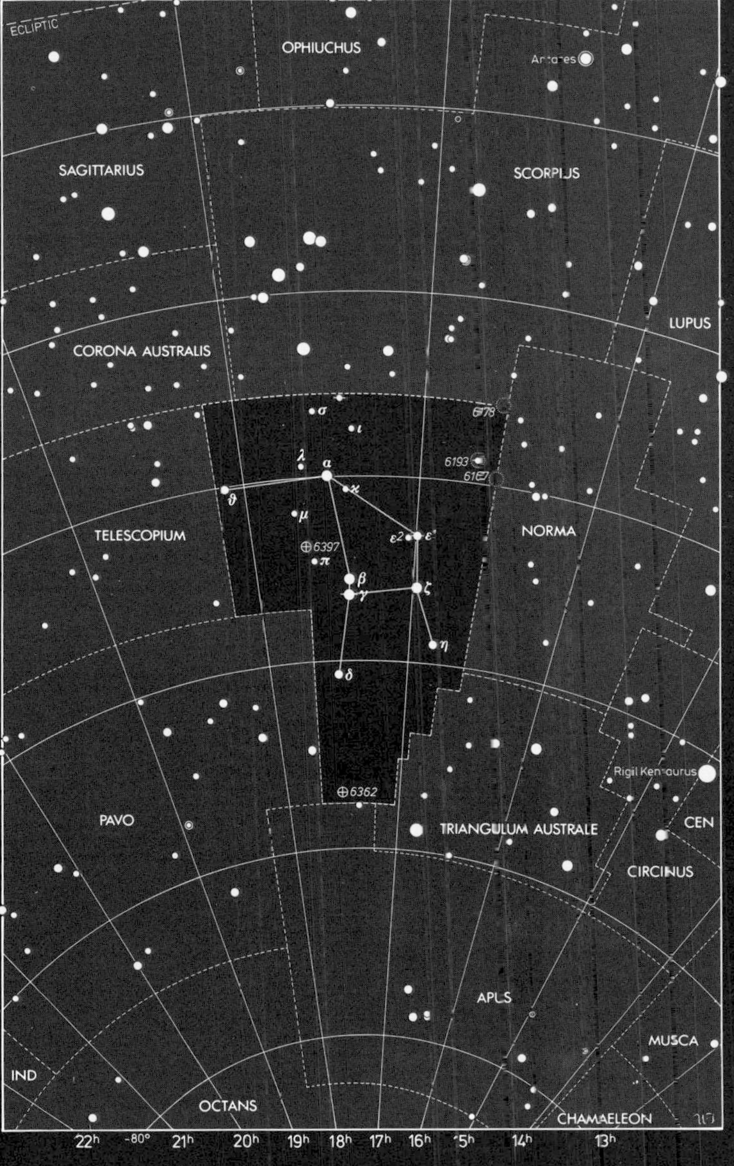

ARIES The Ram

A constellation whose origin dates back to ancient times, lying between Taurus and Andromeda. Aries represents the ram of Greek legend whose golden fleece was sought by Jason and the Argonauts. Despite its faintness, Aries has assumed great importance in astronomy because in Ancient Greek times, over 2000 years ago, it contained the point where the Sun passes from south to north across the celestial equator each year. This point, the vernal equinox, marked the start of northern hemisphere spring, and from it the celestial coordinate known as right ascension is measured; this point is also known as the First Point of Aries. However, it no longer lies in Aries but has moved into neighbouring Pisces, a result of the slight wobble of the Earth in space known as precession (see diagram below).

α (alpha) Arietis, 2h 07m +23°, (Hamal, from the Arabic for 'lamb'), mag. 2.0, is a yellow giant star 75 l.y. away.

β (beta) Ari, 1h 55m +21°, (Sheratan, from the Arabic meaning 'two'), mag. 2.6, is a white star 52 l.y. away.

γ (gamma) Ari, 1h 54m +19°, (Mesarthim), is a striking double consisting of twin white stars of mags. 4.7 and 4.6, clearly visible through small telescopes even under low magnification. The stars lie about 120 l.y. away.

ε (epsilon) Ari, 2h 59m +21°, is a challenging double star for apertures of 100 mm or over. High magnification reveals a tight pair of white stars of mags. 5.2 and 5.5. The distance is uncertain.

λ (lambda) Ari, 1h 58m +24°, 105 l.y. away, is a white star of mag. 4.8 with a yellow mag. 7.3 companion visible in small telescopes or even good binoculars.

π (pi) Ari, 2h 49m +17°, 550 l.y. away, is a blue-white star of mag. 5.2, with a close mag. 8.7 companion, difficult to distinguish in the smallest telescopes.

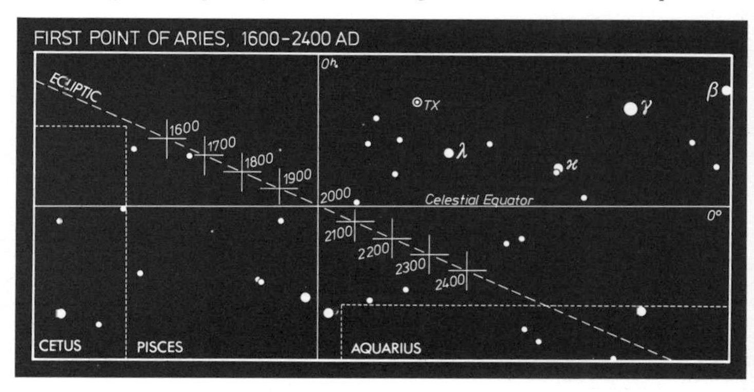

Movement of the so-called First Point of Aries over 800 years. The First Point of Aries currently lies in Pisces and is approaching Aquarius. Wil Tirion.

7
ARIES
Ari · Arietis

Magnitudes: −1 0 1 2 3 4 5 <5

Double or multiple · Variable · Open cluster · Globular cluster · Galaxy · Diffuse neb. · Planetary neb.

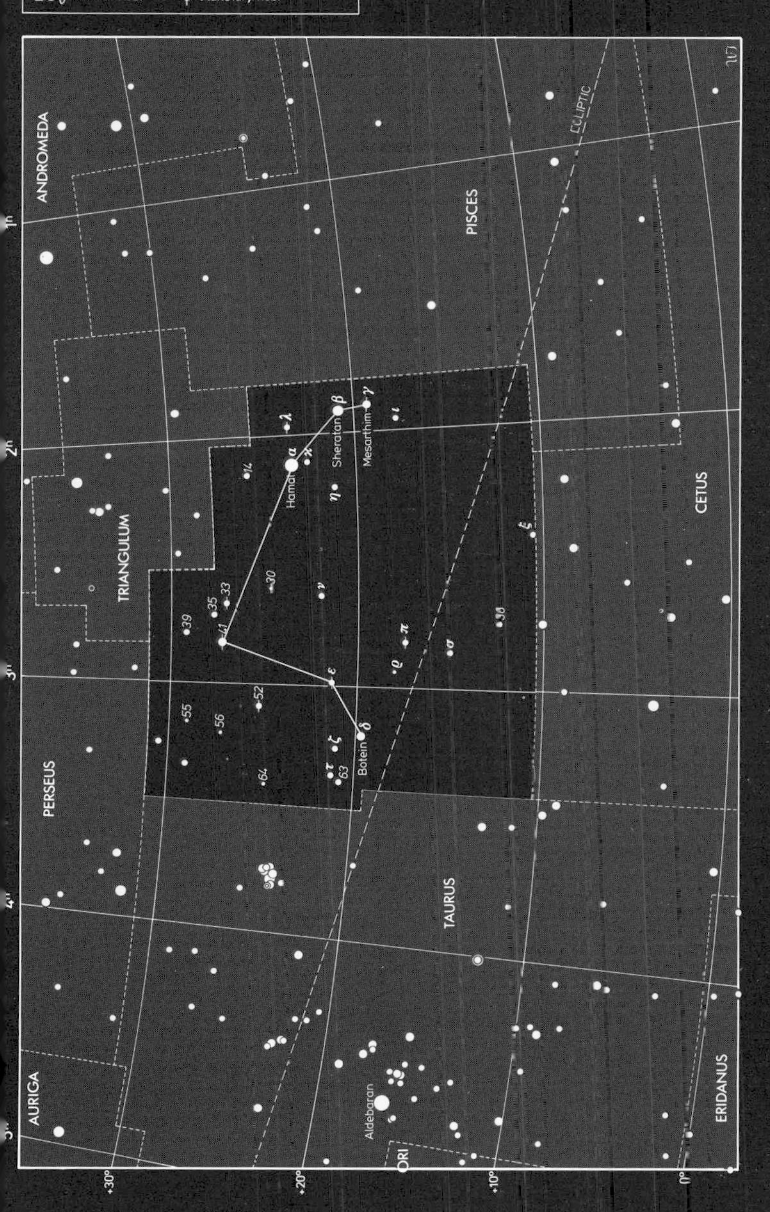

AURIGA The Charioteer

A large and prominent constellation, usually identified by the Ancient Greeks as Erichthonius, a legendary king of Athens and a skilled charioteer. Auriga's leading star is Capella, sixth-brightest in the whole sky. In legend this star represented the she-goat Amaltheia, which suckled the infant Zeus; the stars ζ (zeta) and η (eta) Aurigae are supposedly her kids. The star γ (gamma) Aurigae, marking the charioteer's foot, was once regarded as being shared with Taurus the Bull, but it is now assigned exclusively to Taurus as β (beta) Tauri.

α (alpha) Aurigae, 5h 17m +46°, (Capella, 'she-goat'), mag. 0.08, lies 46 l.y. away. It is a spectroscopic binary, consisting of two yellow giant stars orbiting every 104 days, although they do not eclipse each other.

β (beta) Aur, 6h 00m +45°, (Menkalinan, 'shoulder of the charioteer'), 46 l.y. away, is an eclipsing variable of approximate mag. 1.9 consisting of two white stars that eclipse every 3.96 days, causing the star to vary by about 0.1 mag.

ε (epsilon) Aur, 5h 02m +44°, a white supergiant about 2000 l.y. away, is an eclipsing binary with an exceptionally long period. Normally it shines at mag. 3.0, but every 27 years it sinks to mag. 3.8 as it is eclipsed by a dark companion, remaining at minimum for a year. One theory is that the companion of ε Aurigae is a binary star surrounded by a disk of material. Its next eclipse is due to start in 2009.

ζ (zeta) Aur, 5h 02m +41°, 360 l.y. away, is a famous eclipsing binary of contrasting stars: an orange giant orbited every 972 days by a smaller blue companion. During eclipses, ζ Aurigae's brightness drops from mag. 3.7 to 4.0.

ϑ (theta) Aur, 6h 00m +37°, about 150 l.y. away, is a blue-white star of mag. 2.6. It has a white companion of mag. 7.1 which, because of its closeness and relative faintness, needs at least 100 mm aperture and high magnification to distinguish. This is a tough double for steady nights.

4 Aur, 4h 59m +38°, 175 l.y. away, is a double star of mags. 5.0 and 8.0, visible in small telescopes.

14 Aur, 5h 15m +33°, 190 l.y. away, is a double star of mags. 5.1 and 7.4, split in small telescopes.

UU Aur, 6h 37m +38°, is a deep red semi-regular variable star over 3000 l.y. away. It varies between 5th and 7th mags. with a rough period of 234 days.

M36 (NGC 1960), 5h 36m +34°, is a small, bright cluster of about 60 stars, visible in binoculars and resolvable into stars in small telescopes. M36 lies 4100 l.y. away.

M37 (NGC 2099), 5h 52m +33°, is the largest and richest of the clusters in Auriga, containing about 150 stars. In binoculars the cluster appears as a hazy, unresolved patch, but a 100-mm telescope resolves it into a sparkling field of faint stardust, with a brighter orange star at the centre. It lies 4400 l.y. away. ▶

8
AURIGA
Aur · Aurigae

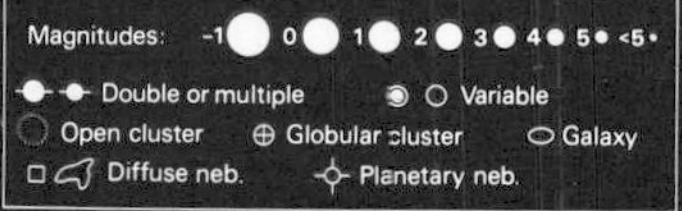

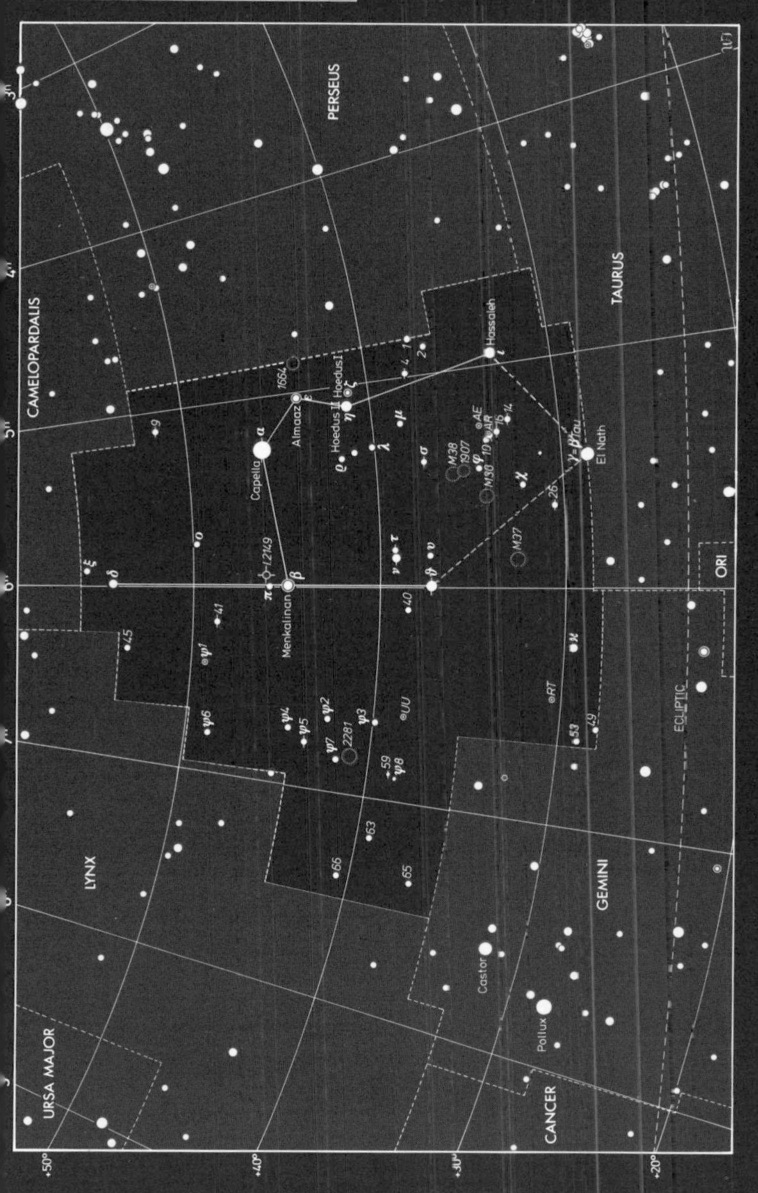

BOÖTES The Herdsman

An ancient constellation, representing a herdsman driving a bear (Ursa Major) around the sky; the herdsman is often depicted holding the leash of the hunting dogs, Canes Venatici. The name of the constellation's brightest star, Arcturus, actually means 'bearkeeper' in Greek. In Greek mythology Boötes represents Arcas, the son of Zeus and the nymph Callisto (neighbouring Ursa Major represents Callisto herself, who was turned into a bear by Zeus's jealous wife Hera). Arcturus is the brightest star in the northern hemisphere of the sky, and is easily identified: the curving handle of the Big Dipper or Plough acts as a pointer to it. Arcturus forms the base of a large 'Y' shape with ε (epsilon) Boötis, γ (gamma) Boötis and α (alpha) Coronae Borealis. The rest of the constellation is much fainter than Arcturus, but contains many double stars of interest. The year's most abundant meteor shower, the Quadrantids, radiates from the northern part of Boötes, an area of sky that was once occupied by the now-abandoned constellation of Quadrans Muralis, the Mural Quadrant (hence the shower's name). The Quadrantid meteors reach a peak of about 100 meteors per hour on January 3–4 each year, although they are not as bright as other rich showers such as the Perseids.

α (alpha) Boötis, 14h 16m +19°, (Arcturus), mag. –0.04, is the fourth-brightest star in the entire sky. It is an orange giant, about 27 times the diameter of the Sun, lying 36 l.y. away; its ruddy colour is noticeable to the naked eye, and is more striking with optical aid. Arcturus has a mass similar to the Sun's, and it is believed that our Sun will swell up to become a red giant like Arcturus in 5000 million years.

β (beta) Boo, 15h 02m +40°, (Nekkar, corrupted from the Arabic for 'ox-driver', referring to the whole constellation), mag. 3.5, is a yellow giant 140 l.y. away.

γ (gamma) Boo, 14h 32m +38°, (Haris or Seginus), mag. 3.0, is a giant white star about 100 l.y. away.

δ (delta) Boo, 15h 16m +33°, mag. 3.5, is a yellow giant 160 l.y. away. It has a wide binocular companion of mag. 7.8. ▶

◀ M38 (NGC 1912), 5h 29m +36°, is a large, scattered cluster of about 100 faint stars, visible in binoculars, with a noticeable cross-shape when seen through a telescope. Its distance is 4300 l.y. Next to it lies the small fuzzy blob of NGC 1907, a much smaller and fainter cluster 4500 l.y. away.

NGC 2281, 6h 49m +41°, is a binocular cluster of about 30 stars, 1600 l.y. away. Through a telescope the stars appear to be arranged in a crescent, four brighter stars forming a diamond shape.

9
BOÖTES
Boo · Boötis

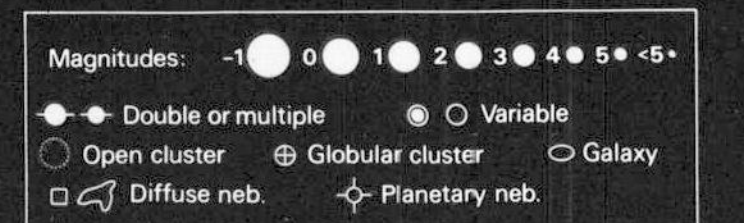

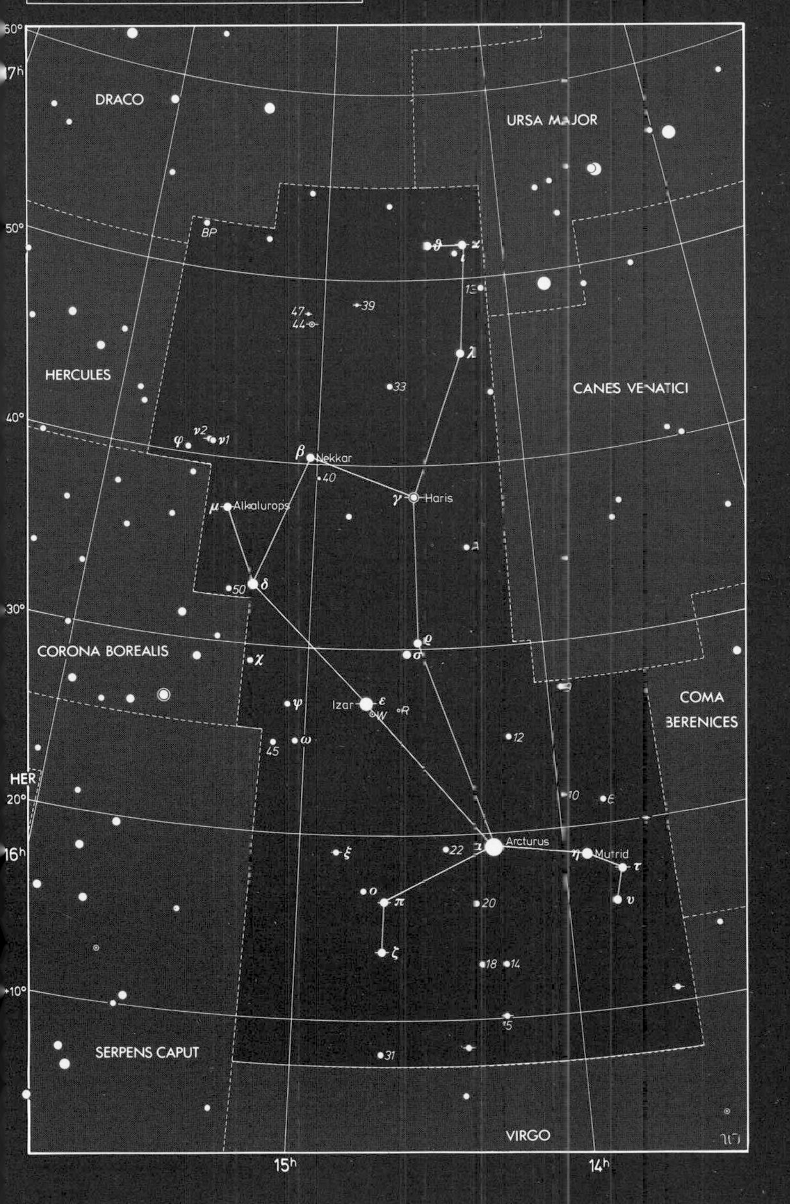

CAELUM The Chisel

An obscure, almost irrelevant constellation at the foot of Eridanus, representing an engraving tool. It was introduced in the 1750s by Nicolas Louis de Lacaille during his mapping of the southern sky.

α (alpha) Caeli, 4h 41m –42°, mag. 4.5, is a white star 62 l.y. away.

β (beta) Cae, 4h 42m –37°, mag. 5.1, is a white star 65 l.y. away.

γ (gamma) Cae, 5h 04m –35°, mag. 4.6, is an orange star 280 l.y. away. It has a close mag. 8.1 companion, difficult to see in the smallest apertures because of the brightness contrast.

δ (delta) Cae, 4h 31m –45°, mag. 5.1, is a blue-white star 1150 l.y. away.

◀ ε (epsilon) Boo, 14h 45m +27°, (Izar, 'girdle' or 'loincloth'), 150 l.y. away, is a celebrated double star: an orange giant primary of mag. 2.7 with a blue companion of mag. 5.1. This close double of contrasting colours requires a telescope of at least 75 mm at ×100 power or more, because the bright primary tends to overwhelm its fainter companion; but its appearance when split has led to the alternative name Pulcherrima, meaning 'most beautiful'.

ι (iota) Boo, 14h 16m +51°, 75 l.y. away, is a wide double star of mags. 4.8 and 7.5.

κ (kappa) Boo, 14h 13m +52°, 160 l.y. away, is an easy double star for small telescopes, consisting of components of mags. 4.5 and 6.7.

μ (mu) Boo, 15h 24m +37°, (Alkalurops, 'club' or 'staff'), 80 l.y. distant, is an attractive triple star. To the naked eye it appears as a blue-white star of mag. 4.3, but binoculars reveal a wide companion of mag. 6.5. Telescopes of 75 mm aperture with high magnification show that this companion actually consists of two close stars of mags. 7.0 and 7.6; they orbit each other every 260 years.

ν (nu) Boo, 15h 31m +41°, is a binocular duo consisting of a white star, mag. 5.0, 125 l.y. distant, and an unrelated orange giant, also of mag. 5.0, 330 l.y. away.

π (pi) Boo, 14h 41m +16°, 165 l.y. away, is a double star with blue-white components of mags. 4.9 and 5.8, visible in small telescopes.

ξ (xi) Boo, 14h 51m +19°, 23 l.y. away, is a showpiece double for small telescopes, consisting of yellow and orange stars of mags. 4.7 and 7.0, orbiting each other every 150 years.

44 Boo, 15h 04m +48°, is a complex double–variable star 39 l.y. away. To the naked eye it appears as a yellow star of mag. 4.8. In fact, it is a binary of twin yellow stars of mags. 5.3 and 6.2 orbiting every 225 years. Until the year 2020 they will be split in apertures of 75 mm but thereafter they will become more difficult as they start to close, reaching their closest in 2033 when they will be indivisible in amateur telescopes. The fainter star is itself an eclipsing binary with a period of 6.4 hours and a range of 0.6 mag.

10 CAELUM
Cae · Caeli

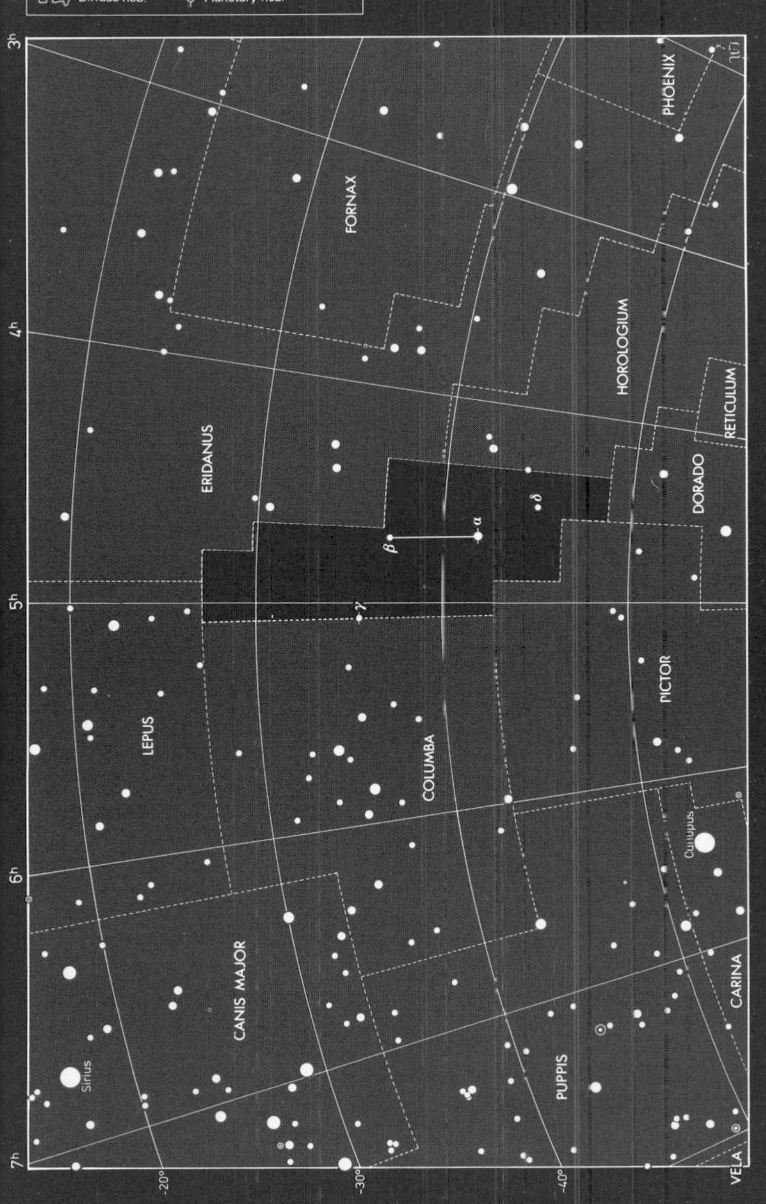

CAMELOPARDALIS The Giraffe

A faint and obscure constellation, sometimes written as Camelopardus, an obsolete variant of the name. It was invented in 1613 by the Dutch theologian and astronomer Petrus Plancius and supposedly represents the animal on which Rebecca rode into Canaan to marry Isaac.

α (alpha) Camelopardalis, 4h 54m +66°, mag. 4.3, is a blue supergiant star 3000 l.y. away.

β (beta) Cam, 5h 03m +60°, at mag. 4.0 the brightest star in the constellation, is a yellow supergiant 1000 l.y. distant. It has a wide mag. 8.6 companion star visible in small telescopes or even good binoculars.

Σ 1694 (Struve 1694), 12h 49m +83°, 460 l.y. away, is a pair of white stars, mags. 5.3 and 5.8, easily split in small telescopes.

NGC 1502, 4h 08m +62°, is a small 6th-mag. star cluster with about 45 members visible in binoculars or a small telescope, somewhat triangular in shape and with two easy double stars at its centre. It lies 3100 l.y. away.

NGC 2403, 7h 37m +66°, is an 8th-mag. spiral galaxy ¼° long, easily seen as an elliptical glow in a 100-mm telescope on a good night. It lies about 12 million l.y. away.

NGC 2523 is a curiously shaped barred spiral galaxy in Camelopardalis. At 13th mag. it is beyond the range of most amateur telescopes. Hale Observatories photograph.

11
CAMELOPARDALIS
Cam · Camelopardalis

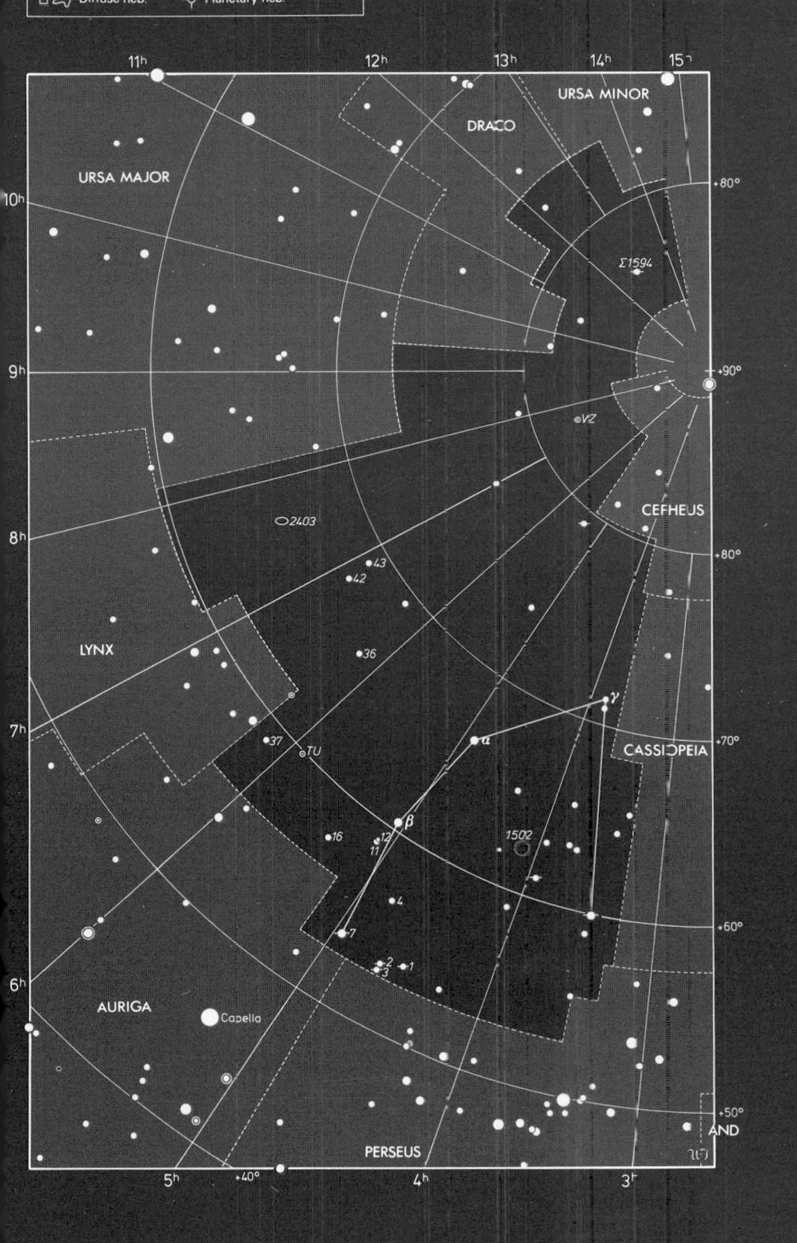

CANCER The Crab

Cancer represents the crab that attacked Hercules when he was fighting the multi-headed Hydra; the luckless crab was crushed underfoot by mighty Hercules but was subsequently elevated to the heavens. In ancient times, the Sun reached its most northerly point in the sky each year while it was in Cancer. The date on which the Sun is farthest north of the Earth's equator, June 21, is known as the summer solstice; on this day the Sun appears overhead at noon at latitude 23½° north on Earth. This latitude came to be known as the Tropic of Cancer, a name it retains today even though, because of the effect of precession, the Sun now lies in the neighbouring constellation of Gemini on the summer solstice. Cancer is the faintest of the 12 constellations of the zodiac through which the Sun passes each year, but it nevertheless contains much of interest, notably the star cluster Praesepe, the Manger. Praesepe is flanked by two stars, Asellus Borealis and Asellus Australis, the northern and southern donkeys, visualized as feeding at the stellar manger.

α (alpha) Cancri, 8h 58m +12°, (Acubens, 'the claw'), mag. 4.2, is a white star about 130 l.y. away. It has a 12th-mag. companion visible with telescopes of 75 mm aperture and over.

β (beta) Cnc, 8h 17m +9°, mag. 3.5, an orange giant 160 l.y. away, is the brightest star in the constellation.

γ (gamma) Cnc, 8h 43m +21°, (Asellus Borealis, 'northern donkey'), mag. 4.7, is a white star 280 l.y. away.

δ (delta) Cnc, 8h 45m +18°, (Asellus Australis, 'southern donkey'), mag. 3.9, is a yellow giant star 160 l.y. away.

ζ (zeta) Cnc, 8h 12m +18°, 82 l.y. away, is an interesting multiple star. A small telescope reveals two yellow stars of mags. 5.1 and 6.2; they form a genuine binary with an estimated orbital period of over 1000 years. Larger telescopes split the brighter component into a tight binary of mags. 5.6 and 6.0, orbital period 60 years. The stars are currently moving apart; until the year 2000 an aperture of 200 mm will be needed to split them, but 150 mm should do it by 2005 and at their widest, around 2020, 100 mm may be sufficient.

ι (iota) Cnc, 8h 47m +29°, 160 l.y. away, is a yellow giant of mag. 4.0 with a blue-white mag. 6.6 companion just visible in binoculars, or easily seen through a small telescope.

M44 (NGC 2632), 8h 40m +20°, Praesepe ('manger'), commonly called the Beehive Cluster, is a swarm of about 50 stars of 6th mag. and fainter, visible as a misty patch to the naked eye and best seen through binoculars. The brightest member, ε (epsilon) Cancri, is mag. 6.3. Praesepe covers 1½° of sky, three times the apparent diameter of the Moon. It lies 520 l.y. away.

M67 (NGC 2682), 8h 50m +12°, is a smaller and denser cluster of stars than M44, visible as a Moon-sized misty ellipse in binoculars or small telescopes and needing apertures of at least 75 mm to resolve the brightest of its 200 or so individual stars of 10th mag. and fainter. It lies 2600 l.y. away.

12
CANCER
Cnc · Cancri

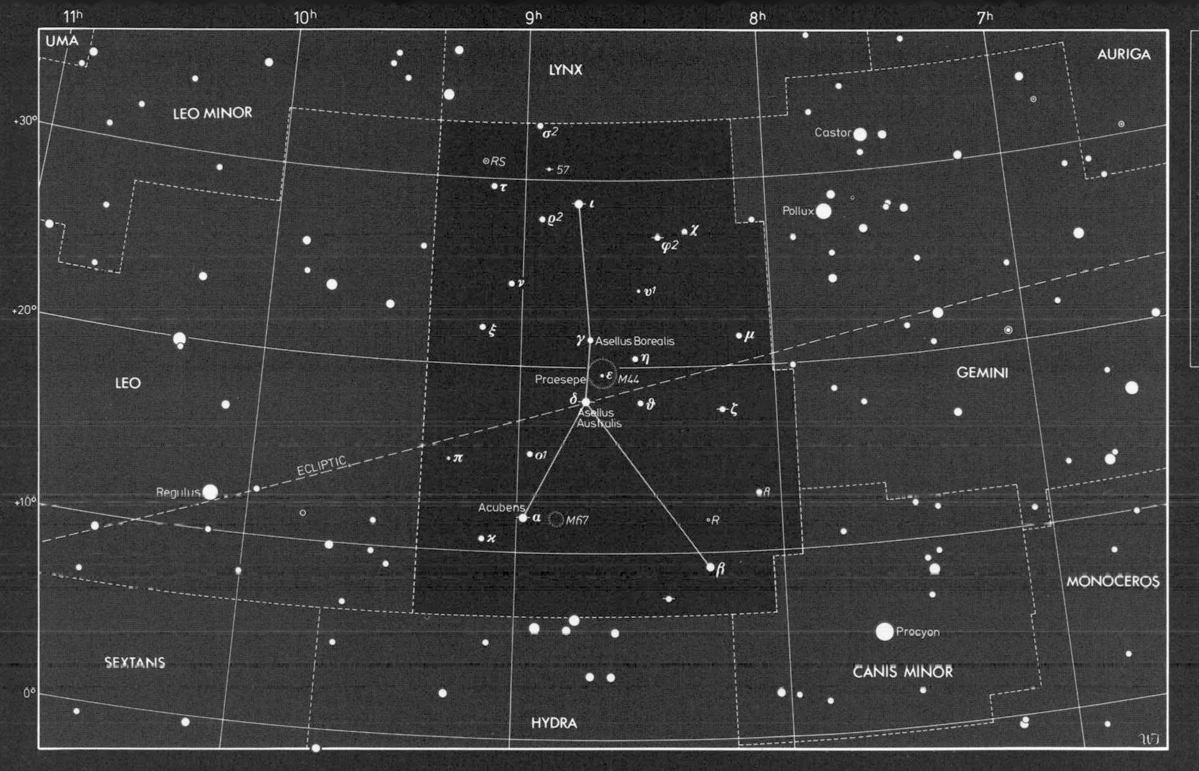

CANES VENATICI The Hunting Dogs

A constellation introduced in 1687 by the Polish astronomer Johannes Hevelius, consisting of a sprinkling of faint stars below Ursa Major. It represents two dogs, Asterion ('little star') and Chara ('joy'), held on a leash by neighbouring Boötes as they pursue the Great Bear around the pole. Canes Venatici contains numerous galaxies, the most famous being M51, the Whirlpool, a beautiful face-on spiral (pictured on page 282). It was the first galaxy in which spiral form was detected, by Lord Rosse in 1845, with his 72-inch (1.8-m) reflector at Birr Castle, Ireland.

α (alpha) Canum Venaticorum, 12h 56m +38°, is popularly called Cor Caroli, meaning 'Charles's heart', a reference to the executed King Charles I of England; it is reputed, doubtless apocryphally, to have shone particularly brightly in 1660 on the arrival of Charles II in England at the Restoration of the monarchy. It is a double star of mags. 2.9 and 5.6, easily split in small telescopes. Both stars are white, but various observers have reported subtle shades of colour when seen through a telescope. The brighter star is the standard example of a rare class of stars with strong and variable magnetic fields; its brightness fluctuates slightly but not enough to be noticeable to the eye. Cor Caroli is 91 l.y. away.

β (beta) CVn, 12h 34m +41°, (Chara), mag. 4.3, is the only other star of any prominence in the constellation. It is a yellow star, 30 l.y. away.

Y CVn, 12h 45m +45°, is a deep-red semi-regular variable star sometimes known as La Superba. It ranges between mags. 5 and 6.5 in approximately 160 days.

M3 (NGC 5272), 13h 42m +28°, is a rich globular cluster located midway between Cor Caroli and Arcturus, regarded as one of the finest globular clusters in the northern sky. It is on the naked-eye limit at 6th mag., but can be picked up easily as a hazy star in binoculars or a small telescope; a 5th-mag. star nearby acts as a guide. In small telescopes M3 appears as a condensed ball of light with a faint outer halo. Apertures of 100 mm or more are needed to resolve individual stars in its outer regions. M3 is 32,000 l.y. away.

M51 (NGC 5194), 13h 30m +47°, the Whirlpool Galaxy, is an 8th-mag. spiral galaxy about 15 million l.y. away with a smaller satellite galaxy, NGC 5195, apparently lying at the end of one of its arms; in fact, this companion lies slightly behind M51, having brushed past it within the last 100 million years or so. The Whirlpool can be seen in binoculars, appearing elongated. It is disappointing in small telescopes, which show a faint milky radiance around the starlike nuclei of the galaxy and its satellite; apertures of at least 250 mm are needed to see the arms. Nevertheless, M51 is well worth hunting for on clear, dark nights. (For a photograph see page 282.)

M94 (NGC 4736), 12h 51m +41°, is a compact spiral galaxy presented nearly face-on. In amateur telescopes it looks like an 8th-mag. comet, with a fuzzy star-like nucleus surrounded by an elliptical halo. M94 is about 15 million l.y. away.

13
CANES VENATICI
CVn · Canum Venaticorum

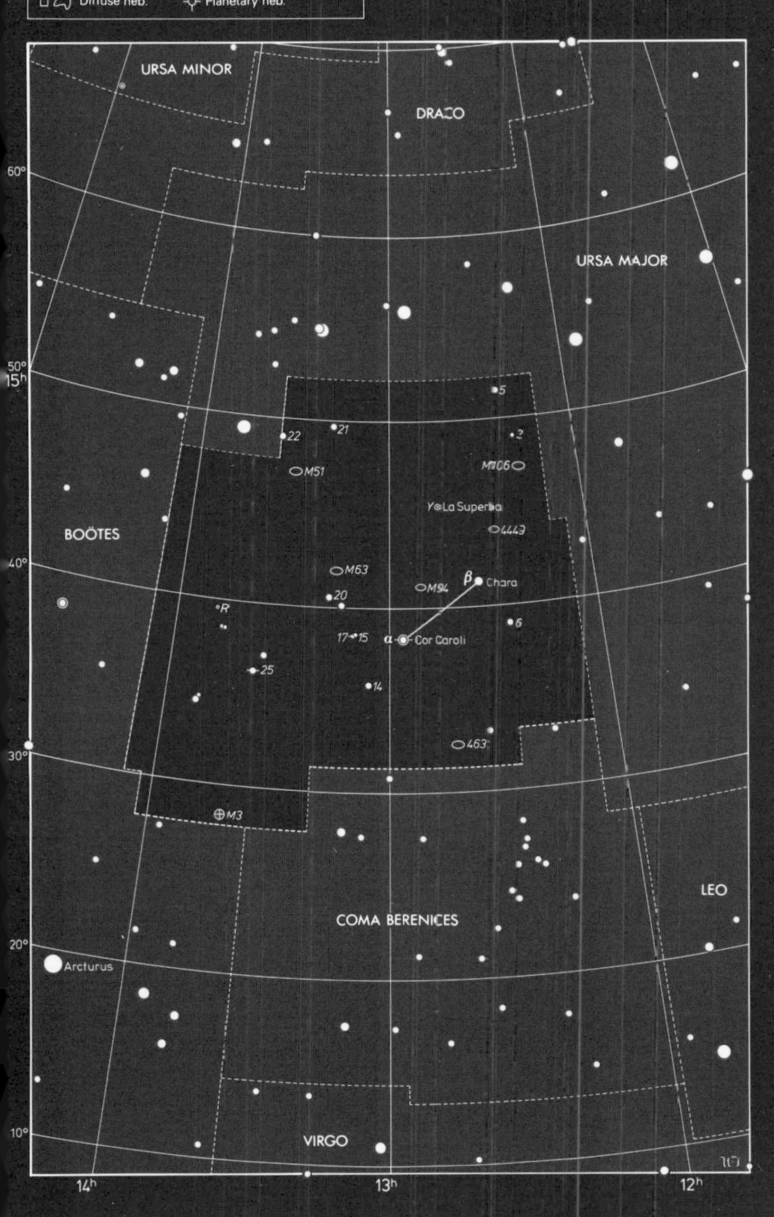

CANIS MAJOR The Greater Dog

An ancient constellation, representing one of the two dogs (the other being Canis Minor) following at the heels of Orion. Canis Major contains many brilliant stars, making it one of the most prominent constellations; its leading star, Sirius, is the brightest in the entire sky. Sirius features in many legends. The Ancient Egyptians based their calendar on its yearly motion around the sky.

α (alpha) Canis Majoris, 6h 45m –17°, (Sirius, from the Greek meaning 'searing' or 'scorching'), mag. –1.47, is a brilliant white star 8.7 l.y. away, one of the Sun's closest neighbours. It has a white dwarf companion of mag. 8.5 that orbits it every 50 years. The brilliance of Sirius overpowers this white dwarf so that, even when the two stars are at their greatest separation, as between the years 2020 and 2025, telescopes of 200 mm aperture or more and steady atmospheric conditions are required to see it. During the 1990s, when the two stars are at their closest, the companion will be impossible to see in any amateur telescope.

β (beta) CMa, 6h 23m –18°, (Mirzam, 'the announcer'), mag. 2.0, is a blue giant 850 l.y. away. It is a pulsating star whose variations, of a few hundredths of a magnitude every 6 hours, are undetectable to the naked eye.

δ (delta) CMa, 7h 08m –26°, (Wezen, 'the weight'), mag. 1.8, is a yellow supergiant star 2300 l.y. away.

ε (epsilon) CMa, 6h 59m –29°, (Adhara, 'the virgins'), mag. 1.5, is a blue giant 490 l.y. away. It has a mag. 7.4 companion which is difficult to see in small telescopes because of the glare from the primary.

η (eta) CMa, 7h 24m –29°, (Aludra), mag. 2.4, is a blue supergiant 2300 l.y. away.

μ (mu) CMa, 6h 56m –14°, mag. 5.0, is a yellow giant with a close 9th-mag. blue-white companion difficult to pick up in the smallest apertures because of the magnitude contrast.

ν^1 (nu^1) CMa, 6h 36m –19°, mag. 5.7, is a yellow star 110 l.y. away with an unrelated mag. 7.7 companion, estimated to be three times more distant, visible in small telescopes.

M41 (NGC 2287), 6h 47m –21°, is a large and bright open cluster of about 80 stars, easily visible through binoculars or a small telescope and, with a total mag. of 4.5, detectable by the naked eye under good conditions – it was known to the Ancient Greeks. A low-power view in a small telescope shows the individual stars grouped in bunches and curves, covering an area of sky equivalent to the apparent diameter of the Moon. The brightest star in the cluster is a 7th-mag. orange giant. M41 is 2300 l.y. away.

NGC 2362, 7h 19m –25°, is a compact cluster surrounding the mag. 4.4 blue supergiant star τ (tau) Canis Majoris. Small telescopes show about 60 stars in the cluster. τ CMa is a member of the cluster, which lies about 5000 l.y. away.

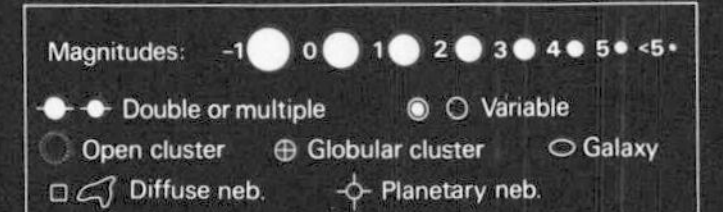

14
CANIS MAJOR
CMa · Canis Majoris

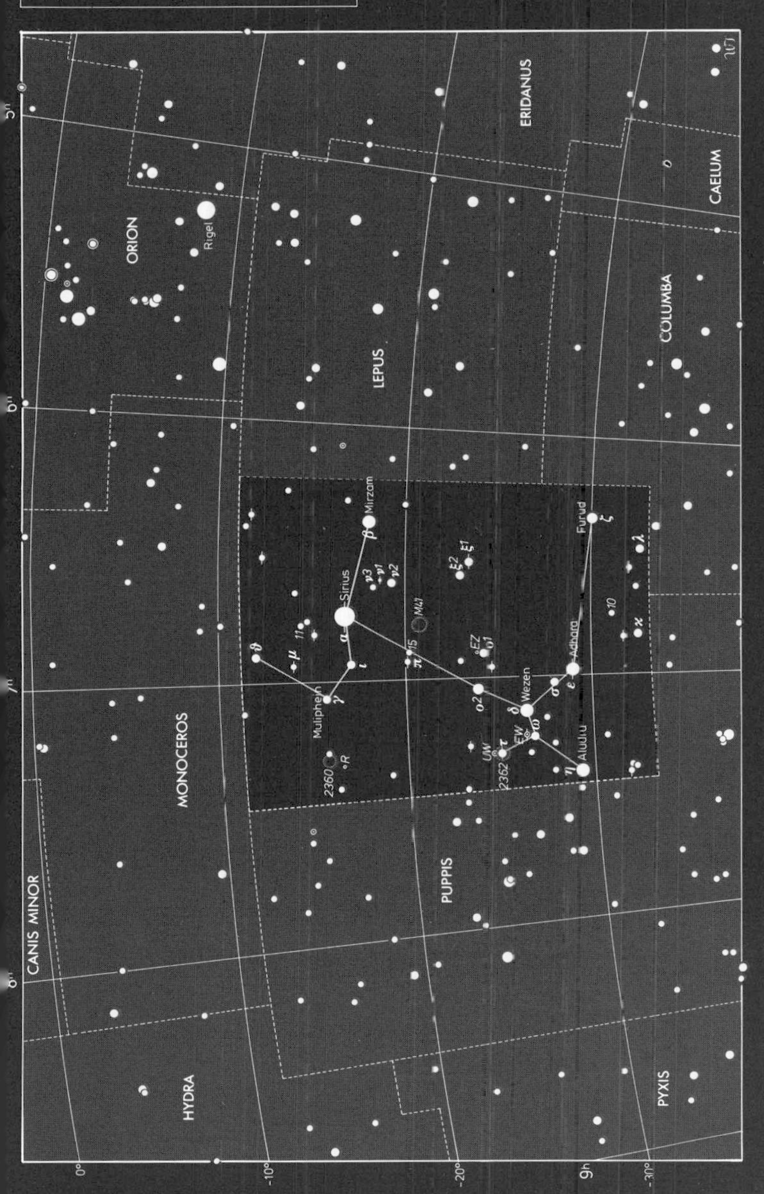

CANIS MINOR The Lesser Dog

The second of the two dogs of Orion, the other being Canis Major. Apart from its leading star Procyon, the eighth-brightest star in the sky, there are few objects of importance in Canis Minor. Procyon forms a prominent equilateral triangle with the bright stars Sirius (in Canis Major) and Betelgeuse (in Orion).

α (alpha) Canis Minoris, 7h 39m +5°, (Procyon, from the Greek meaning 'before the dog', referring to its rising before Canis Major), mag. 0.34, is a yellow-white star 11.4 l.y. away, and therefore among the nearest stars to the Sun. Like Sirius, Procyon has a white dwarf companion but this star, of mag. 10.3, is even more difficult to see than the companion of Sirius, requiring the use of large professional telescopes. Procyon's companion orbits it every 41 years.

β (beta) CMi, 7h 27m +8°, (Gomeisa), mag. 2.9, is a blue-white star about 150 l.y. away.

White dwarfs

By a remarkable coincidence both Sirius, the brightest star in Canis Major, and Procyon, the brightest star in Canis Minor, are accompanied by tiny, faint stars known as white dwarfs. The existence of companion stars to Sirius and Procyon was predicted in 1844 by the German astronomer Friedrich Wilhelm Bessel, who detected a wobble in the proper motions (page 12) of Sirius and Procyon. Bessel realized that this wobble was most probably caused by the presence of unseen companions orbiting around the visible stars. The companion of Sirius, called Sirius B, was first seen in 1862 by the American astronomer Alvan G. Clark using a 47-cm (18½-inch) refractor, and the companion to Procyon (Procyon B) was first seen in 1896 by John M. Schaeberle with the 91-cm (36-inch) refractor at Lick Observatory. But not until 1915 was the truly extraordinary nature of these stars realized. Observations showed that Sirius B was very hot, very small and very dense. In fact Sirius B has the mass of the Sun packed into a sphere only one-fiftieth of the Sun's diameter. The resulting density of Sirius B is over 100,000 times that of water. A white dwarf is a star at the end of its life; it is the shrunken remnant of a once-proud star like the Sun whose central nuclear fires have gone out. The cause of the immense densities of white dwarfs is the inexorable pull of gravity, which squeezes the electrons of the dying star as closely together as is physically possible.

15
CANIS MINOR
CMi · Canis Minoris

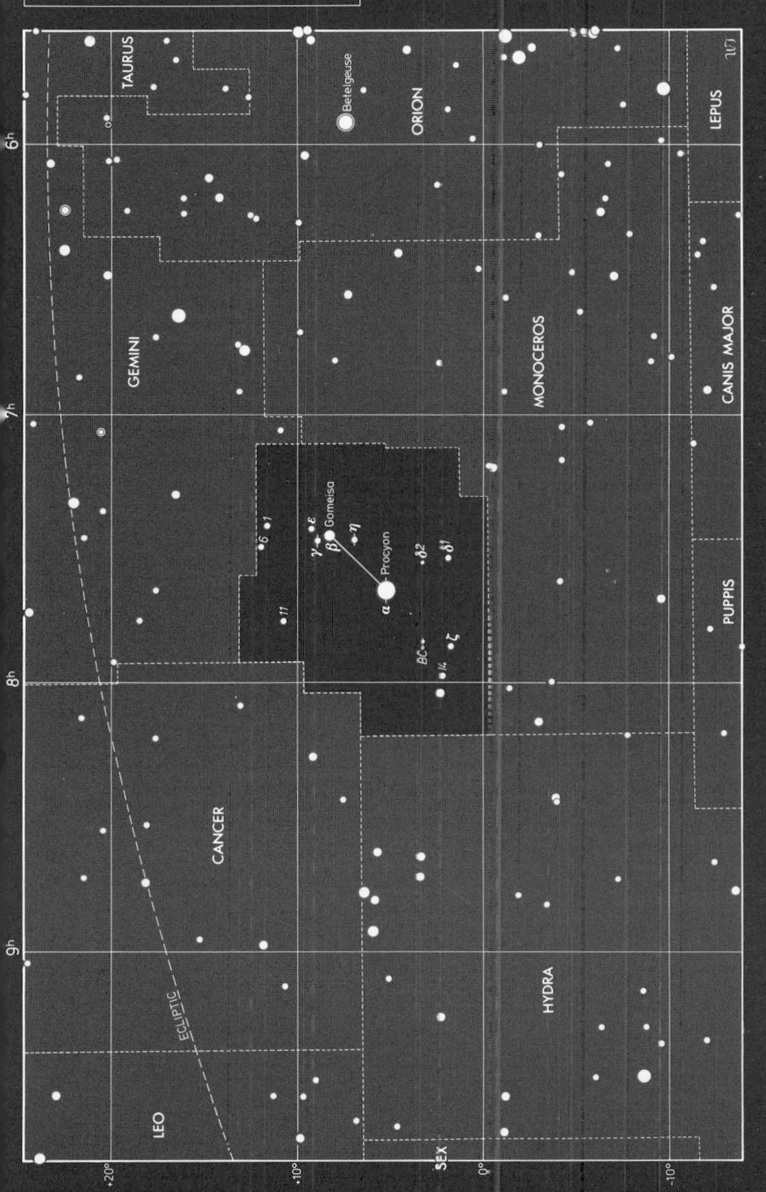

CAPRICORNUS The Sea Goat

Capricornus is depicted as a goat with a fish's tail. Amphibious creatures feature prominently in ancient legends, and the origin of Capricornus certainly dates back to ancient times. In Greek legend, the constellation represented the goat-headed god Pan who jumped into a river to escape the approach of the monster Typhon, turning his lower half into a fish. About 2500 years ago the Sun was in Capricornus when it reached its farthest point south of the equator at the winter solstice, December 22. The latitude on Earth, 23½° south, at which the Sun appears overhead at noon on the winter solstice therefore became known as the Tropic of Capricorn. Because of precession, the winter solstice has now moved into the neighbouring constellation of Sagittarius, but the Tropic of Capricorn retains its name. The Sun is in the constellation from late January to mid-February.

α (alpha) Capricorni, 20h 18m –13°, (Algedi or Giedi, meaning 'goat' or 'ibex') is a multiple star, consisting of unrelated yellow supergiant and orange giant stars 1300 and 150 l.y. away, of mags. 4.2 and 3.6 respectively, visible separately with the naked eye or binoculars. Telescopes reveal that each star is itself double. The fainter of the pair, α^1, has a wide mag. 9.2 companion visible in small telescopes; α^2 has its own companion of mag. 11. Telescopes of at least 100 mm aperture show that this faint companion is itself composed of two 11th-mag. stars. α Capricorni is therefore a fascinating hybrid system.

β (beta) Cap, 20h 21m –15°, (Dabih, 'lucky one of the slaughterers'), mag. 3.1, is a golden-yellow giant star 330 l.y. away. It has a wide, unrelated blue-white giant companion of mag. 6.1 and 850 l.y. away, visible through binoculars or small telescopes.

γ (gamma) Cap, 21h 40m –17°, (Nashira), mag. 3.7, is white star about 100 l.y. away.

δ (delta) Cap, 21h 47m –16°, (Deneb Algedi, 'the goat's tail'), mag. 2.9, is the brightest star in the constellation. It is an eclipsing binary, varying by a barely perceptible 0.2 mag. over 24.5 hours. It lies 49 l.y. away.

π (pi) Cap, 20h 27m –18°, 1050 l.y. away, is a blue-white giant of mag. 5.3 with a close mag. 8.9 companion visible with a small telescope.

M30 (NGC 7099), 21h 40m –23°, is a mag. 7.5 globular cluster 27,000 l.y. away, visible in small telescopes and resolvable in 100 mm aperture, notably condensed at the centre with chains of stars leading outwards.

16 CAPRICORNUS
Cap · Capricorni

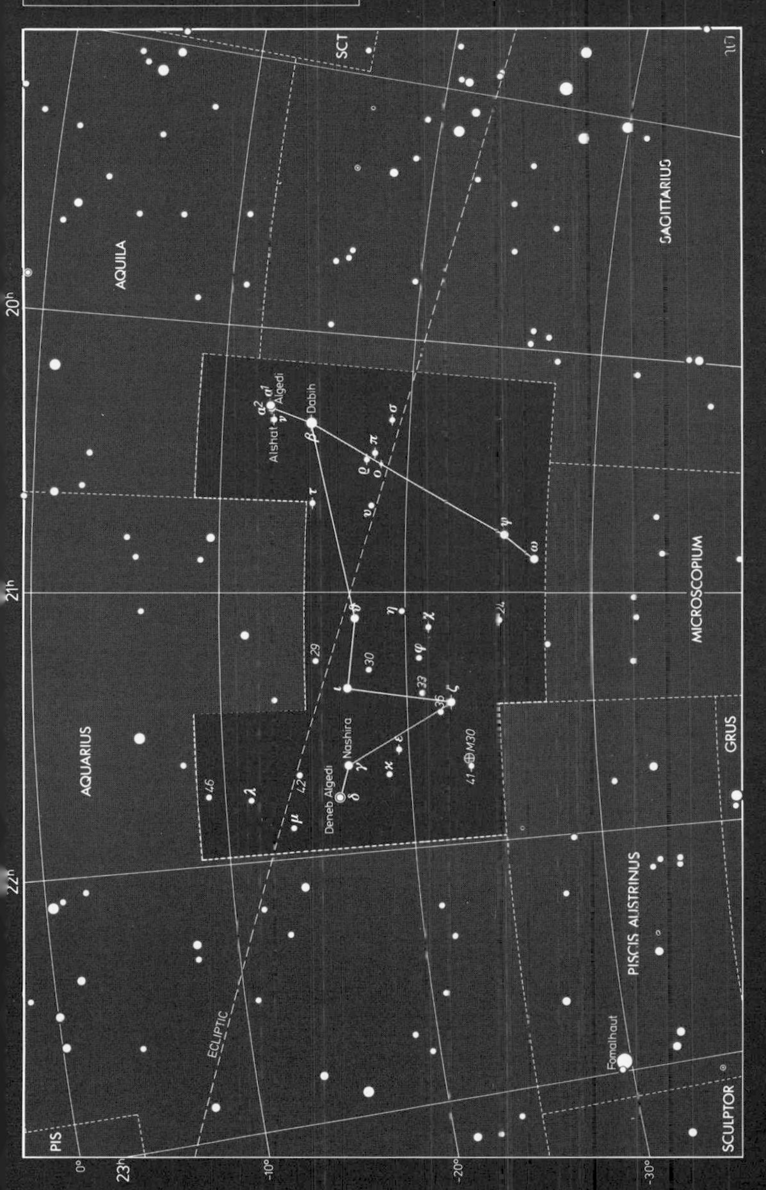

CARINA The Keel

This constellation was originally part of the extensive Argo Navis, the Ship of the Argonauts, until subdivided in 1763 by the French celestial cartographer Nicolas Louis de Lacaille. As a part of Argo Navis, Carina originated in Greek times and is associated with the legend of Jason and the Argonauts and their quest for the Golden Fleece. Carina lies in the Milky Way, providing rich starfields and clusters for binoculars. The stars ι (iota) and ε (epsilon) Carinae, together with $\varkappa$ (kappa) and δ (delta) Velorum, form the False Cross, sometimes confused with the real Southern Cross.

α (alpha) Carinae, 6h 24m –53°, (Canopus), mag. –0.72, the second-brightest star in the sky, is a white supergiant 205 l.y. away. It is named after the helmsman of the Greek King Menelaus, and appropriately enough is now used by spacecraft as a guide for navigation.

β (beta) Car, 9h 13m –70°, (Miaplacidus), mag. 1.7, is a blue-white star 55 l.y. away.

ε (epsilon) Car, 8h 23m –60°, mag. 1.9, is a yellow giant star 29 l.y. away.

η (eta) Car, 10h 45m –60°, 9000 l.y. away, is a peculiar nova-like variable star embedded in a nebula called NGC 3372 (see page 102). η Carinae in the past has fluctuated erratically in brightness, reaching a maximum of mag. –1 in 1843 when it was temporarily the second-brightest star in the sky, but has now settled at around 6th mag. The star is estimated to be over 100 times more massive and 4 million times brighter than the Sun. It is surrounded by a shell of dust thrown off in the 1843 outburst; further variations can come at any time, and the star is thought to be a candidate for a future supernova.

ϑ (theta) Car, 10h 43m –64°, mag. 2.8, is a blue-white star 490 l.y. away at the centre of the cluster IC 2602 (see page 102).

ι (iota) Car, 9h 17m –59°, mag. 2.2, is a white supergiant 750 l.y. away.

υ (upsilon) Car, 9h 47m –65°, 320 l.y. away, is a double star consisting of two white giants of mags. 3.0 and 6.0, divisible in small telescopes.

R Car, 9h 32m –63°, is a red giant long-period variable star of uncertain distance that ranges between 4th and 10th mags. with a period of 309 days.

S Car, 10h 09m –62°, is a red giant variable similar to R Car, varying from 5th to 10th mag. over 150 days.

NGC 2516, 7h 58m –61°, is a naked-eye cluster of 80 stars, 1300 l.y. away, as large as the full Moon and resolvable in binoculars. Small telescopes show a mag. 5.2 red giant near its centre, along with three double stars of 8th and 9th mags.

NGC 3114, 10h 03m –60°, is a naked-eye cluster of 7th-mag. stars and fainter, the size of the full Moon, 2900 l.y. away, a poorer version of NGC 2516. ▶

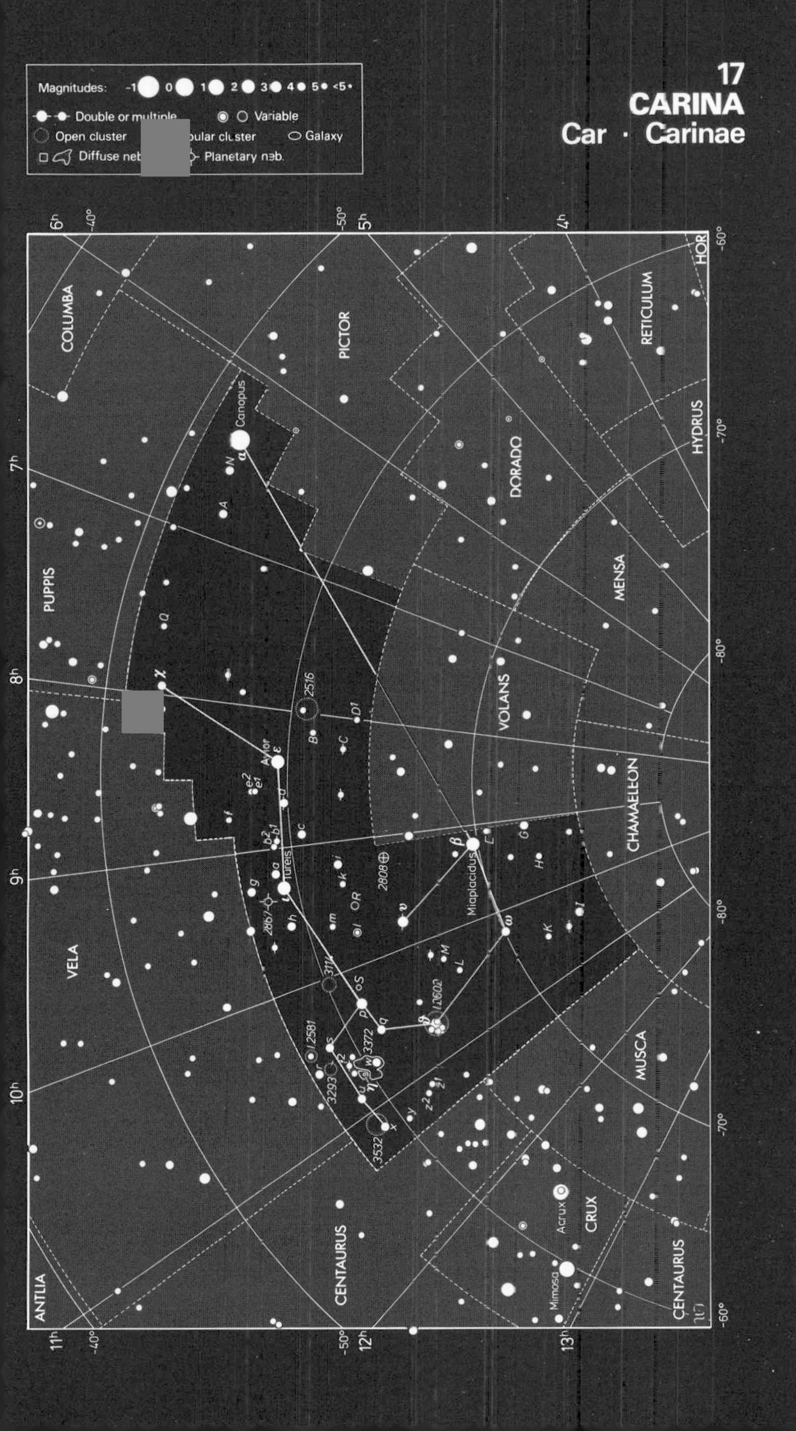
17
CARINA
Car · Carinae
Magnitudes: -1 0 1 2 3 4 5 <5
Double or multiple
Variable
Open cluster
ular cluster
Galaxy
Diffuse neb
Planetary neb.
COLUMBA
PICTOR
RETICULUM
HOR
HYDRUS
DORADO
PUPPIS
MENSA
VOLANS
CHAMAELEON
VELA
MUSCA
CRUX
CENTAURUS
ANTLIA
Canopus
Avior
Tureis
Miaplacidus
Acrux
Mimosa
2516
2808
2867
3114
3293
3372
3532
I.2581
I.2602
6h
7h
8h
9h
10h
11h
12h
13h
5h
4h
-40°
-50°
-60°
-70°
-80°

CASSIOPEIA

In Greek legend, Cassiopeia was the beautiful but boastful Queen of Ethiopia, wife of King Cepheus and mother of Andromeda. In the sky she is depicted sitting in a chair. The constellation is easily identifiable by the distinctive W-shape of its five brightest stars. Cassiopeia lies on the opposite side of the Pole Star from Ursa Major, in a rich part of the Milky Way. At 0h 25.3m, +64° 09′, near the star $\varkappa$ (kappa) Cassiopeiae, occurred the famous supernova outburst of 1572, observed by Tycho Brahe. The remains of this supernova are now a strong radio source, about 20,000 l.y. away. The remains of another supernova, which erupted around 1660 but which went unseen at the time, form the strongest radio source in the sky, Cassiopeia A; it lies at 23h 23.4m, +58° 50′, and is 10,000 l.y. away.

α (alpha) Cassiopeiae, 0h 41m +57°, (Schedar, 'the breast'), mag. 2.2, is a yellow giant star 120 l.y. away. It has a wide mag. 8.9 companion, unrelated.

β (beta) Cas, 0h 09m +59°, (Caph), mag. 2.3, is a white star 46 l.y. away.

γ (gamma) Cas, 0h 57m +61°, 780 l.y. away, is a remarkable blue-white variable of the type known as a *shell star*. It throws off rings of gas at irregular intervals, apparently because its high-speed rotation makes it unstable, causing it to vary unpredictably between mags. 3.0 and 1.6. Usually it hovers around mag. 2.5.

δ (delta) Cas, 1h 26m +60°, (Ruchbah, 'the knee'), mag. 2.7, is a blue-white star 52 l.y. away. It varies by about 0.1 mag. with a period of 2 years, and is thought to be an eclipsing binary. ▶

◀ NGC 3372, 10h 44m –60°, is a celebrated diffuse nebula easily visible to the naked eye as a brilliant patch of the Milky Way four Moon diameters wide, surrounding the erratic variable star η (eta) Carinae. The nebula shines from the light of brilliant young stars born within it. Binoculars show jewelled star clusters and swirls of glowing gas alternating with dark lanes in this nebula. Most famous is a dark notch, called the Keyhole because of its distinctive shape, silhouetted against the nebula's brightest central portion near η Carinae itself. The whole NGC 3372 nebula lies about 9000 l.y. from us, the same distance as η Carinae. (For photograph see page 270.)

NGC 3532, 11h 06m –59°, 1300 l.y. away, is an outstanding cluster of 150 stars of 7th mag. and fainter, covering nearly 1° of sky. It is visible to the naked eye as a fuzzy 3rd-mag. star and is glorious in binoculars, set among rich Milky Way starfields. It is elliptical in shape with stars arranged in chains and with several orange giants. The mag. 3.9 yellow supergiant x Carinae at the cluster's edge is not a member but a background object over four times as far away.

IC 2602, 10h 43m –64°, is a large and brilliant cluster of 60 stars, similar to the Pleiades, lying 490 l.y. away. Its brightest members are visible to the naked eye, notably the 3rd-mag. ϑ Carinae (see page 100). The whole cluster appears twice as wide as the full Moon.

18
CASSIOPEIA
Cas · Cassiopeiae

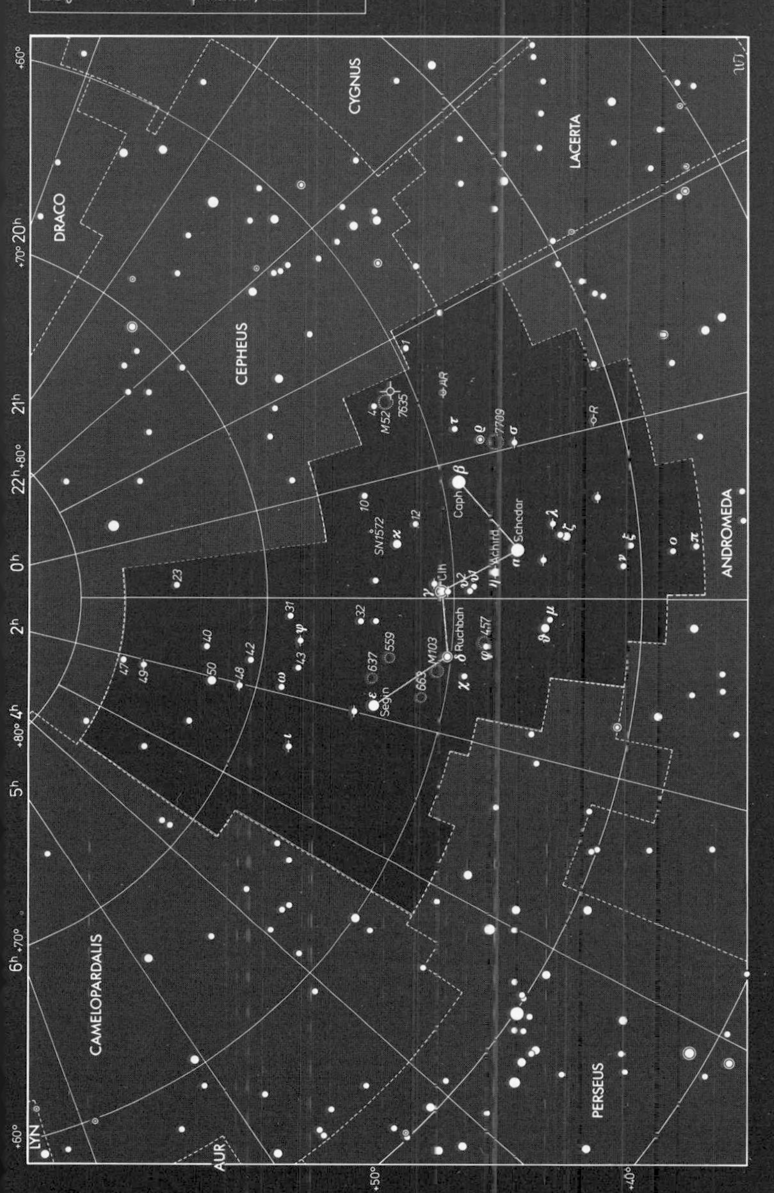

CENTAURUS The Centaur

A large and rich constellation representing a centaur, the mythical beast that was half man, half horse. Reputedly Centaurus depicts the scholarly centaur Chiron, the tutor of many Greek gods and heroes, who was raised to the sky after being accidentally struck by a poisoned arrow from Hercules. Centaurus is of particular interest because it contains the closest star to the Sun, α (alpha) Centauri, which is actually a family of three stars linked by gravity. A line from α through β (beta) Centauri points to Crux, the Southern Cross. One of the strongest radio sources in the sky, Centaurus A, is associated with the galaxy NGC 5128. Centaurus lies in a prominent part of the Milky Way.

α (alpha) Centauri, 14h 40m –61°, (Rigil Kentaurus, 'foot of the centaur', or Toliman) lies 4.3 l.y. away. To the naked eye it shines at mag. –0.27, the third-brightest star in the sky, but the smallest of telescopes reveals that it consists of twin yellow stars of mags. –0.01 and 1.33. The brighter of these is very similar ▶

◀ ε (epsilon) Cas, 1h 54m +64°, mag. 3.4, is a blue giant star 520 l.y. away.

η (eta) Cas, 0h 49m +58°, 19 l.y. away, is a beautiful double star with yellow and red components of mags. 3.5 and 7.5, visible in small telescopes. They form a true binary with a period of 480 years.

ι (iota) Cas, 2h 29m +67°, 160 l.y. away, is a mag. 4.5 white star with a wide mag. 8.4 companion visible through a 60-mm telescope. With an aperture of 100 mm and high magnification, the brighter star is seen to have a closer mag. 6.9 yellow companion, making this an impressive triple.

σ (sigma) Cas, 23h 59m +56°, 1400 l.y. away, is a close pair of mags. 5.0 and 7.1 appearing green and blue, in striking contrast to the warmer hues of η Cas. An aperture of 75 mm and high power will split the pair, but 150 mm is needed to show the colours.

ψ (psi) Cas, 1h 26m +68°, 250 l.y. away, is a mag. 4.7 yellow giant with a wide 9th-mag. companion visible in a small telescope. High powers reveal that this companion is itself a close binary.

M52 (NGC 7654), 23h 24m +62°, is a cluster of about 100 stars, 5200 l.y. away, visible as a misty patch in binoculars. It is somewhat kidney-shaped, with a mag. 5.0 yellow-orange star, 4 Cas, embedded at one edge, like a poorer version of the celebrated Wild Duck Cluster (M11 in Scutum); however, this star may be a foreground object. M52 can be resolved into stars with 75 mm aperture.

M103 (NGC 581), 1h 33m +61°, is a scattering of about 25 faint stars with a central diamond shape, 8500 l.y. away.

NGC 457, 1h 19m +58°, is a loose cluster of about 80 stars, 9000 l.y. away, seemingly arranged in chains. The mag. 5.0 yellow supergiant φ (phi) Cas on its outskirts is a member.

NGC 663, 1h 46m +61°, is a binocular cluster of about 80 stars, 7200 l.y. away.

19

CENTAURUS

Cen · Centauri

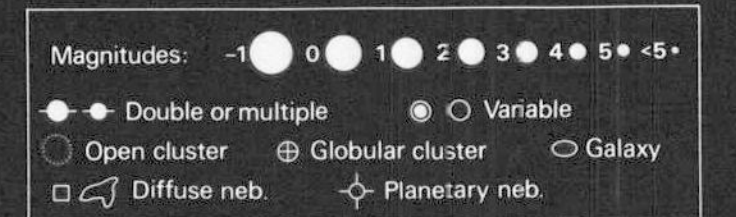

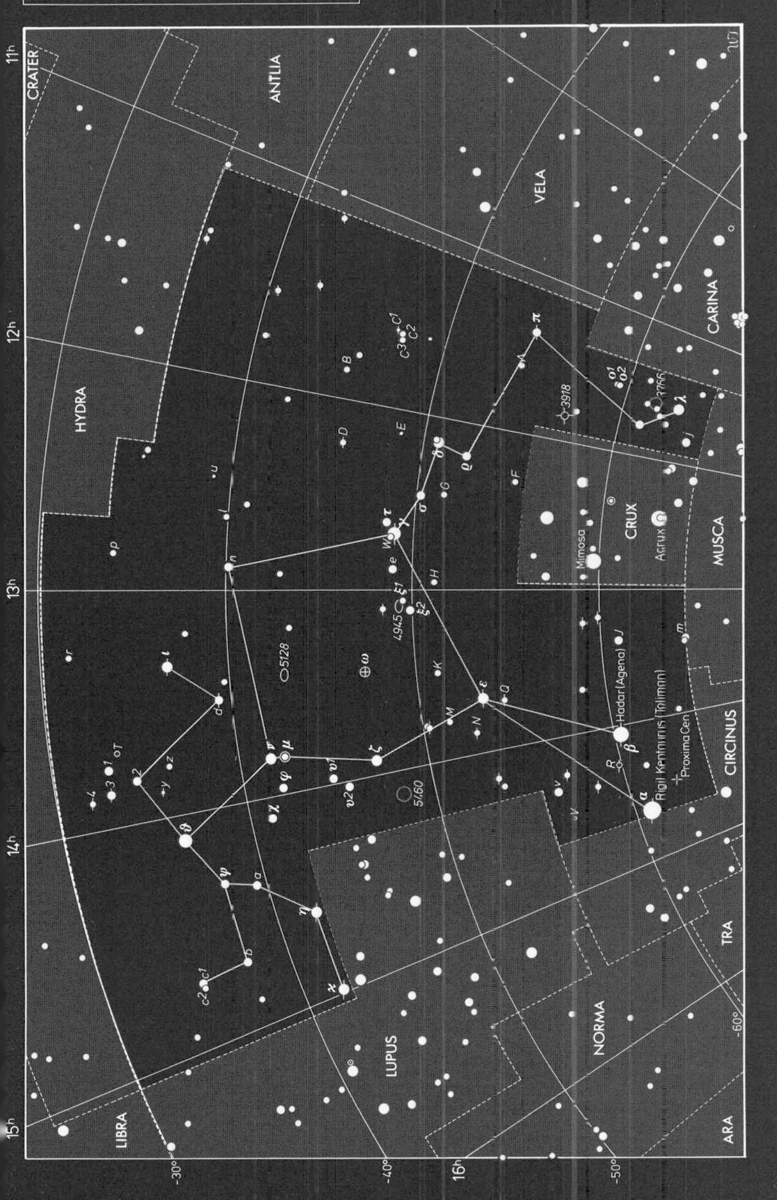

ω (omega) Centauri, the finest of all globular star clusters, appears noticeably elliptical in this photograph. Royal Observatory Edinburgh.

◀ in nature to the Sun. They orbit each other every 80 years and are always divisible in amateur telescopes, although at their closest around the years 2037–8 they will need 75 mm aperture. Also associated with α Centauri is an 11th-mag. red dwarf called Proxima Centauri, lying 2° away and therefore not even in the same telescopic field of view. (See the finder chart on page 107.) This star is estimated to take as long as a million years to orbit its two brilliant companions. At present, Proxima Centauri is about 0.1 l.y. closer to us than the two other members of α Centauri. Proxima Centauri is a flare star, suddenly increasing in brightness by as much as one magnitude for several minutes.

β (beta) Cen, 14h 04m –60°, (Hadar or Agena), mag. 0.6, is a blue giant 360 l.y. away.

γ (gamma) Cen, 12h 42m –49°, 330 l.y. away, is a close double with blue-white components each of mag. 2.9, orbiting each other every 85 years. Together they shine as a star of mag. 2.2. They are divisible in apertures of 150 mm until the

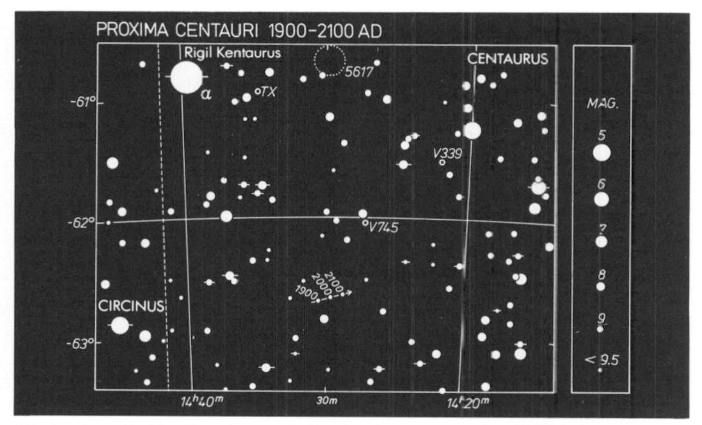

Finder chart for Proxima Centauri, showing its proper motion. Wil Tirion.

year 2000, but 220 mm will be needed by 2005. When at their closest, between the years 2010 and 2020, amateur telescopes will be unable to split them, but they become divisible in 220 mm again by 2025, and 150 mm from 2030.

3 Cen, 13h 52m –33°, is a striking pair of unrelated blue-white stars of mags. 4.6, 820 l.y. away, and 6.1, 590 l.y. away, divisible in small telescopes.

ω (omega) Cen (NGC 5139), 13h 27m –47°, is the largest and brightest globular cluster in the sky – so prominent that it was labelled as a star on early charts. It appears of mag. 3.7 to the naked eye, noticeably elliptical in shape and as large as the full Moon. Small telescopes or even binoculars begin to resolve its outer regions into stars, and it is a showpiece for all apertures. Its brilliance and large apparent size are due in part to its relative closeness of 17,000 l.y., which makes it among the nearest globular clusters to us.

NGC 3766, 11h 36m –62°, is a naked-eye open cluster of about 100 stars, 5500 l.y. away.

NGC 3918, 11h 50m –57°, is an 8th-mag. planetary nebula 2600 l.y. away, discovered by John Herschel and called by him the Blue Planetary. It is similar in appearance to the planet Uranus, but three times the apparent diameter. Its central star, of 11th mag., should be detectable in modest amateur instruments.

NGC 5128, 13h 26m –43°, is a peculiar 7th-mag. galaxy known to radio astronomers as Centaurus A. In long-exposure photographs it appears as a giant elliptical galaxy with an encircling band of dust, apparently the result of a merger between an elliptical and a spiral galaxy. Under good skies it is visible in binoculars, but at least 100 mm aperture is necessary to trace its outline and the dark bisecting lane of dust. NGC 5128 is about 15 million l.y. away.

NGC 5460, 14h 08m –48°, is a large, 6th-mag. cluster of about 40 stars visible in binoculars or small telescopes. It lies 1600 l.y. away.

CEPHEUS

An ancient constellation representing the mythological King Cepheus of Ethiopia, husband of Cassiopeia and father of Andromeda, themselves depicted by nearby constellations. Cepheus is situated on the edge of the Milky Way and, although it contains no spectacular star clusters or nebulae, it is replete with double and variable stars, including the celebrated δ (delta) Cephei, prototype of the Cepheid variables, used as 'standard candles' for distance-finding in space. This star's fluctuations in light output were discovered in 1784 by the English amateur astronomer John Goodricke, a deaf-mute who died in 1786 at the age of 21.

α (alpha) Cephei, 21h 19m +63°, (Alderamin), mag. 2.4, is a white star 52 l.y. away.

β (beta) Cep, 21h 29m +71°, (Alfirk, meaning 'flock', i.e. of sheep), 815 l.y. away, is both a double and a variable star. A small telescope shows that this blue giant star, of mag. 3.2, has an unrelated mag. 7.9 companion, 590 l.y. away. β Cephei is the prototype of a class of pulsating variable stars (also known as β Canis Majoris stars) with periods of a few hours and tiny light fluctuations. Over 4.6 hours or so β Cephei varies by 0.1 mag., an amount indistinguishable to the naked eye but which can be detected by sensitive instruments.

γ (gamma) Cep, 23h 39m +78°, (Errai, 'the shepherd'), mag. 3.2, is a yellow star 52 l.y. away.

δ (delta) Cep, 22h 29m +58°, 1300 l.y. away, is a famous pulsating variable star, the prototype of the classic Cepheid variables. This yellow supergiant varies between mags. 3.5 and 4.4 in 5 days 9 hours, changing in size between about 40 and 46 times the Sun's diameter as it does so. Less well known is that δ Cephei is also an attractive double star for binoculars or the smallest telescopes, with a wide, bluish mag. 6.3 companion.

μ (mu) Cep, 21h 44m +59°, 2800 l.y. away, is a famous red star, called the Garnet Star by William Herschel because of its striking tint, which is notable in binoculars. μ Cephei is a red supergiant, the prototype of a class of variable stars known as semi-regular variables. It varies between mags. 3.4 and 5.1 with a period of around 2 years.

ξ (xi) Cep, 22h 04m +65°, 110 l.y. away, is a double star of mags. 4.4 and 6.5 visible in small telescopes. The components are blue-white and yellow and form a true binary with an estimated orbital period of nearly 4000 years.

ο (omicron) Cep, 23h 19m +68°, 140 l.y. away, is a yellow giant of mag. 4.9 with a close mag. 7.1 companion for telescopes of 60 mm aperture and above. They orbit each other every 800 years.

CEPHEUS
Cep · Cephei

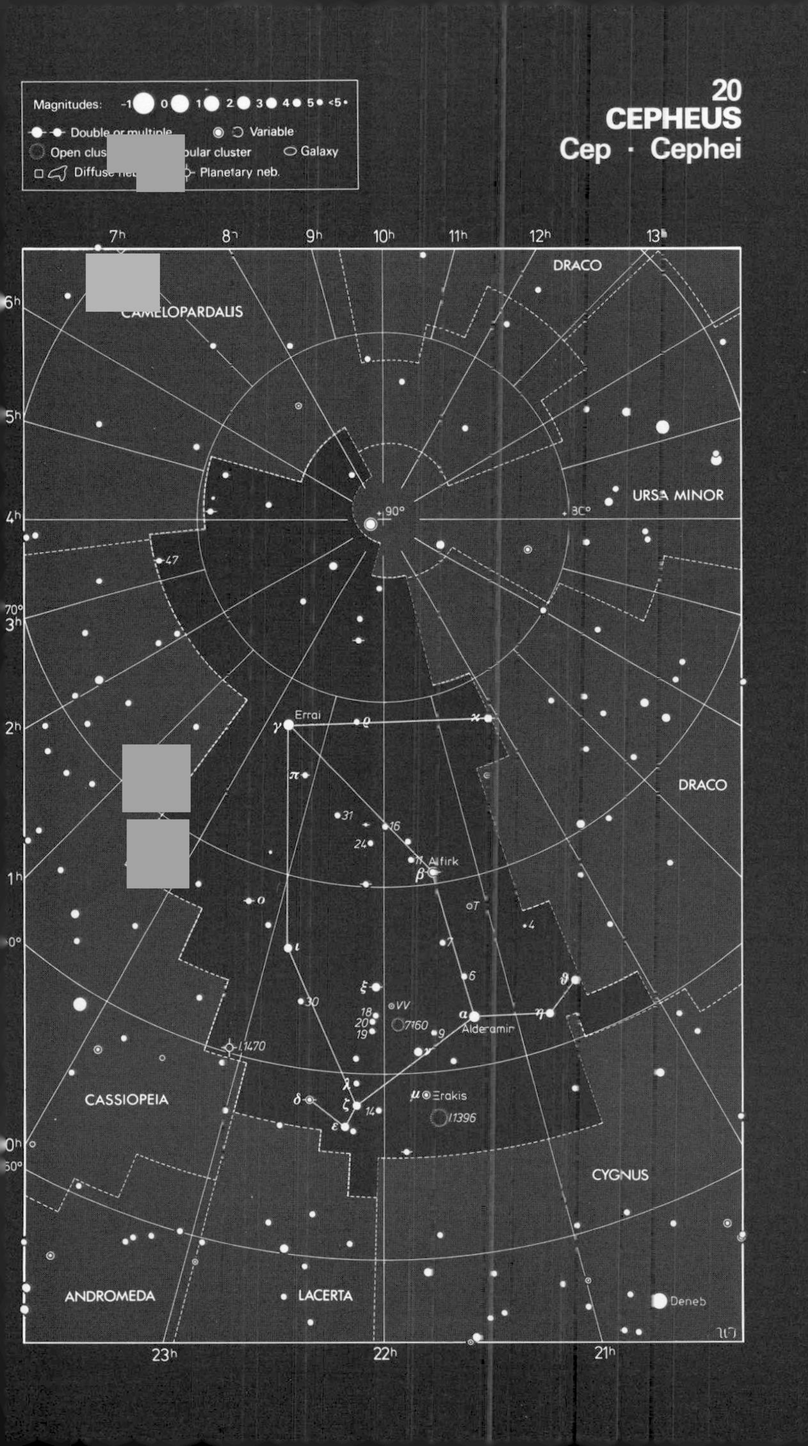

CETUS The Whale

An ancient constellation depicting the sea monster that threatened to devour Andromeda before she was rescued by Perseus. Cetus is found in the sky basking on the banks of Eridanus, the River. The constellation is large but faint; nevertheless it contains several stars of particular interest, notably o (omicron) Ceti and τ (tau) Ceti. One faint but famous star is UV Ceti, position 1h 38.8m, –17° 58′, which consists of a pair of 13th-mag. red dwarfs 8.4 l.y. away, one of which is the prototype of a class of erratic variables known as *flare stars*; these are red dwarfs that undergo sudden increases in light output lasting only a few minutes. The outbursts of the flare star component of UV Ceti take it from its normal level of 13th mag. to as bright as 7th mag. ▶

Finder chart for the variable star Mira, also known as o (omicron) Ceti. The numbers by the surrounding stars are their magnitudes with the decimal point omitted. This convention prevents confusion between faint stars and decimal points. The magnitude of Mira may be estimated by comparison with the surrounding stars. Wil Tirion.

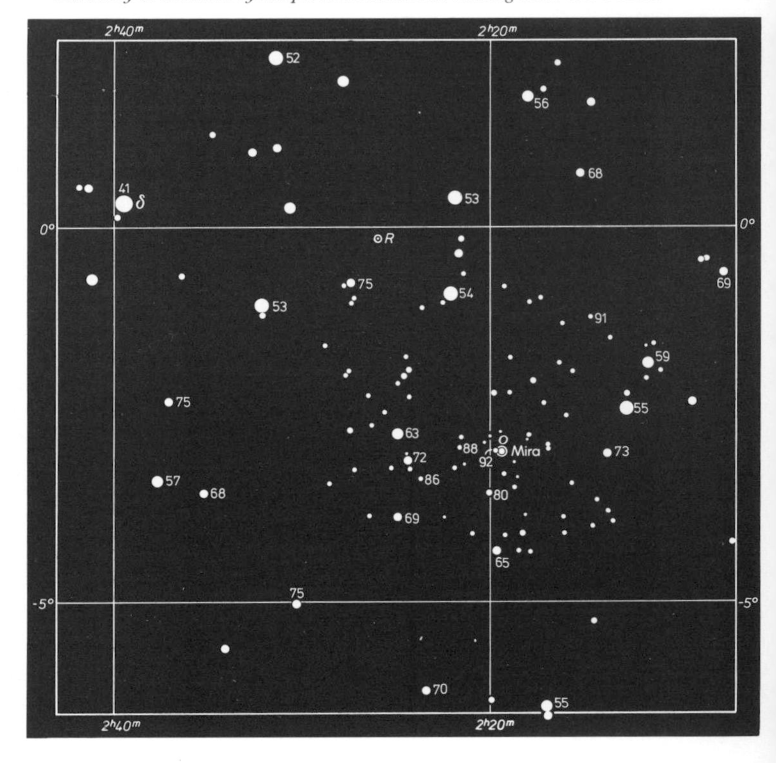

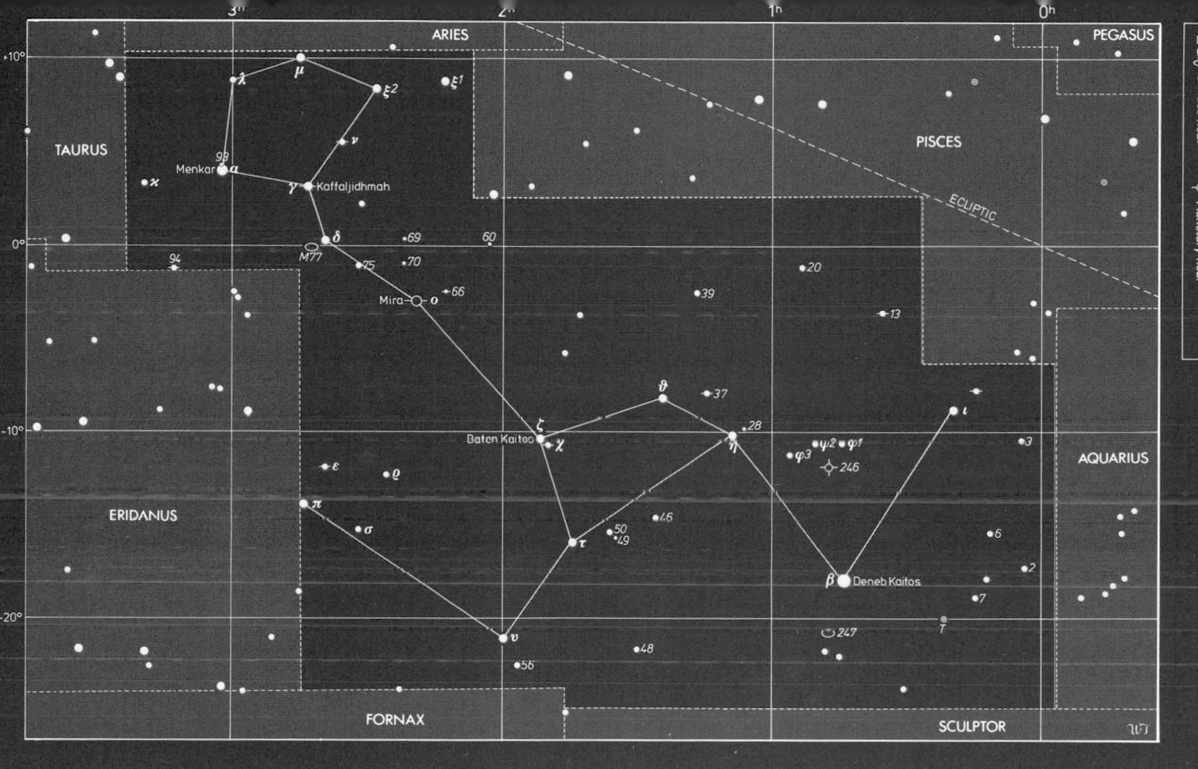

21
CETUS
Cet · Ceti

Magnitudes: -1 0 1 2 3 4 5 <5
Double or multiple
Variable
Open cluster
Globular cluster
Galaxy
Diffuse neb.
Planetary neb.

CHAMAELEON The Chameleon

A faint and unremarkable constellation, introduced into the southern skies by the Dutch navigators Pieter Dirkszoon Keyser and Frederick de Houtman at the end of the 16th century.

α (alpha) Chamaeleontis, 8h 19m –77°, mag. 4.1, is a white star 72 l.y. away.

β (beta) Cha, 12h 18m –79°, mag. 4.3, is a blue-white star 390 l.y. away.

γ (gamma) Cha, 10h 35m –79°, mag. 4.1, is a red giant 280 l.y. away.

δ (delta) Cha, 10h 45m –80°, consists of a wide pair of unrelated stars, both clearly seen in binoculars: δ^1 Cha, mag. 5.5, is an orange giant 360 l.y. away and δ^2 Cha, mag. 4.4, is a blue star 780 l.y. away.

ε (epsilon) Cha, 12h 00m –78°, 285 l.y. away, consists of a very close pair of stars of mags. 5.4 and 6.0, needing apertures of 150 mm or above with high magnification to split.

NGC 3195, 10h 09m –81°, is a faint planetary nebula of similar apparent size to the planet Jupiter, needing at least 100 mm aperture to be seen well.

◀ α (alpha) Ceti, 3h 02m +4°, (Menkar, 'nose'), mag. 2.5, is a red giant star 160 l.y. away. Binoculars show a wide mag. 5.6 blue-white companion, 93 Ceti, which is unrelated, lying over three times as far away.

β (beta) Cet, 0h 44m –18°, (Deneb Kaitos, 'tail of the whale'), mag. 2.0, the brightest star of the constellation, is a yellow giant 62 l.y. away.

γ (gamma) Cet, 2h 43m +3°, 75 l.y. away, is a close double star needing telescopes of at least 60 mm aperture and high power to split. The stars are of mags. 3.5 and 7.3, the colours yellow and bluish.

o (omicron) Cet, 2h 19m –3°, (Mira, 'the amazing one'), 250 l.y. away, is the prototype of a famous class of red giant long-period variable stars. Mira itself varies between about 3rd and 9th mags. (although it can become as bright as 2nd mag.) in an average of 332 days, changing in diameter from 300 to 400 times the size of the Sun as it does so. Mira's light variations were first noted in 1596 by the Dutch astronomer David Fabricius, making it the first variable star (other than novae) to be discovered. (See the finder chart on page 110.)

τ (tau) Cet, 1h 44m –16°, mag. 3.5, a yellow dwarf, is one of the nearest stars to the Sun, lying 11.7 l.y. away. Its main claim to fame is that, of all the nearby single stars, it is the one most like the Sun, although we do not as yet know whether it also has planets. Radio astronomers have listened for possible alien radio messages coming from the star, but without success.

M77 (NGC 1068), 2h 43m 0°, is a small, softly glowing 9th-mag. face-on spiral galaxy requiring a telescope of 150 mm aperture or above to be seen well. It lies roughly 50 million l.y. away, and is perhaps the most remote of the objects on Messier's list. M77 is classified as a Seyfert galaxy, a type of spiral with a brilliant nucleus, and it is a radio source.

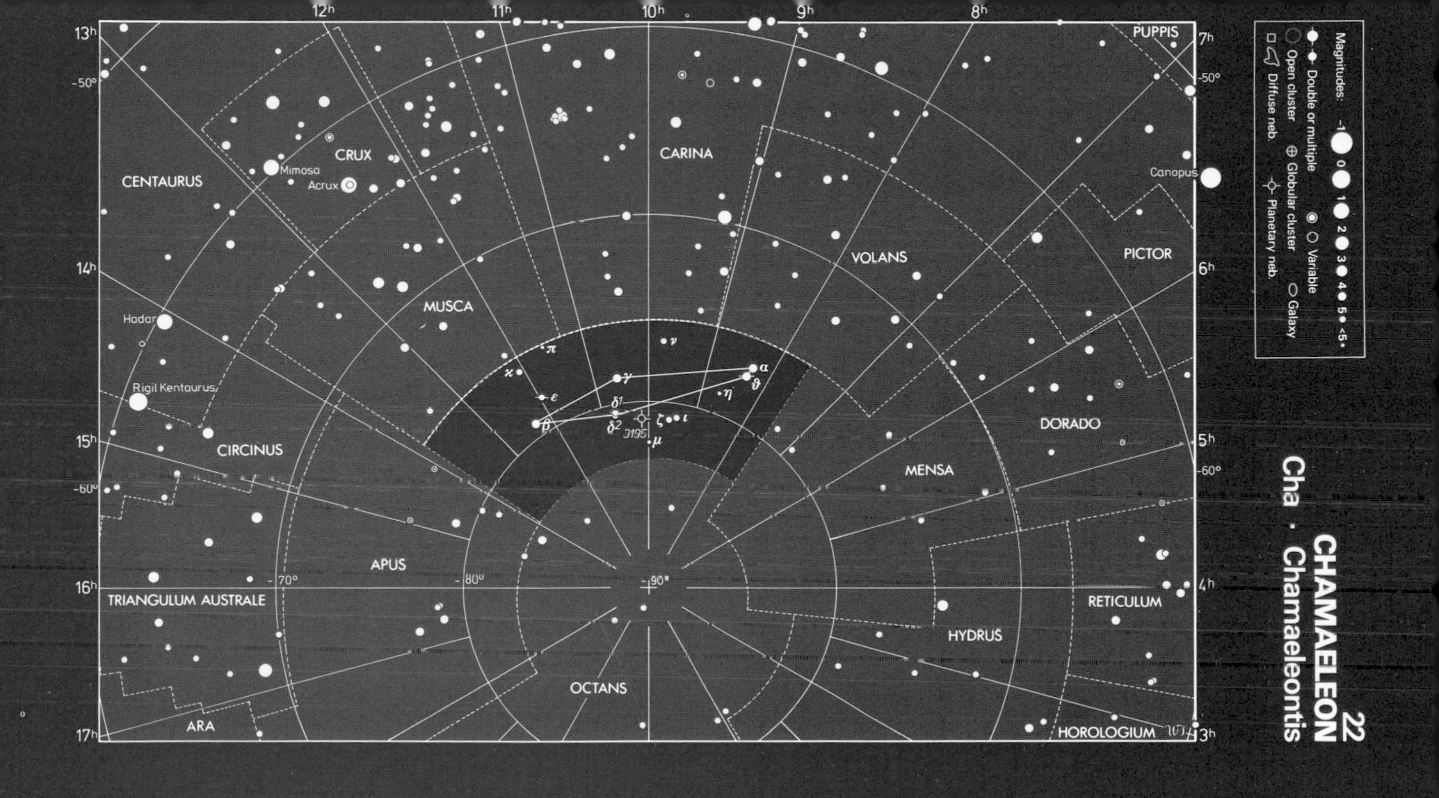
22
CHAMAELEON
Cha · Chamaeleontis
Magnitudes: -1 0 1 2 3 4 5 <5
Double or multiple
Variable
Open cluster
Globular cluster
Galaxy
Diffuse neb.
Planetary neb.
PUPPIS
PICTOR
DORADO
RETICULUM
HOROLOGIUM
HYDRUS
MENSA
VOLANS
CARINA
OCTANS
MUSCA
APUS
CRUX
CIRCINUS
CENTAURUS
TRIANGULUM AUSTRALE
ARA
Canopus
Acrux
Mimosa
Hadar
Rigil Kentaurus
-50°
-60°
-70°
-80°
-90°
7h
6h
5h
4h
3h
8h
9h
10h
11h
12h
13h
14h
15h
16h
17h

CIRCINUS The Compasses

Another of the small and obscure southern constellations introduced in 1756 by the French astronomer Nicolas Louis de Lacaille. It represents a pair of compasses as used by surveyors, and is appropriately placed in the sky next to Norma, the Level. It is overshadowed by the brilliance of neighbouring Centaurus.

α (alpha) Circini, 14h 43m –65°, 65 l.y. away, is a white star of mag. 3.2 with a mag. 8.6 companion easily visible in small telescopes.

γ (gamma) Cir, 15 23m –59°, consists of a very close pair of blue and yellow stars of mags. 5.1 and 5.5, probably unrelated, divisible in a telescope of 150 mm aperture using high magnification.

In the neighbouring constellation of Norma at 15h 52m –52° lies the satisfyingly symmetrical 13th-mag. planetary nebula known variously as Shapley 1 (Sp 1), PK 329+02.1 or RCW 100, surrounding a 14th-mag. central star. Anglo-Australian Telescope Board.

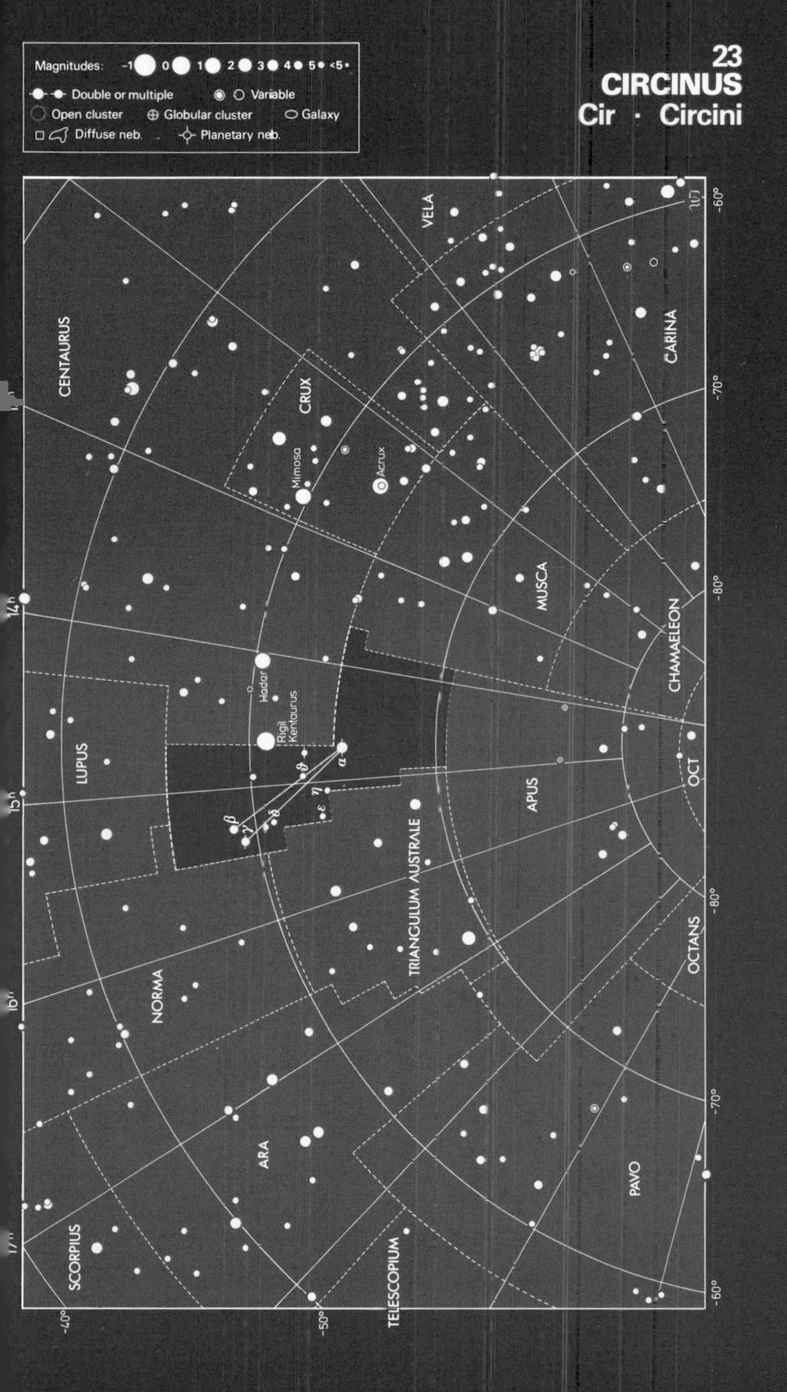
23
CIRCINUS
Cir · Circini
Magnitudes: -1 0 1 2 3 4 5 <5
Double or multiple
Variable
Open cluster
Globular cluster
Galaxy
Diffuse neb.
Planetary neb.
CENTAURUS
CRUX
VELA
CARINA
Mimosa
Acrux
MUSCA
CHAMAELEON
Hadar
Rigil Kentaurus
LUPUS
APUS
OCT
TRIANGULUM AUSTRALE
NORMA
OCTANS
ARA
PAVO
SCORPIUS
TELESCOPIUM
14h
15h
16h
17h
-40°
-50°
-60°
-70°
-80°

COLUMBA The Dove

A constellation representing the dove that followed Noah's Ark, or possibly the dove that the Argonauts sent ahead to help them pass safely between the Symplegades, the Clashing Rocks, at the mouth of the Black Sea; fittingly Columba is placed next to Puppis, the stern of the ship Argo. Columba originated in 1592 when the Dutchman Petrus Plancius formed it from some stars adjacent to Canis Major that had not previously been part of any constellation. It contains little of interest for amateur telescopes.

α (alpha) Columbae, 5h 40m –34°, (Phakt, 'ring dove'), mag. 2.6, is a blue-white star 170 l.y. away.

β (beta) Col, 5h 51m –36°, mag. 3.1, is a yellow giant 130 l.y. away.

NGC 1851, 5h 14m –40°, is a 7th-mag. globular cluster for small telescopes, 35,000 l.y. away.

In Coma Berenices lies the classically elegant NGC 4565, a spiral galaxy seen exactly edge-on (see page 120). Note the central bulge of stars and the dark lanes of dust in the galaxy's plane. Hale Observatories photograph.

24 COLUMBA
Col · Columbae

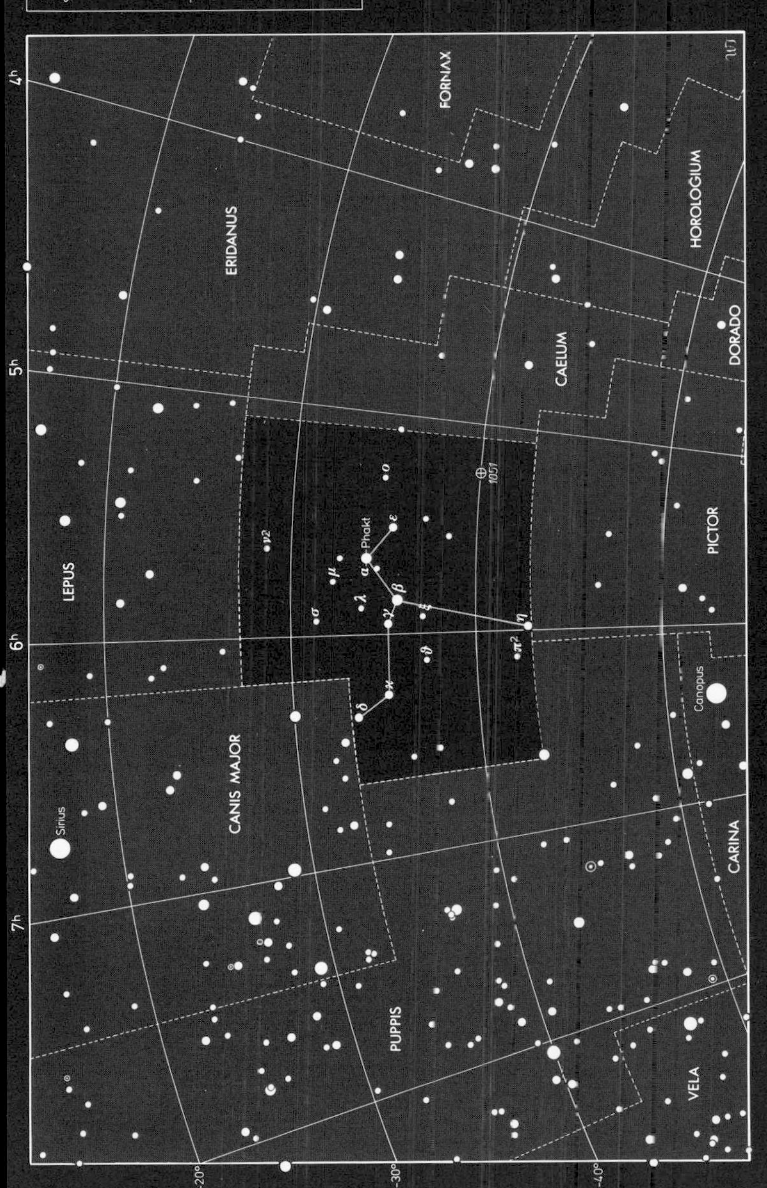

COMA BERENICES Berenice's Hair

This faint constellation represents the flowing locks of Queen Berenice of Egypt, who cut off her hair in gratitude to the gods for the safe return of her husband Ptolemy III Euergetes from battle. Although this legend dates from Greek times, this group of stars was regarded as part of Leo until 1551 when the Dutch cartographer Gerard Mercator made them into a separate constellation. The main part of the Queen's severed tresses is represented by a scattered group of about 50 stars known as the Coma Star Cluster or Melotte 111, covering several degrees of sky near γ (gamma) Comae Berenices and best seen in binoculars. The brightest members of the cluster, of 4th and 5th mags., form a noticeable V-shape. The cluster is 260 light years distant. Coma Berenices also contains another type of cluster – a cluster of galaxies. The Coma Cluster of galaxies (not to be confused with the Coma Star Cluster) lies over 250 million l.y. away, so its members are too faint for all but the largest amateur telescopes. But the constellation also contains some brighter galaxies, members of the nearer Virgo Cluster, the brightest of which are visible in amateur telescopes. The north pole of our Galaxy lies in Coma Berenices.

α (alpha) Comae Berenices, 13h 10m +18°, (Diadem), mag. 4.3, is a tight binary 64 l.y. away, consisting of twin yellow-white stars of mag. 5.1. that orbit each other every 26 years. However, even at their widest, around the years 1993 and 2010, they are at the limit of resolution of a 220-mm telescope.

β (beta) Com, 13h 12m +28°, mag. 4.3, is a yellow star 27 l.y. away.

γ (gamma) Com, 12h 27m +28°, mag. 4.4, is an orange giant 260 l.y. away, the brightest member of the Coma Star Cluster.

24 Com, 12h 35m +18°, 280 l.y. away, is a beautiful coloured double star for small telescopes, consisting of an orange giant of mag. 5.0 and a mag. 6.6 blue-white companion.

M53 (NGC 5024), 13h 13m +18°, is an 8th-mag. globular cluster 56,000 l.y. away, visible in small telescopes as a rounded, hazy patch.

M64 (NGC 4826), 12h 57m +22°, is a famous spiral galaxy, called the Black Eye Galaxy because of a dark cloud of dust silhouetted against its nucleus. This dark dust lane shows up well in telescopes above 150 mm aperture; observers with smaller instruments must content themselves simply with locating this 9th-mag. galaxy 20 million l.y. away, closer than the Virgo Cluster and not a member of it.

M85 (NGC 4382), 12h 25m +18°, is a 9th-mag. elliptical galaxy in the Virgo Cluster, 50 million l.y. away. Small telescopes show a bright, star-like centre.

M88 (NGC 4501), 12h 32m +14°, is a 10th-mag. spiral galaxy in the Virgo Cluster, 50 million l.y. away. It is presented nearly edge-on, so that it appears elliptical. ▶

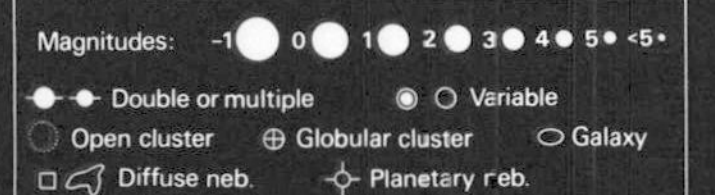

25
COMA BERENICES
Com · Comae Berenices

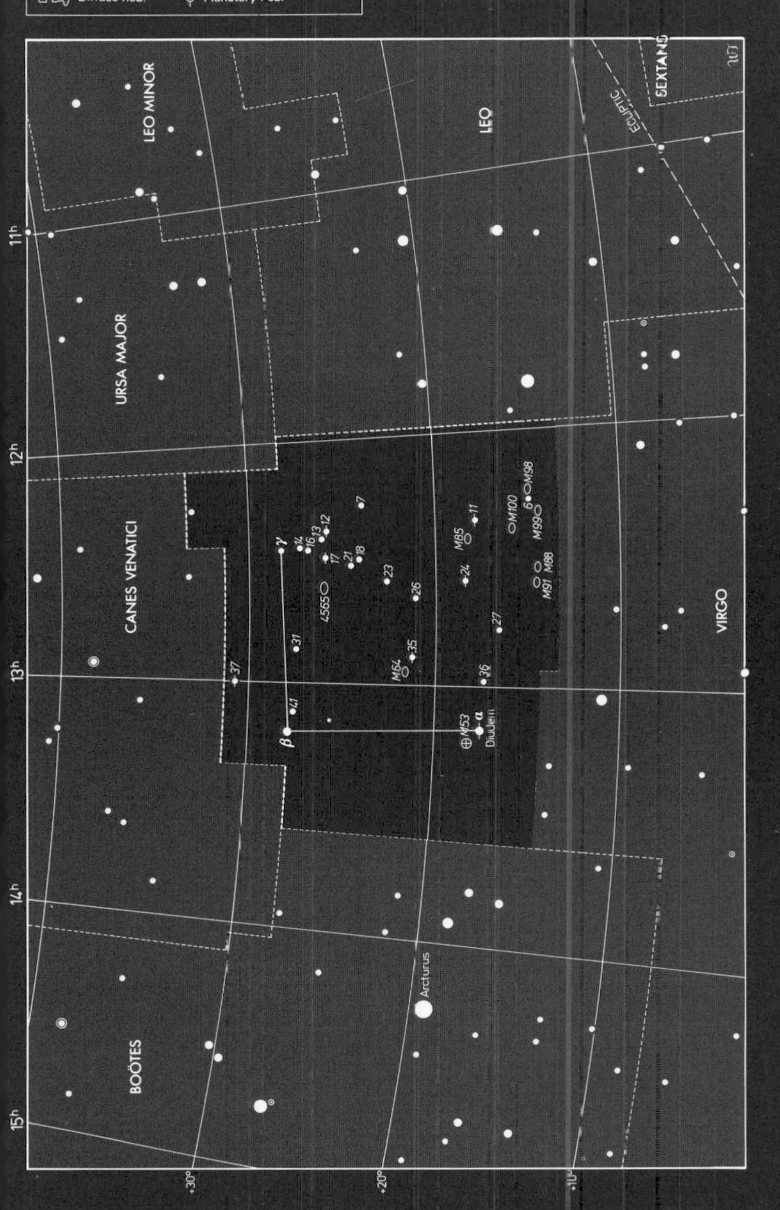

CORONA AUSTRALIS The Southern Crown

The southern counterpart of the Northern Crown (Corona Borealis). Corona Australis has been known since the time of the Greek astronomer Ptolemy in the 2nd century AD who knew it not as a crown but as a wreath. In one legend it represents the crown placed in the sky by Bacchus when he rescued his dead mother from the Underworld; alternatively, it could simply have slipped from the head of the centaur Sagittarius, at whose feet it lies. Although faint, it is a distinctive figure, and is situated on the edge of the Milky Way.

α (alpha) Coronae Australis, 19h 09m –38°, mag. 4.1, is a white star 140 l.y. away.

β (beta) CrA, 19h 10m –39°, mag. 4.1, is a yellow giant 270 l.y. away.

γ (gamma) CrA, 19h 06m –37°, 68 l.y. away, consists of a pair of near-identical yellow stars, mags. 4.8 and 5.1, orbiting every 120 years, forming a tight double for small telescopes. At their closest together during the 1990s the two stars need 100 mm aperture to split them, but after the year 2000 they become progressively easier.

$\varkappa$ (kappa) CrA, 18h 33m –39°, 420 l.y. away, is a pair of blue-white stars of mags 5.7 and 6.3 easily divisible in small telescopes.

NGC 6541, 18h 08m –44°, is a 7th-mag. globular cluster 22,000 l.y. away, visible in binoculars or small telescopes.

◀ M99 (NGC 4254), 12h 19m +14°, is a 10th-mag. spiral galaxy 50 million l.y. away in the Virgo Cluster, presented face-on so that it appears almost circular.

M100 (NGC 4321), 12h 23m +16°, 50 million l.y. away, is a 9th-mag. Virgo Cluster spiral seen face-on, similar to M99 but larger.

NGC 4565, 12h 36m +26°, is a 10th-mag. spiral galaxy seen edge-on. It is the most famous of the edge-on spirals, and is pictured on page 116. Instruments of 200 mm aperture and above show it as a cigar-shaped body split by a dark band (actually a dust lane) and a noticeable central bulge with a star-like core. It is not a member of the Virgo Cluster. NGC 4565 lies about 20 million l.y. away.

Many of the brightest galaxies in Coma Berenices are members of the Virgo Cluster, the centre of which lies just south of the Coma / Virgo border. At the cluster's heart lies the massive elliptical galaxy M87, which photographs show to be shooting out a long jet of hot gas. For a description see page 252. Lick Observatory.

CORONA AUSTRALIS

CrA · Coronae Australis

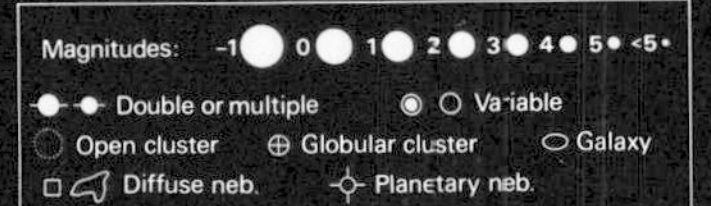

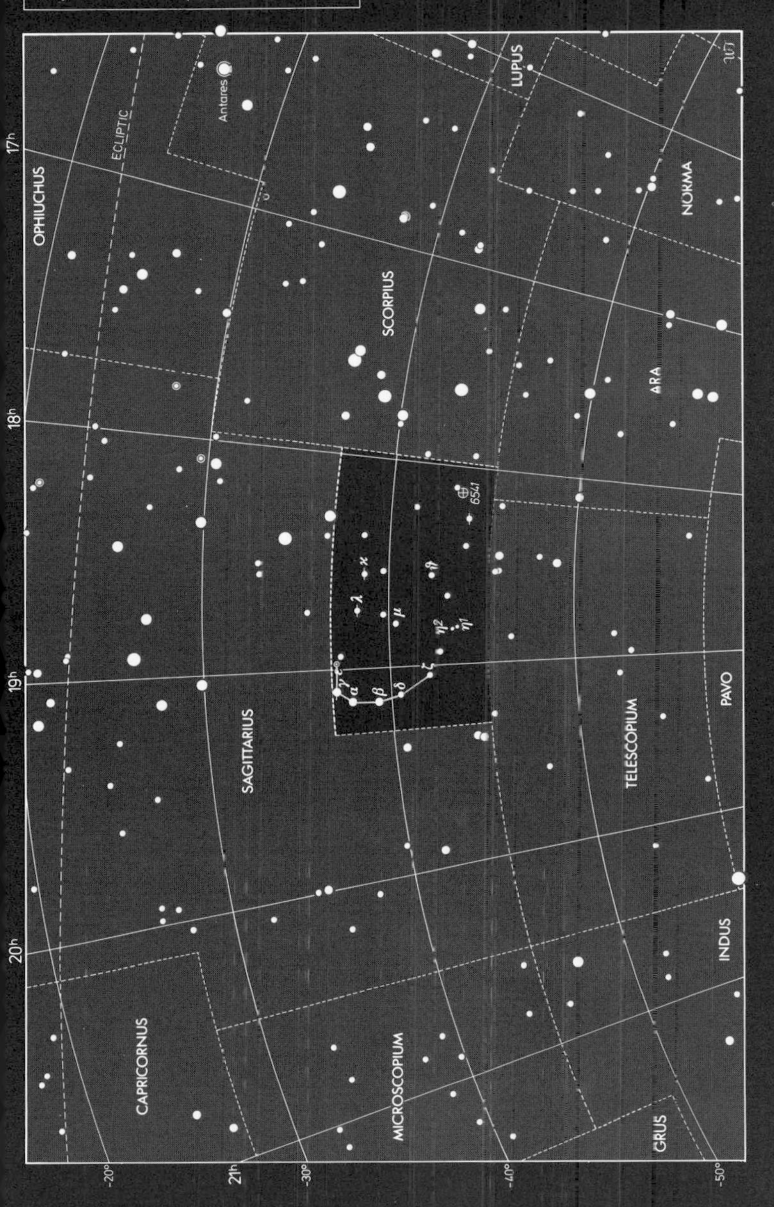

CORONA BOREALIS The Northern Crown

An ancient constellation, representing the jewelled crown worn by Ariadne when she married Bacchus and cast by him into the sky to mark the happy event. It consists of an arc of seven stars, all but one of 4th magnitude. The exception is 2nd-magnitude Gemma, which is set in the crown like a central gem, as its name implies; it is alternatively known as Alphecca, from the Arabic name for the constellation. Corona Borealis contains a famous cluster of about 400 galaxies more than 1000 million light years away. Being so very distant, the galaxies are no brighter than 16th magnitude and are thus far beyond the reach of amateur telescopes.

α (alpha) Coronae Borealis, 15h 35m +27°, (Gemma or Alphecca), mag. 2.2, is a blue-white star 75 l.y. distant. It is an eclipsing binary of the Algol type, but its variation every 17.4 days is only 0.1 mag., too slight to be noticeable to the naked eye.

ζ (zeta) CrB, 15h 39m +37°, 420 l.y. away, is a pair of blue-white stars, mags. 5.0 and 6.0, visible in small telescopes.

ν (nu) CrB, 16h 22m +34°, is a wide binocular pair of orange giants of mags. 5.2 and 5.4, but unrelated – one is 590 l.y. distant and the other is 460 l.y. away.

σ (sigma) CrB, 16h 15m +34°, 78 l.y. away, is a pair of yellow stars for small telescopes, of mags. 5.6 and 6.6. They form a genuine binary with an estimated orbital period of about 1000 years.

R CrB, 15h 49m +28°, is a remarkable yellow supergiant star lying within the arc of the Crown, halfway between the stars α (alpha) and ι (iota). It usually appears about 6th mag., but occasionally and unpredictably drops in a matter of weeks to as faint as mag. 15, from where it may take many months to regain its former brightness. Recent catastrophic declines in the star's brilliance occurred in 1962, 1972 and 1977, and another could occur at any time. These sudden dips in the light of R Coronae Borealis are believed to be due to the accumulation of carbon particles (i.e. soot) in its atmosphere. R Coronae Borealis is estimated to lie over 7000 l.y. away.

T CrB, 16h 00m +26°, is another spectacular variable star, known as the Blaze Star, which performs in almost the opposite way to R Coronae Borealis. It is a recurrent nova, usually slumbering at around mag. 11 but which can suddenly and unpredictably brighten to mag. 2. Its last recorded outburst was in 1946, and the previous one was 80 years before that. It is not known when it may erupt again.

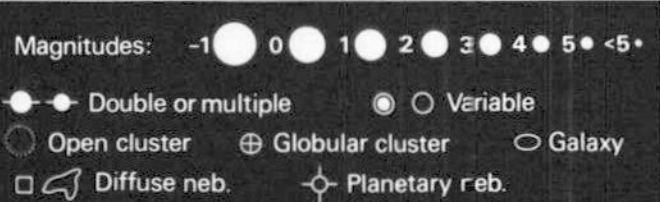

27
CORONA BOREALIS
CrB · Coronae Borealis

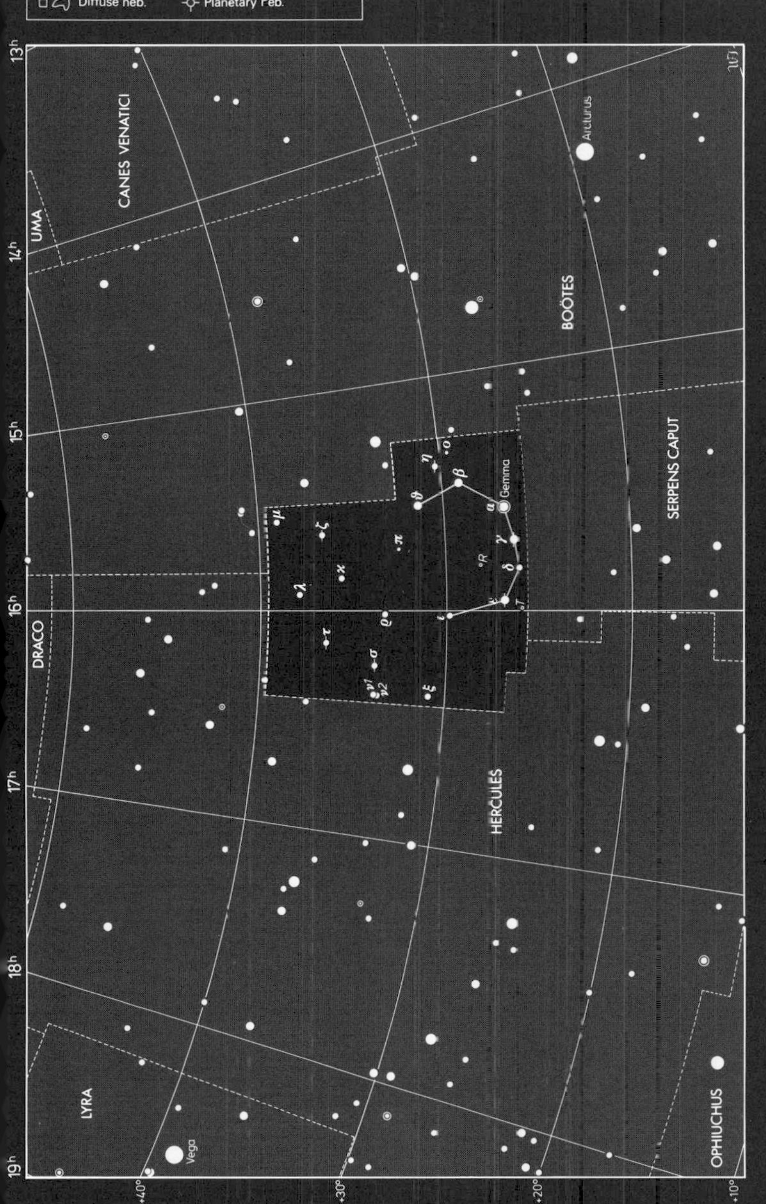

CORVUS The Crow

In Greek legend, Corvus is associated with the neighbouring constellations of Crater, the Cup, and Hydra, the Water Snake. The crow is said to have been sent by Apollo to fetch water in a cup, but along the way it dallied to eat figs. When the crow returned to Apollo it carried the water snake in its claws, claiming that this creature had been blocking the spring and was the cause of its delay. Apollo, seeing through the lie, banished the the trio to the sky, where the Crow and the Cup lie on the back of Hydra. For its misdeed the crow was condemned to suffer from eternal thirst, which is why crows croak so harshly; in the sky, the cup is just out of the thirsty crow's reach. In another legend, a snow-white crow brought Apollo the bad news that his lover Coronis had been unfaithful to him. In his anger, Apollo turned the crow black. Apollo and crows are closely linked in legend, for during the war waged by the giants on the gods, Apollo turned himself into a crow.

α (alpha) Corvi, 12h 08m –25°, (Alchiba), mag. 4.0, is a white star 52 l.y. away.

β (beta) Crv, 12h 34m –23°, mag. 2.7, is a yellow giant star 265 l.y. away.

γ (gamma) Crv, 12h 16m –18°, mag. 2.6, is a blue-white giant 220 l.y. away.

δ (delta) Crv, 12h 30m –17°, is a wide double star for small telescopes. The brightest component, visible to the naked eye, is a mag. 3.0 blue-white star 115 l.y. away, which is accompanied by a mag. 9.2 star often described as purplish in colour.

Σ 1669 (Struve 1669), 12h 41m –13°, is a neat pair of white stars 230 l.y. away, appearing to the naked eye as a single star of mag. 5.3 but divisible by small telescopes into components of mags. 6.0 and 6.1.

Just over the border between Corvus and Virgo lies the Sombrero Galaxy, M104, looking rather like the planet Saturn. It is a spiral galaxy with a large nucleus and a dense lane of dust. See page 252. Anglo-Australian Telescope Board.

28
CORVUS
Crv · Corvi
Magnitudes: -1 0 1 2 3 4 5 <5
Double or multiple
Variable
Open cluster
Globular cluster
Galaxy
Diffuse neb.
Planetary neb.
SEXTANS
LEO
CRATER
HYDRA
ANTLIA
VIRGO
ECLIPTIC
Spica
Algorab
Gienah
Minkar
Alchiba
Kru7
R
Σ1669
LIBRA
CENTAURUS
11h
12h
13h
14h
0°
-10°
-20°
-30°

CRATER The Cup

An ancient constellation representing the chalice of Apollo; it is associated in legend with neighbouring Corvus (see page 124). It contains no objects of particular interest.

α (alpha) Crateris, 11h 00m –19°, mag. 4.1, is a yellow giant star 165 l.y. away.

β (beta) Crt, 11h 12m –23°, mag. 4.5, is a blue-white star 72 l.y. away.

γ (gamma) Crt, 11h 25m –18°, mag. 4.1, is a white star 75 l.y. away. It has a mag. 9.6 companion visible in small telescopes.

δ (delta) Crt, 11h 19m –15°, mag. 3.6, the constellation's brightest star, is an orange giant lying 62 l.y. away.

In neighbouring Corvus, near the border with Crater, lies the peculiar pair of 11th-mag. galaxies NGC 4038–9, known as the Antennae because of their long tails. The two galaxies seem to have nearly collided. Royal Observatory Edinburgh.

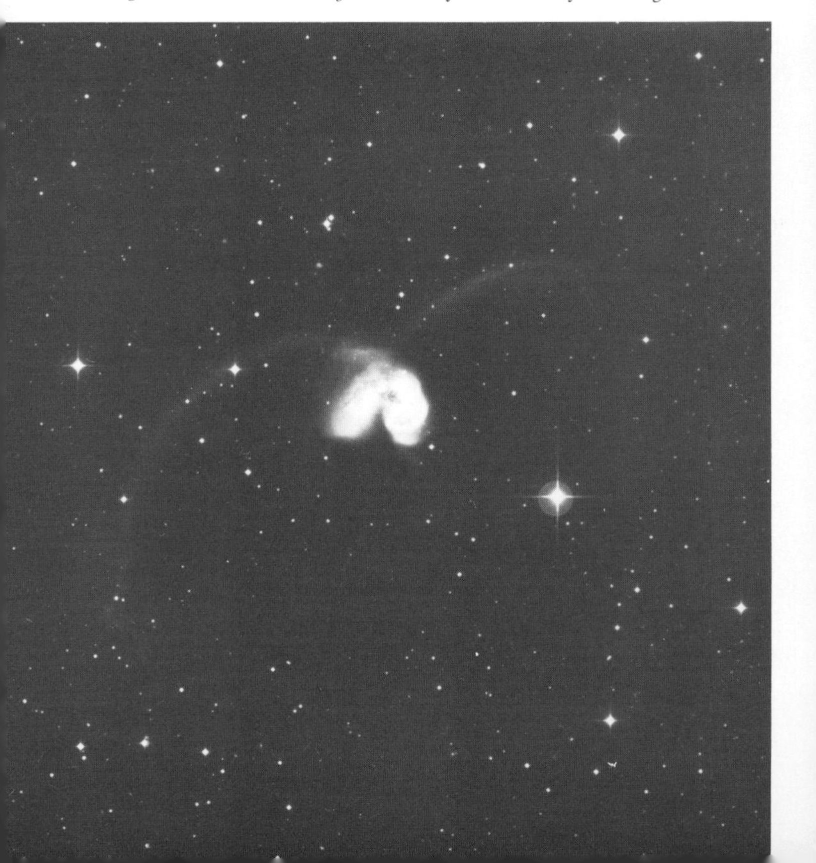

29
CRATER
Crt · Crateris

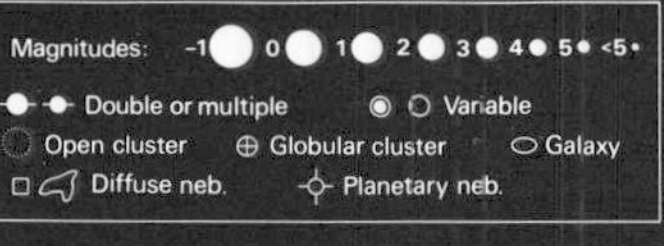

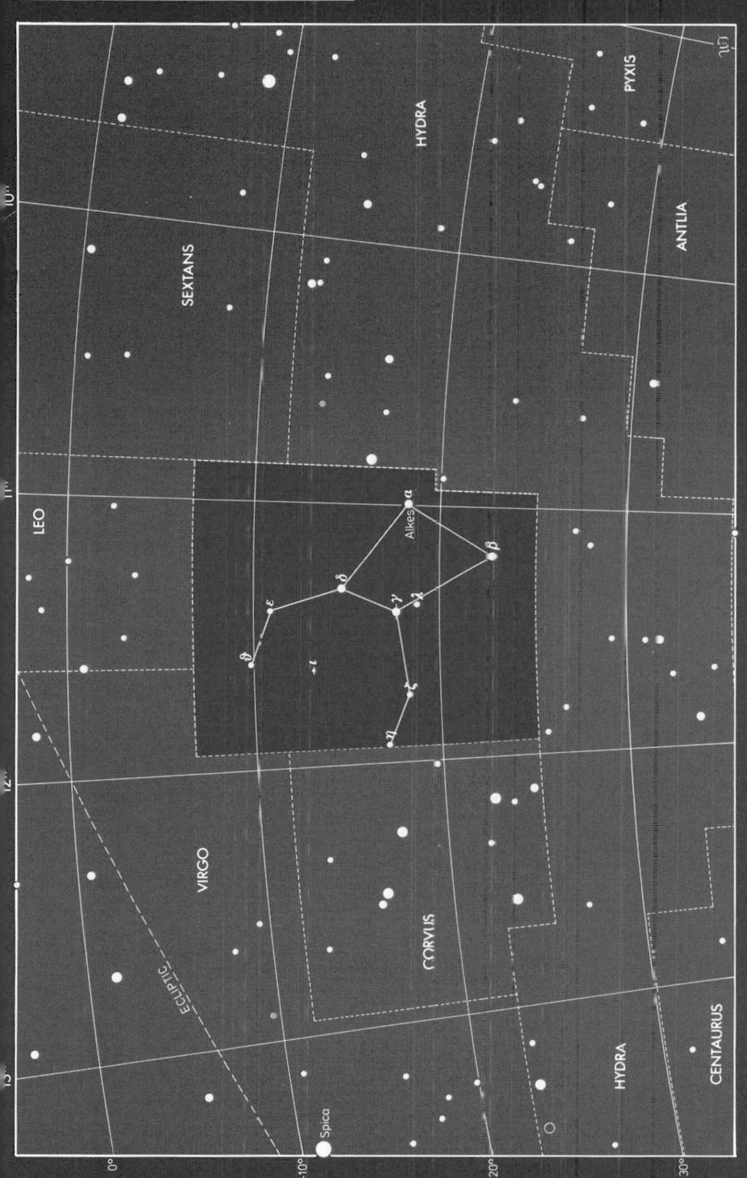

CRUX The Southern Cross

The smallest constellation in the sky, but one of the most celebrated and distinctive. It was formed from some of the stars of Centaurus by various seamen and astronomers in the 16th century. Its long axis points towards the south celestial pole. Crux lies in a dense and brilliant part of the Milky Way, which makes the famous dark nebula known as the Coalsack seem even more striking in silhouette against the starry background. The Coalsack Nebula, which spills over into neighbouring Centaurus and Musca, lies an estimated 400 l.y. away and covers nearly 7° by 5° of sky. Crux, like Centaurus, was visible from the Mediterranean area in ancient times, so its stars were known to Greek astronomers; the effects of precession have since carried it below the horizon from such northerly latitudes.

α (alpha) Crucis, 12h 27m –63°, (Acrux), 520 l.y. away, appears to the naked eye as a star of mag. 0.8. A small telescope shows it to be a sparkling double star with blue-white components of mags. 1.3 and 1.7.

β (beta) Cru, 12h 48m –60°, (Mimosa), mag. 1.2, is a blue-white giant 490 l.y. away. It is a variable star of the β Cephei type, fluctuating by less than 0.1 mag. every 6 hours.

γ (gamma) Cru, 12h 31m –57°, mag. 1.6, is a red giant star 105 l.y. away. There is a very wide mag. 6.5 unrelated companion, 275 l.y. away, visible in binoculars.

δ (delta) Cru, 12h 15m –59°, mag. 2.8, the faintest of the four main stars in the Cross, is a blue-white star 460 l.y. away.

ε (epsilon) Cru, 12h 21m –60°, mag. 3.6, is an orange giant 160 l.y. away.

ι (iota) Cru, 12h 46m –61°, 240 l.y. away, is a yellow giant of mag. 4.7 with a mag. 9.5 companion visible in small telescopes.

μ (mu) Cru, 12h 55m –57°, is a wide pair of blue-white stars of mags. 4.0 and 5.2, distances 680 and 550 l.y., splittable by the smallest telescopes or even good binoculars.

NGC 4755, 12h 54m –60°, the Jewel Box or κ (kappa) Crucis Cluster, is one of the finest star clusters in the sky, visible to the naked eye as a hazy 4th-mag. star. A small telescope shows at least 50 stars of various colours, mostly blue; the brightest individual members are blue supergiants of 6th and 7th mag., with an 8th-mag. red supergiant at the centre. The star κ Crucis itself is a blue supergiant of mag. 5.9. John Herschel gave this cluster its popular name when he likened it to a piece of multicoloured jewellery. The distance of the Jewel Box is 7600 l.y. (For a photograph see page 258.)

30
CRUX
Cru · Crucis

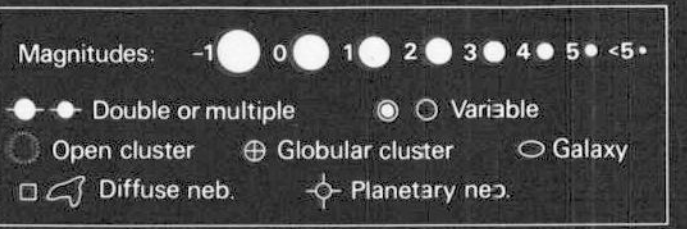

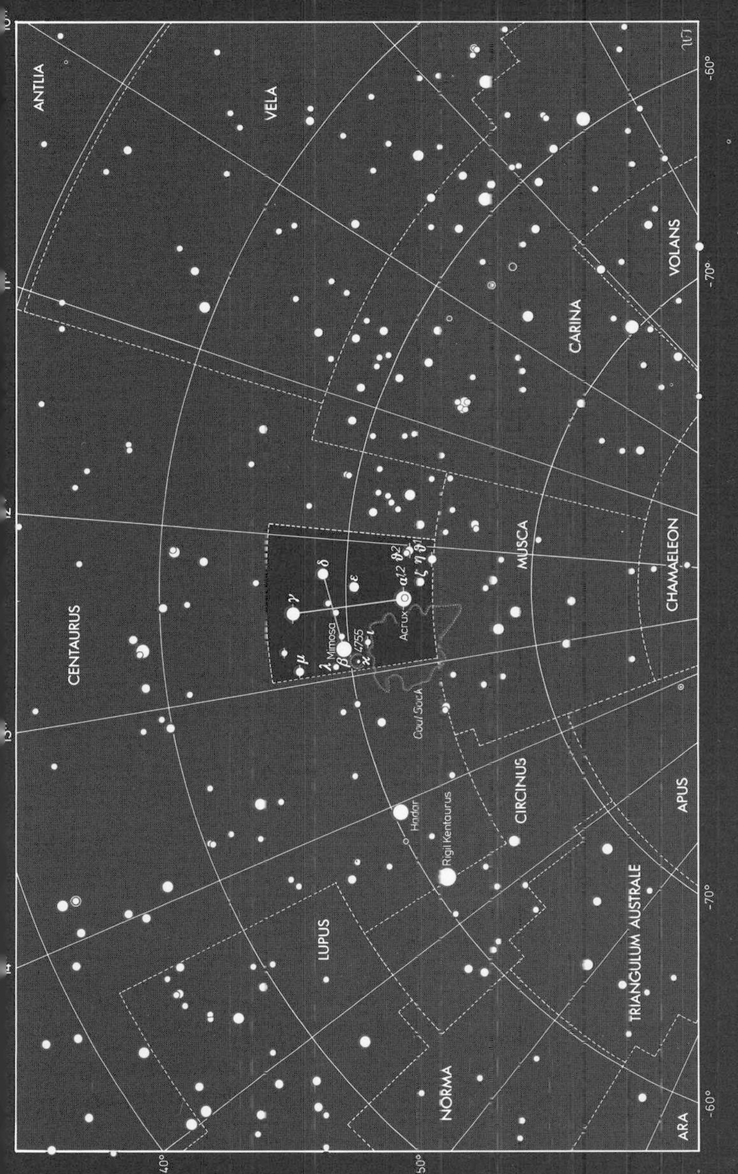

CYGNUS The Swan

Cygnus represents a swan flying down the Milky Way. In Greek mythology, the god Zeus visited Leda, wife of King Tyndareus of Sparta, in the guise of a swan; the result of their union was Pollux, one of the heavenly twins. The flying swan's tail is marked by the star Deneb, its beak by Albireo and its wings by δ (delta) and ε (epsilon) Cygni. These stars form a distinctive cross-shape, and so the constellation is also referred to as the Northern Cross; it is far larger than the Southern Cross. Cygnus lies in a rich part of the Milky Way, which is split here by a dark lane of dust known as the Cygnus Rift or the Northern Coalsack. Deneb, the constellation's brightest star, forms one corner of the Summer Triangle, completed by Altair and Vega.

Among the fascinating objects in Cygnus is an X-ray source called Cygnus X-1, believed to mark the position of a black hole; it lies at 19h 58.4m, +35° 12′, near η (eta) Cygni. Cygnus A, at 19h 59.5m, +40° 44′, near γ (gamma) Cygni, is a powerful radio source, believed to be two galaxies colliding. Long-exposure photographs of the region between ε (epsilon) Cygni and the border with Vulpecula reveal beautiful swirls of gas, known as the Cygnus Loop (see page 133).

α (alpha) Cygni, 20h 41m +45°, (Deneb, 'tail'), mag. 1.3, is a blue-white supergiant star 1700 l.y. away.

β (beta) Cyg, 19h 31m +28°, (Albireo), 390 l.y. away, is one of the sky's showpiece doubles. It consists of gloriously contrasting yellow and blue-green stars, like a celestial traffic light. The brighter star, of mag. 3.1, is a yellow giant, and its blue-green companion is mag. 5.1. They can be separated through good binoculars and are a beautiful sight in any amateur telescope.

γ (gamma) Cyg, 20h 22m +40°, (Sadr, 'breast'), mag. 2.2, is a yellow-white supergiant 550 l.y. away.

δ (delta) Cyg, 19h 45m +45°, 190 l.y. away, is a blue-white giant of mag. 2.9 with a close mag. 6.3 companion, visible in a telescope of 100 mm aperture or above at high magnification. The stars have an orbital period of over 800 years.

ε (epsilon) Cyg, 20h 46m +34°, (Gienah, 'wing'), mag. 2.5, is a yellow giant 75 l.y. away.

μ (mu) Cyg, 21h 44m +29°, 72 l.y. away, is a pair of white stars of mags. 4.8 and 6.1 orbiting each other about every 500 years, requiring an aperture of 150 mm with high magnification to be split; they are currently closing and will probably require 220 mm by the year 2010, although they start to open again after 2020. A wide mag. 6.9 binocular companion is unrelated.

o^1 (omicron^1) Cyg, 20h 14m +47°, is perhaps the most beautiful binocular double in the heavens, consisting of stars of orange and turquoise, mags. 3.8 and 4.9, 200 l.y. and 270 l.y. away, like a wider version of Albireo. A small telescope, or binoculars held steadily, shows a closer blue companion of mag. 7.0 to the brighter (orange giant) star. ▶

31
CYGNUS
Cyg · Cygni

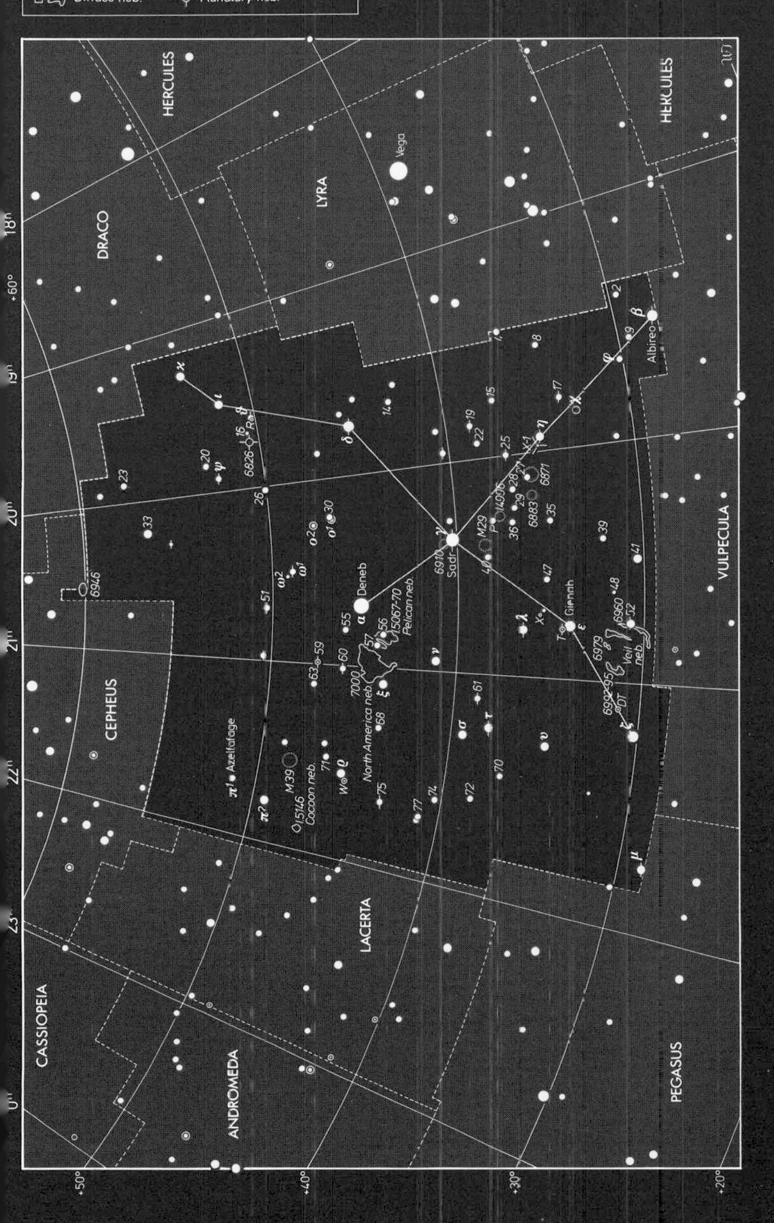

◀ χ (chi) Cyg, 19h 51m +33°, 82 l.y. away, is a red giant long-period variable that varies between 3rd and 14th mag. every 400 days or so.

ψ (psi) Cyg, 19h 56m +52°, 130 l.y. away, is a pair of white stars of mags. 4.9 and 7.4 for small to moderate apertures.

61 Cyg, 21h 07m +39°, 11.1 l.y. away, is a showpiece pair of orange dwarf stars of mags. 5.2 and 6.0, orbiting each other in about 650 years, divisible in a small telescope or even binoculars. In addition to being among the closest stars to Earth, 61 Cygni was the first star to have its parallax measured, by the German astronomer Friedrich Wilhelm Bessel in 1838.

P Cyg, 20h 18m +38°, is an erratically variable blue supergiant that normally resides around 5th mag., although in the year 1600 it reached a peak of 3rd mag. Evidently the star is so large and luminous that it is close to instability. It has brightened gradually since the 18th century as it evolves into a red supergiant. Its distance is uncertain, but is probably several thousand light years.

M39 (NGC 7092), 21h 32m +48°, is a large, loose cluster of about 30 stars of 7th mag. and fainter, visible in binoculars. It lies 880 l.y. away.

NGC 6826, 19h 45m +51°, is an 8th-mag. planetary nebula about 3200 l.y. away, known as the Blinking Planetary because it appears to blink on and off. It is visible as a pale blue disk in 75-mm telescopes, but 150 mm is needed to show it to advantage. At its centre is a 10th-mag. star. Looking alternately at this star and away from it produces the 'blinking' effect. It lies within 1° of the double star 16 Cygni, which consists of two 6th-mag. creamy-white stars 53 l.y. away.

NGC 7000, the North America Nebula, can be seen as a hook-shaped brightening in the Milky Way with the naked eye or binoculars in good skies. Despite its large size, 2° across at its widest, it can be difficult to detect because of its low surface brightness. Long-exposure photographs show its shape most clearly, like the continent of North America. The nebula is about 1500 l.y. away, similar to the distance of the star Deneb. However, the nebula is believed to be lit up not by Deneb but by an extremely hot 6th-mag. star that lies within it.

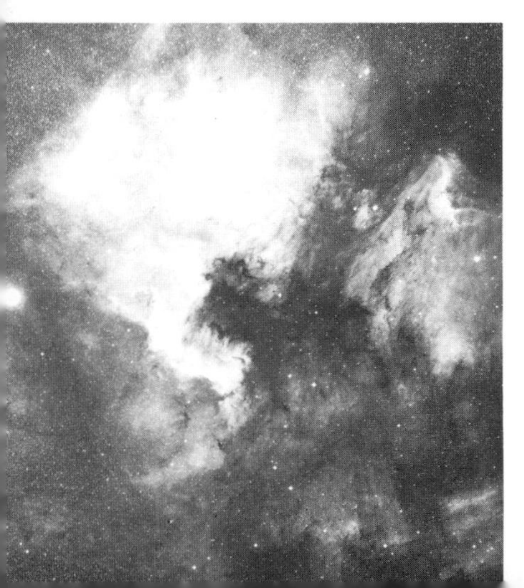

The North America Nebula is a cloud of glowing gas shaped like the continent of North America, its particularly distinctive Gulf of Mexico being formed by an obscuring cloud of dark dust. It lies near the star ξ (xi) Cygni, an orange supergiant of mag. 3.7. Hale Observatories photograph.

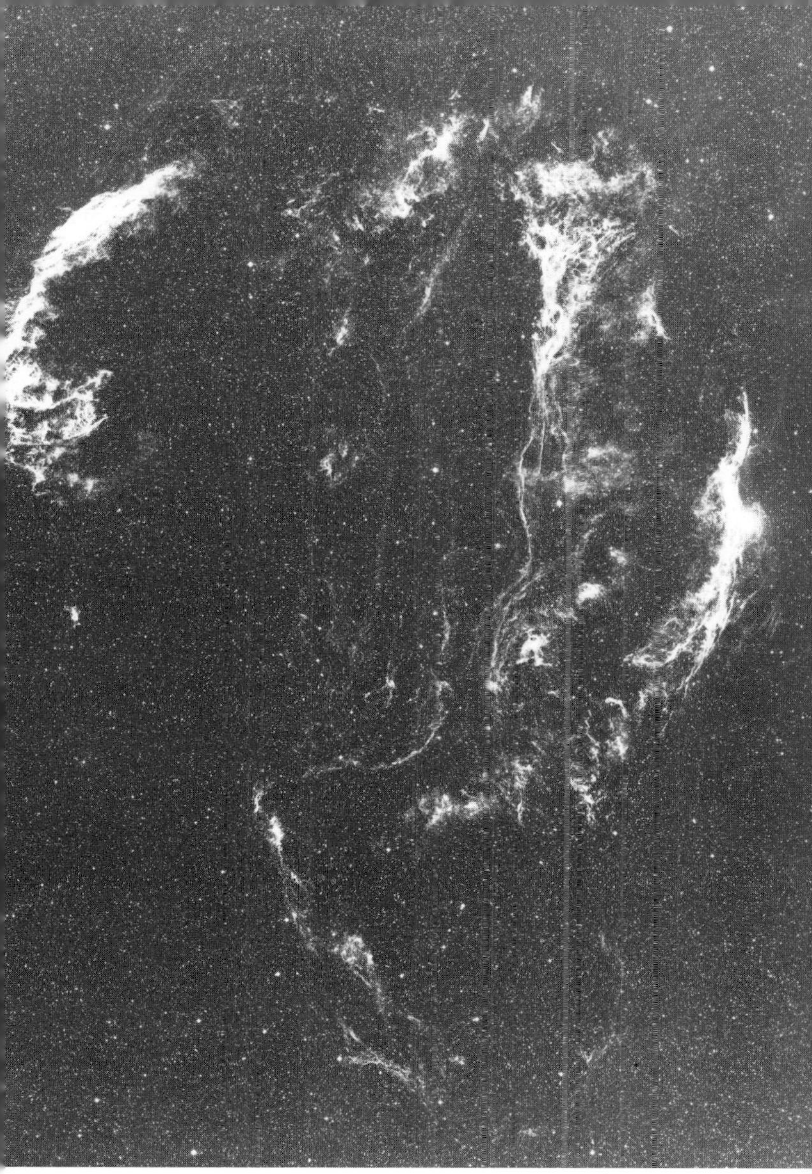

Drifting apart over thousands of years, the delicate traceries of the Cygnus Loop are the tenuous remains of a star that shattered itself in a supernova explosion some 60,000 years ago. At top left is its brightest part, NGC 6992, the Veil Nebula, which can be glimpsed in binoculars under clear, dark skies. The fainter arc on the right is NGC 6960, next to the 4th-mag. star 52 Cygni. The Cygnus Loop lies 4200 l.y. away. Hale Observatories photograph.

DELPHINUS The Dolphin

A constellation which originated in Greek times, celebrating the long-standing relationship between humans and this intelligent sea creature. In legend, dolphins were the messengers of the sea-god Poseidon. They were credited with saving the life of Arion, a musician and poet, who was attacked on a ship. In this legend, the nearby constellation Lyra represents Arion's lyre. Delphinus has a distinctive shape, its four main stars forming a rectangle known as Job's Coffin. Its two brightest stars are called Sualocin and Rotanev, which backwards read Nicolaus Venator, the Latinized form of the name Niccolò Cacciatore, who was assistant and successor to the Italian astronomer Giuseppe Piazzi at Palermo Observatory. Like the small neighbouring constellations Vulpecula and Sagitta, Delphinus lies in a rich area of the Milky Way and has become a favourite hunting-ground for novae.

α (alpha) Delphini, 20h 40m +16°, (Sualocin), mag. 3.8, is a blue-white star 190 l.y. away.

β (beta) Del, 20h 38m +15°, (Rotanev), mag. 3.6, is a white star 72 l.y. away. It is a binary with an orbital period of 27 years, but the components are too close to split in amateur telescopes.

γ (gamma) Del, 20h 47m +16°, 125 l.y. away, is a showpiece double star consisting of golden and yellow-white stars of mags. 4.3 and 5.1, neatly separated in a small telescope. In the same telescopic field of view appears a faint binary, Struve 2725, consisting of stars of mags. 7.6 and 8.4.

The Naming of Variable Stars

In addition to the normal system of naming stars in a constellation (page 6), stars that vary in brightness have their own system of nomenclature. Stars that were already named when their variability was discovered, such as δ (delta) Cephei, β (beta) Persei or o (omicron) Ceti, retain their existing designation. Other variable stars are denoted by a system of one or two letters or, where that is insufficient, by the letter V and a number. The first nine variables in a constellation are given the letters R to Z. Then double letters are given, from RR to RZ. Next the lettering runs from SS to SZ, and so on until ZZ is reached. Then the sequence goes from AA to AZ, BB to BZ, ending with QZ, when 334 variable stars have been named. (The letter J is omitted.) Further variables are given the designation V335, V336, and so on. Novae, too, are given variable-star designations. Hence the nova that erupted in Delphinus in 1967 became known as HR Delphini.

32
DELPHINUS
Del · Delphini

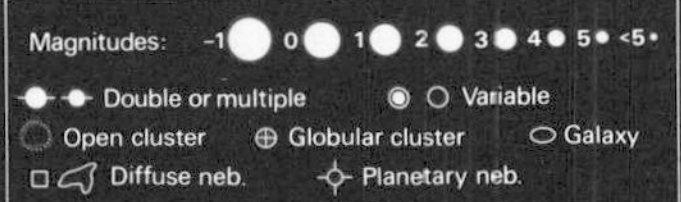

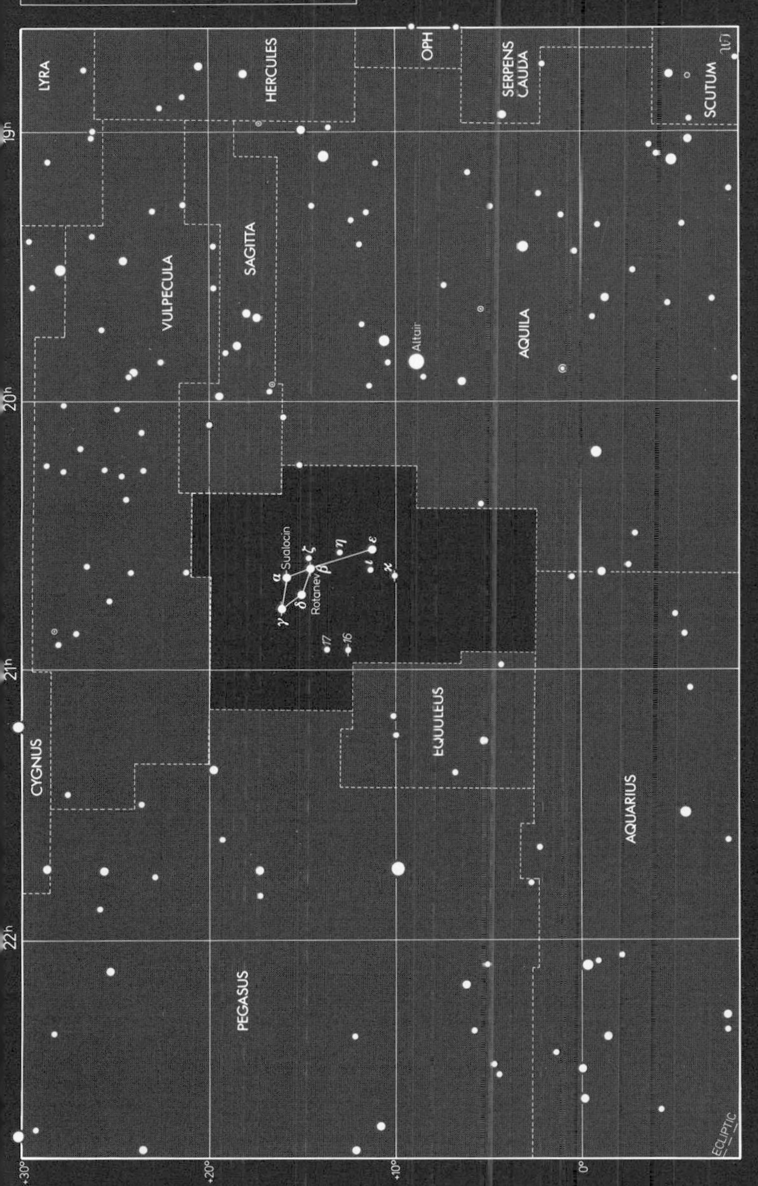

DORADO The Goldfish

This constellation, also known as the Swordfish, was introduced at the end of the 16th century by the Dutch navigators Pieter Dirkszoon Keyser and Frederick de Houtman. Its most notable feature is the Large Magellanic Cloud, the larger of the two satellite galaxies that accompany our Milky Way. In 1987 the first supernova visible to the naked eye since 1604 erupted in the Large Magellanic Cloud, at 5h 35m –69°.

α (alpha) Doradus, 4h 34m –55°, mag. 3.3, is a blue-white giant 190 l.y. away.

β (beta) Dor, 5h 34m –62°, a yellow supergiant 5000 l.y. away, is one of the brightest Cepheid variables, ranging from mag. 3.5 to 4.1 every 9 days 20 hours.

Large Magellanic Cloud (LMC) is a mini-galaxy 169,000 l.y. away, a satellite of the Milky Way, containing perhaps 10,000 million stars. To the naked eye it appears as a fuzzy patch 6° in diameter, 12 times the apparent size of the Moon; its actual diameter is 30,000 l.y. Binoculars and telescopes show individual stars, nebulae (notably the Tarantula Nebula, described below) and clusters.

NGC 2070, 5h 39m –69°, the Tarantula Nebula, is a looping cloud of hydrogen gas about 1000 l.y. in diameter in the Large Magellanic Cloud. To the naked eye the nebula appears as a fuzzy star, also known as 30 Doradus. The nebula's popular name, the Tarantula, comes from its spider-like shape. At the centre of the nebula is a cluster of supergiant stars, called R136, the light from which makes the nebula glow. The Tarantula is larger and brighter than any nebula in the Milky Way. If it were as close to us as the Orion Nebula, the Tarantula would fill the whole constellation of Orion, and would cast shadows.

The elongated shape of the Large Magellanic Cloud, showing many loops of glowing gas in addition to star swarms. The large knot of bright gas at centre left is the Tarantula Nebula, NGC 2070. Royal Observatory Edinburgh.

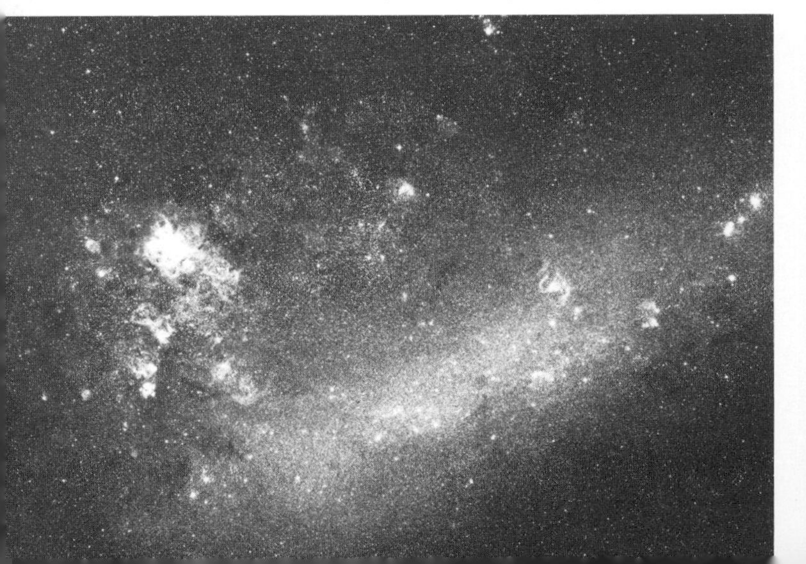

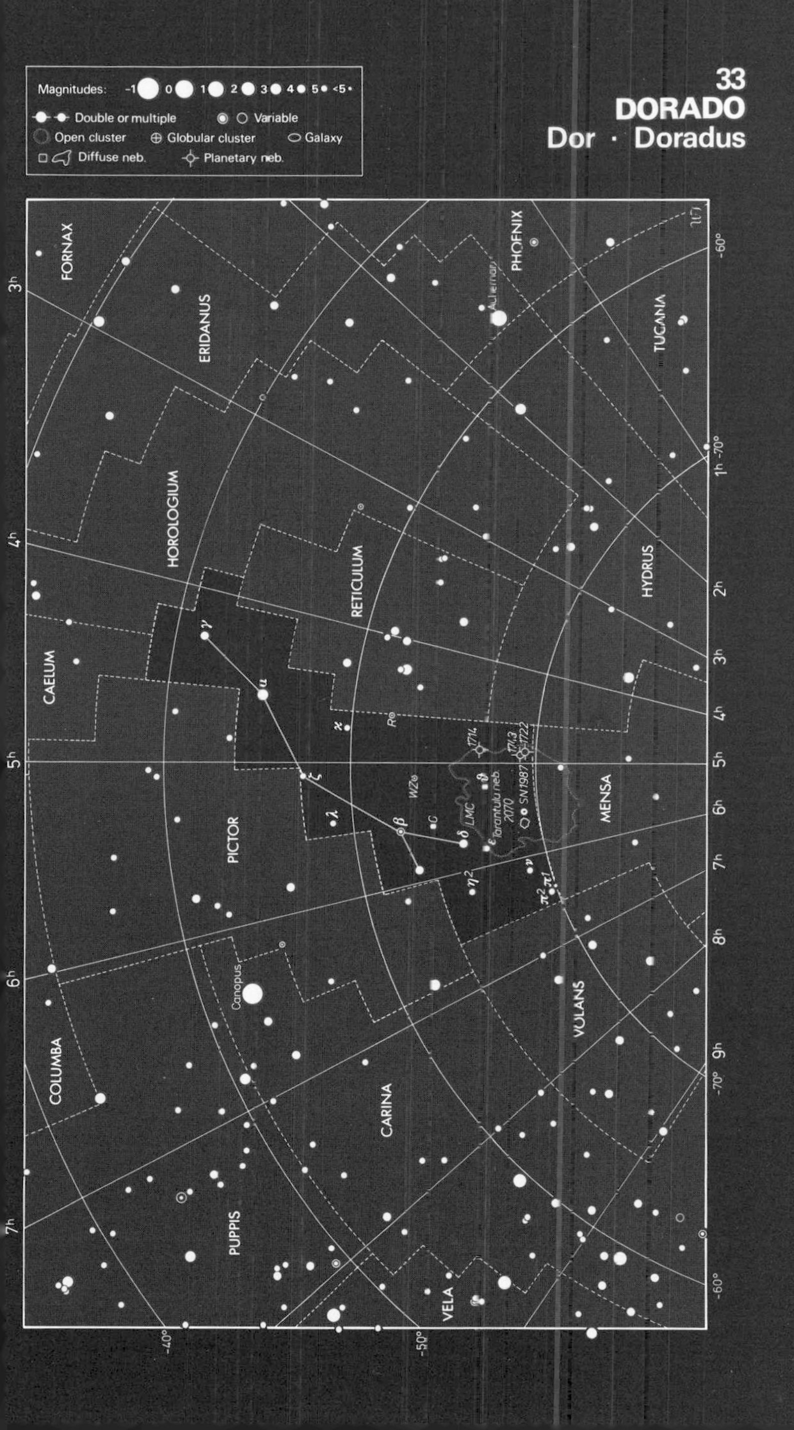
33
DORADO
Dor · Doradus
Magnitudes: -1 0 1 2 3 4 5 <5
Double or multiple
Variable
Open cluster
Globular cluster
Galaxy
Diffuse neb.
Planetary neb.
FORNAX
ERIDANUS
PHOENIX
Achernar
TUCANA
HOROLOGIUM
RETICULUM
HYDRUS
CAELUM
PICTOR
LMC
Tarantula neb.
2070
SN1987
1714
1743
1722
MENSA
COLUMBA
Canopus
CARINA
VOLANS
PUPPIS
VELA
3h
4h
5h
6h
7h
8h
9h
1h
2h
-40°
-50°
-60°
-70°

DRACO The Dragon

Dragons feature in many ancient legends, so it is not surprising to find such a monster in the sky. This one is said to be Ladon, the dragon slain by Hercules so he could steal the golden apples from the garden of the Hesperides. In the sky, one of the feet of Hercules rests upon the dragon's head, while its body lies coiled around the north celestial pole. Although Draco is one of the largest and most ancient constellations it is indistinct, containing no stars brighter than 2nd magnitude. Draco contains the north pole of the ecliptic, i.e. one of two points 90° from the plane of the Earth's orbit.

α (alpha) Draconis, 14h 04m +64°, (Thuban, 'serpent's head'), mag. 3.7, is a blue-white giant star 290 l.y. away. It was the Pole Star in about 2800 BC, but has lost that place to Polaris because of the effect of precession (see page 11).

β (beta) Dra, 17h 30m +52°, (Rastaban, 'serpent's head'), mag. 2.8, is a yellow giant 250 l.y. distant.

γ (gamma) Dra, 17h 57m +51°, (Eltanin, 'the serpent'), mag. 2.2, is an orange giant 105 l.y. away, and the brightest star in the constellation. From observations of this star the English astronomer James Bradley discovered the effect known as the aberration of starlight in 1728.

μ (mu) Dra, 17h 05m +54°, 76 l.y. away, is a close double star with matching cream-coloured components each of mag. 5.7. A telescope of 75 mm aperture with high magnification is needed to separate the stars, which orbit every 480 years. After the year 2000 the two become progressively easier to split.

ν (nu) Dra, 17h 32m +55°, is a pair of identical white stars of mag. 4.9, easily visible in the smallest of telescopes and regarded as one of the finest binocular pairs. They lie about 100 l.y. away.

o (omicron) Dra, 18h 51m +59°, 155 l.y. away, is a mag. 4.7 yellow giant star with a mag. 7.8 companion visible in small telescopes.

ψ (psi) Dra, 17h 42m +72°, 67 l.y. away, is a mag. 4.6 yellow-white star with a yellow mag. 5.8 companion visible in small telescopes or even binoculars.

16–17 Dra, 16h 36m +53°, is a wide pair of blue-white stars of mags. 5.1 and 5.5, 330 l.y. away, easily found in binoculars. Telescopes of 60 mm aperture with high magnification split the brighter star into a binary of mags. 5.5 and 6.5, making this a striking triple system.

39 Dra, 18h 24m +59°, 175 l.y away, is an impressive triple system, the two brightest members of which, at mags. 5.0 and 7.4, appear in binoculars as a wide blue-and-yellow pair. A telescope of 60 mm aperture or more with high magnification reveals that the brighter star has a closer mag. 8.0 companion.

NGC 6543, 17h 59m +67°, is a 9th-mag. planetary nebula 3500 l.y. away, one of the brightest of the class, showing in amateur telescopes as an irregular blue-green disk like an out-of-focus star.

34 DRACO
Dra · Draconis

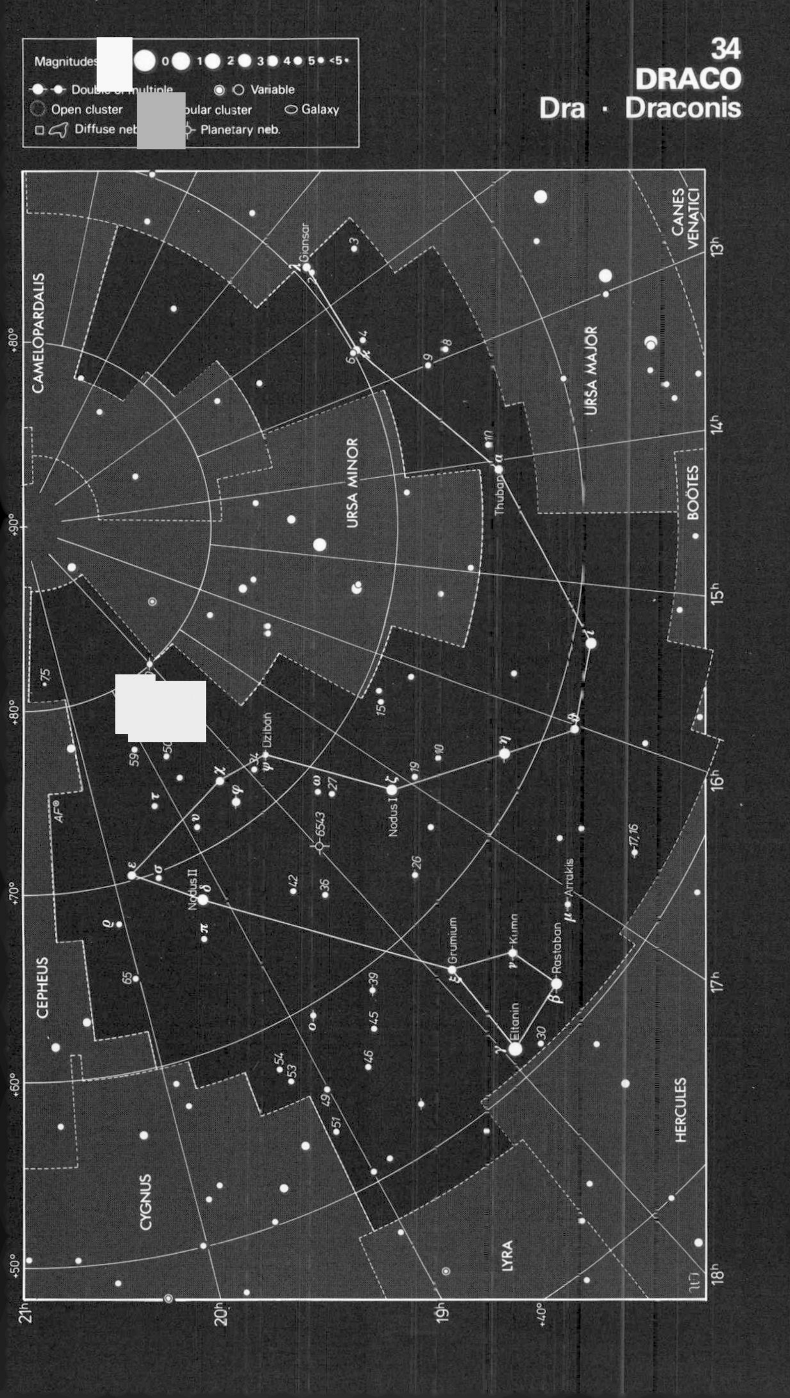

EQUULEUS The Little Horse

The second-smallest constellation in the sky, Equuleus seems to have originated with the Greek astronomer Ptolemy in the second century AD. Only the head of the horse is shown, next to the much larger horse Pegasus, and there are no legends that identify it.

α (alpha) Equulei, 21h 16m +5°, (Kitalpha, 'section of the horse'), mag. 3.9, is a yellow giant 300 l.y. away.

γ (gamma) Equ, 21h 10m +10°, is a mag. 4.7 white star about 120 l.y. away; a mag. 6.1 binocular companion, 6 Equ, is unrelated.

ε (epsilon) Equ, 20h 59m +4°, also known as 1 Equ, is a triple star. In small telescopes it appears as a white-and-yellow pair of mags. 5.4 and 7.4, distances 200 and 125 l.y. The brighter component is also a close binary, of mags. 6.0 and 6.3; these stars orbit each other every 101 years. Currently they are rapidly closing together, and by 2015 will be too close to separate in amateur telescopes.

NGC 1300 in Eridanus, 50 million l.y. away, is a classic barred spiral galaxy with prominent lanes of dust along its central bar. Hale Observatories photograph.

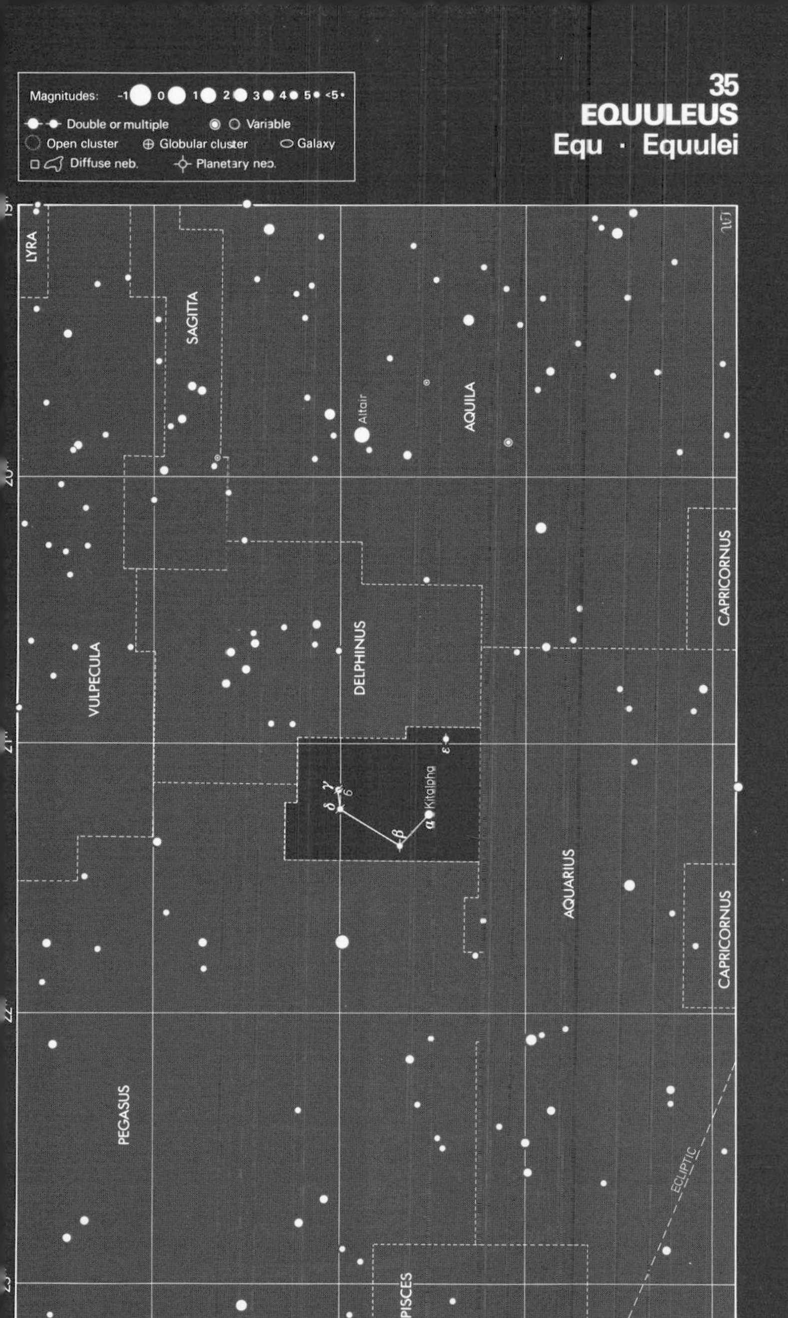
35
EQUULEUS
Equ · Equulei
Magnitudes: -1 0 1 2 3 4 5 <5
Double or multiple
Variable
Open cluster
Globular cluster
Galaxy
Diffuse neb.
Planetary neb.
LYRA
SAGITTA
VULPECULA
DELPHINUS
AQUILA
Altair
Kitalpha
CAPRICORNUS
AQUARIUS
CAPRICORNUS
PEGASUS
PISCES
ECLIPTIC
+20°
+10°
0°
-10°

ERIDANUS The River

An extensive constellation, the sixth-largest in the sky, but often overlooked because of its faintness. It meanders from Taurus in the north to Hydrus in the south. In mythology, Eridanus was the river into which Phaethon fell after trying to drive the chariot of his father Helios, the Sun god. But it also supposedly represents some real river. Early mythologists identified it with the Nile, but later Greek writers said it was the Po in Italy. Originally Eridanus included the stars of what is now Fornax and it stretched only as far as ϑ (theta) Eridani, which was then known as Achernar, from the Arabic meaning 'river's end' (its present name, Acamar, comes from the same Arabic original as Achernar). In recent times Eridanus has been extended to nearly 60° south (below the horizon from Greece), and another star has been given the title Achernar. The constellation contains several interesting galaxies, all too distant and too faint to be picked up easily in amateur telescopes. One celebrated example is NGC 1300, located at 3h 20m –19°, a beautiful 10th-magnitude barred spiral pictured on page 140.

α (alpha) Eridani, 1h 38m –57°, (Achernar, 'river's end'), mag. 0.5, is a blue-white star 91 l.y. away.

β (beta) Eri, 5h 08m –5°, (Cursa, 'the footstool', referring to its position under the foot of Orion), mag. 2.8, is a blue-white giant 78 l.y. away.

ε (epsilon) Eri, 3h 33m –9°, mag. 3.7, is one of the most Sun-like of the nearby stars, lying 10.7 l.y. away. A yellow dwarf, it is believed to be accompanied by a large planet or very small star. ε Eridani has been the subject of several attempts to listen for interstellar radio messages, but nothing has been heard.

ϑ (theta) Eri, 2h 58m –40°, (Acamar), 55 l.y. away, is a striking pair of blue-white stars of mags. 3.2 and 4.3, divisible in small telescopes.

o^2 (omicron^2) Eri, 4h 15m –8°, 16 l.y. away, also known as 40 Eridani, is a remarkable triple star. A small telescope shows that the mag. 4.4 yellow primary, a star similar to the Sun, has a wide mag. 9.5 white dwarf companion, the most easily seen white dwarf in the sky. Small telescopes reveal that the white dwarf has an 11th-mag. companion which is a red dwarf, thereby completing a most interesting trio. The white dwarf and red dwarf orbit each other every 250 years, and remain easy to split until the latter half of the 21st century.

32 Eri, 3h 54m –3°, 290 l.y. away, is a beautiful double star for small telescopes, consisting of an orange giant of mag. 4.8 and a blue-green mag. 6.1 companion.

p Eri, 1h 40m –56°, 22 l.y. away, is a beautiful wide duo of orange stars, mags. 5.8 and 6.0, with an orbital period of about 500 years.

NGC 1535, 4h 14m –13°, is a small 9th-mag. planetary nebula about 2000 l.y. away. Small telescopes show it, but apertures of 150 mm are needed to appreciate its blue-grey disk.

ERIDANUS
Eri · Eridani

FORNAX The Furnace

A barren constellation introduced in the 1750s by Nicolas Louis de Lacaille, originally under the name of Fornax Chemica, the Chemical Furnace. The constellation contains a dwarf elliptical galaxy, a member of our Local Group of galaxies, approximately 400,000 l.y. from the Milky Way but too faint to see in amateur telescopes. Fornax also contains a compact cluster of galaxies containing the 9th-mag. peculiar spiral NGC 1316, also known as the radio source Fornax A, about 75 million l.y. away, at 3h 22.7m, –37° 12′. Another member of the cluster is the 10th-mag. NGC 1365, 3h 33.6m, –36° 08′, a classic barred spiral often pictured in books.

α (alpha) Fornacis, 3h 12m –29°, 46 l.y. away, is a binary consisting of a yellow star of mag. 4.0 with a deeper yellow mag. 6.5 companion of suspected variability. The pair orbit each other every 300 years and will remain divisible in small telescopes throughout the 21st century.

β (beta) For, 2h 49m –32°, mag. 4.5, is a yellow giant star 210 l.y. away.

Among the many galaxies in Fornax is this 9th-mag. barred spiral, NGC 1097, 60 million l.y. away, located at 2h 46m –30°. Anglo-Australian Telescope Board.

37
FORNAX
For · Fornacis

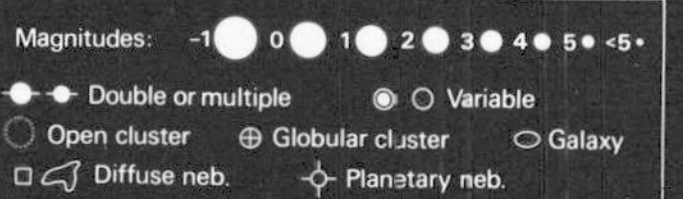

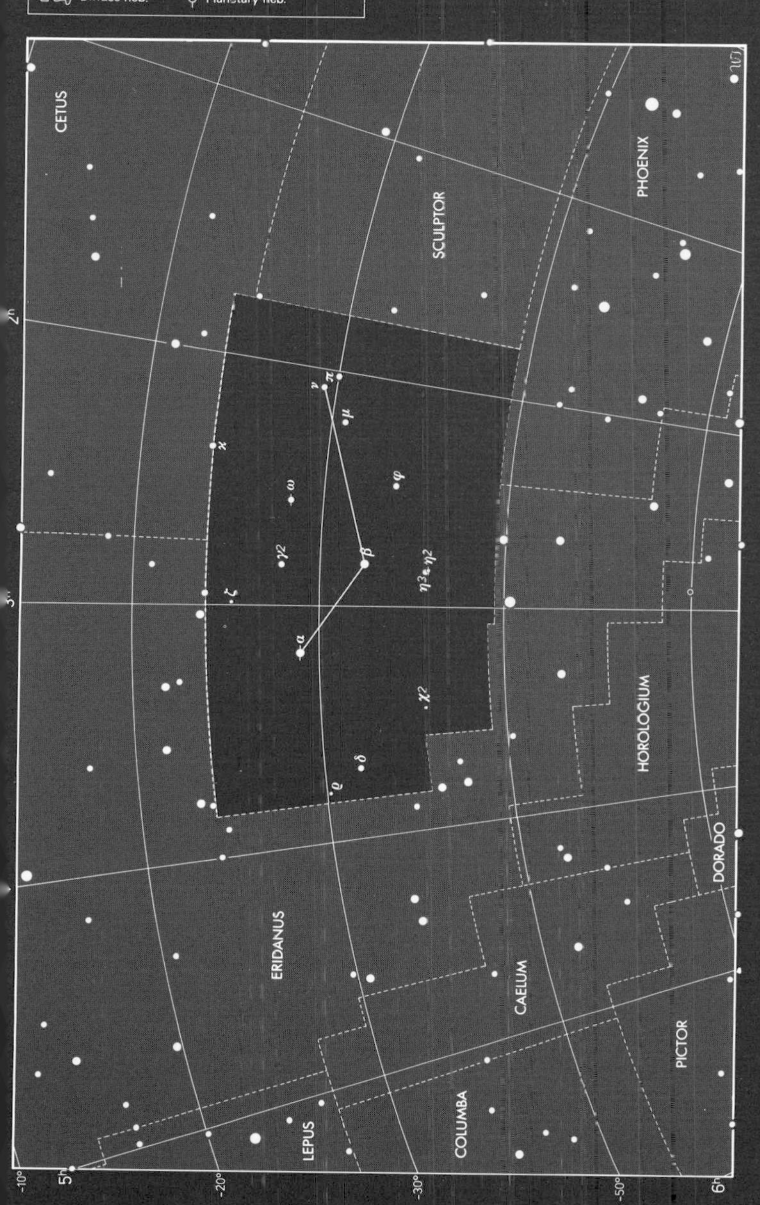

GEMINI The Twins

A constellation that dates from ancient times, representing a pair of twins. We know them as Castor and Pollux, members of the Argonauts' crew and of mixed parentage: both were sons of Queen Leda of Sparta, but Castor's father was her husband King Tyndareus, while the father of Pollux was the god Zeus. The twins were the protectors of mariners, appearing in ships' rigging as the electrical phenomenon now known as St Elmo's fire. In the sky, the stars Castor and Pollux provide a useful yardstick for measuring angular distances – they are exactly 4.5° apart. Gemini is a member of the zodiac and the Sun passes through the constellation from late June to late July. Each year the Geminid meteors, one of the year's richest and brightest showers, radiate from a point near Castor. They reach a maximum around December 13–14, when up to 75 meteors per hour may be seen.

α (alpha) Geminorum, 7h 35m +32°, (Castor), 47 l.y. away, is an astounding multiple star, consisting of six separate components. To the naked eye it appears as a blue-white star of mag. 1.6. A 60-mm telescope with high magnification splits Castor into two components of mags. 1.9 and 2.9, which orbit each other every 470 years. Separation between these stars is increasing, so they will become progressively easier to split until the end of the 21st century, when they will start to close up again. Both these stars are spectroscopic binaries. Small telescopes also show a wide red dwarf companion to Castor; this is itself an eclipsing binary of Algol type, varying between mags. 9.3 and 9.8 in 19.5 hours, completing the six-star system.

β (beta) Gem, 7h 45m +28°, (Pollux), mag. 1.1, is the brightest star in the constellation. Some astronomers have speculated that Pollux, being labelled β Geminorum, was once fainter than Castor and has since brightened, or Castor has faded. But the truth is that Johann Bayer, who allotted the stars Greek letters in 1603, did not distinguish carefully which star was the brighter, and so caused unnecessary confusion. Pollux is a yellow giant 36 l.y. away.

γ (gamma) Gem, 6h 38m +16°, mag. 1.9, is a blue-white star 85 l.y. away.

δ (delta) Gem, 7h 20m +22°, 55 l.y. away, is a creamy-white star of mag. 3.6 with an orange dwarf companion of mag. 8.2. The brightness contrast makes the pair difficult in telescopes below about 75 mm aperture. Their estimated orbital period is over 1000 years.

ε (epsilon) Gem, 6h 44m +25°, mag. 3.0, is a yellow supergiant 680 l.y. away. Powerful binoculars, or a small telescope, reveal a wide mag. 9.2 companion.

ζ (zeta) Gem, 7h 04m +21°, 1700 l.y. away, is both a variable star and a binocular double. A yellow supergiant, it is a Cepheid variable, fluctuating between mags. 3.7 and 4.2 every 10.2 days. Binoculars or small telescopes reveal a wide mag. 7.6 companion, which is unrelated.

η (eta) Gem, 6h 15m +23°, 250 l.y. away, is another double–variable. A red giant star, it fluctuates in semi-regular manner between mags. 3.3 and 3.9 in ►

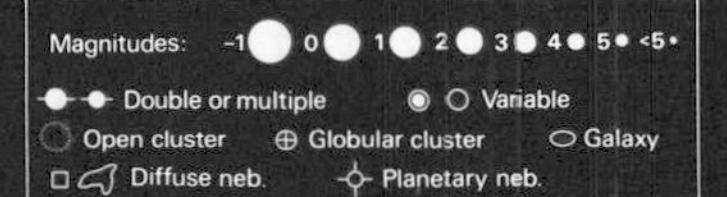

38 GEMINI

Gem · Geminorum

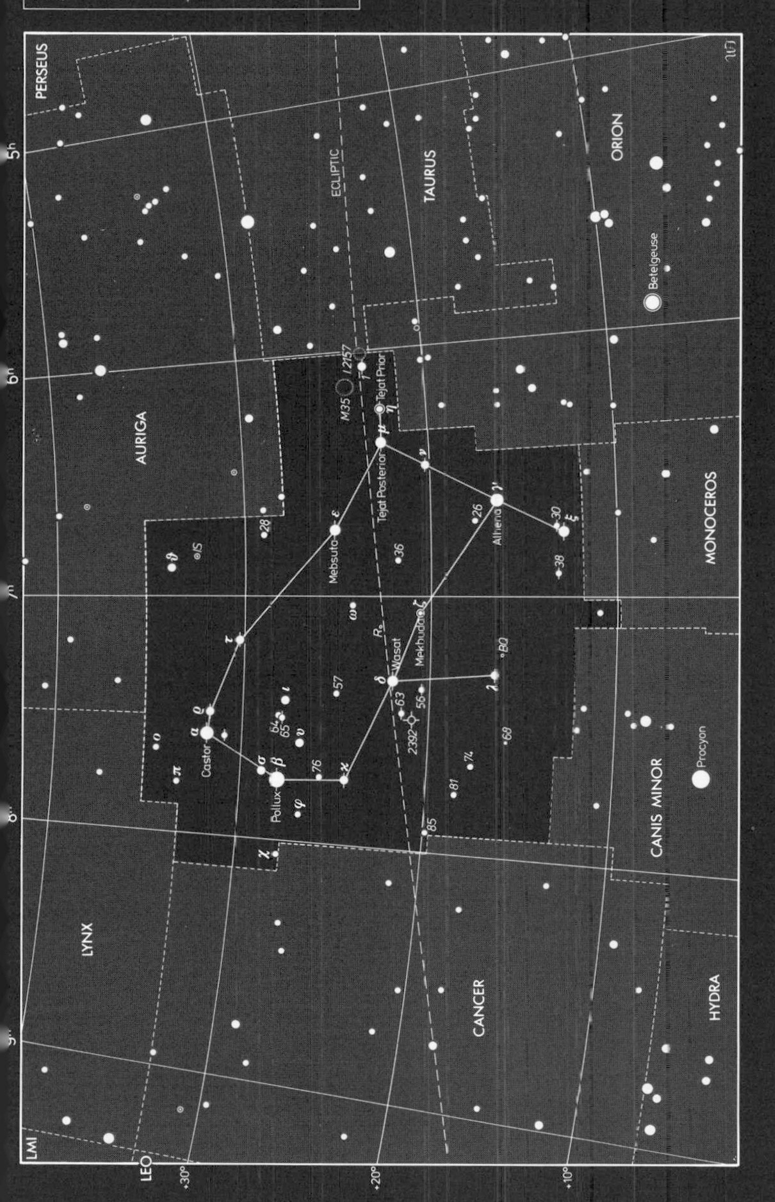

GRUS The Crane

One of the 12 constellations introduced at the end of the 16th century by the Dutch navigators Pieter Dirkszoon Keyser and Frederick de Houtman. It represents a water bird, the long-necked Crane, although it has also been depicted as a flamingo. The stars δ (delta) and μ (mu) Gruis are striking naked-eye doubles.

α (alpha) Gruis, 22h 08m –47°, (Alnair, 'the bright one'), mag. 1.7, is a blue-white star 65 l.y. away.

β (beta) Gru, 22h 43m –47°, is a red giant 115 l.y. away that varies between about mags. 2.0 and 2.3.

γ (gamma) Gru, 21h 54m –37°, mag. 3.0, is a blue-white giant 230 l.y. away.

δ (delta) Gru, 22h 29m –43°, is a naked-eye pairing of two unrelated stars: δ^1, mag. 4.0, is a yellow giant 150 l.y. away; δ^2 is a mag. 4.1 red giant 420 l.y. away.

μ (mu) Gru, 22h 16m –41°, is another naked-eye double of unrelated yellow giants that appear in the same line of sight by chance: μ^1 is of mag. 4.8, 260 l.y. away; μ^2 is of mag. 5.1, 360 l.y. away.

π (pi) Gru, 22h 23m –46°, is a binocular duo of unrelated stars. π^1 is a deep-red semi-regular variable that ranges between mags. 5.4 and 6.7 every 150 days or so; its distance is uncertain. π^2 is a white giant of mag. 5.6, 250 l.y. away.

◀ about 230 days. It has a close mag. 8.8 companion that orbits it in about 500 years and requires a large telescope to distinguish it from the primary's glare.

κ (kappa) Gem, 7h 44m +24°, 145 l.y. away, is a mag. 3.6 yellow giant, with a mag. 8.1 companion made difficult in small telescopes because of the extreme brightness difference.

ν (nu) Gem, 6h 29m +20°, mag. 4.1, is a blue-white giant 620 l.y. away with a wide mag. 8.7 companion visible in binoculars or small telescopes.

38 Gem, 6h 55m +13°, 100 l.y. away, is a double star for small telescopes, with white and yellow components of mags. 4.7 and 7.7.

M35 (NGC 2168), 6h 09m +24°, is a large, 5th-mag. star cluster visible to the naked eye or through binoculars, covering the same area of sky as the Moon. It consists of about 200 stars, and is 2800 l.y. away. Even a small telescope at low magnification will show the members of this outstanding cluster to be arranged in curving chains. Nearby in the sky, but actually about 13,000 l.y. farther off, is NGC 2158, a very rich open cluster that appears as a small, faint patch of light requiring at least 100 mm aperture to distinguish.

NGC 2392, 7h 29m +21°, is an 8th-mag. planetary nebula sometimes known as the Eskimo or Clown Face Nebula because it looks somewhat like a face surrounded by a fringe when seen through a large telescope. A small telescope shows it as a blue-green ellipse about the same size as the disk of Saturn. Its central star is of 10th mag. NGC 2392 lies 3000 l.y. away.

39 GRUS
Gru · Gruis

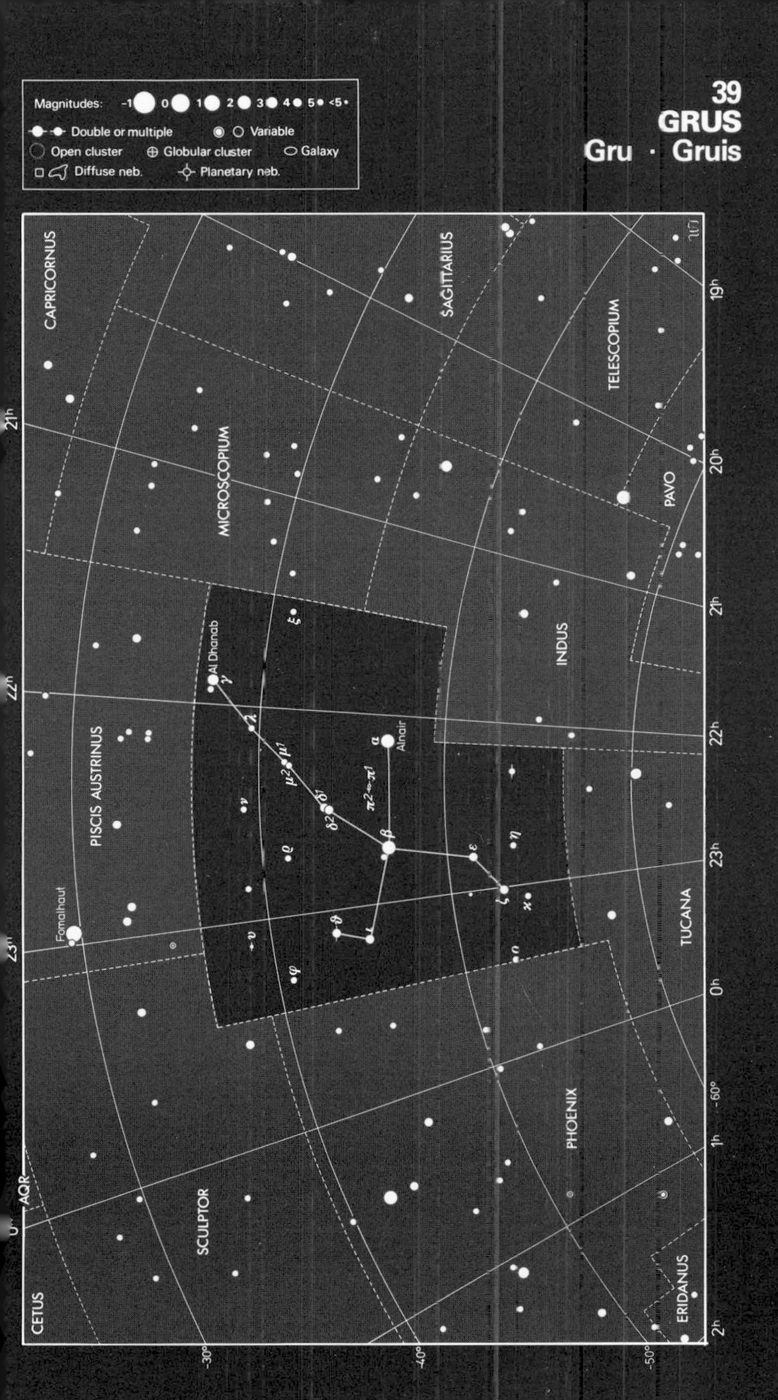

HERCULES

Hercules represents the Greek mythological hero famous for his twelve labours. Originally, though, the constellation was visualized as an anonymous kneeling man, with one foot on the head of the celestial dragon, Draco, which lies adjacent. Some legends identify the constellation with the ancient Sumerian superman, Gilgamesh. Hercules is not one of the most prominent constellations, despite being fifth-largest in the sky. But it is stocked with an abundance of double stars for users of small telescopes, plus one of the brightest and richest globular clusters in the sky, M13, easily located on one side of the central Keystone of four stars that marks the pelvis of Hercules.

α (alpha) Herculis, 17h 15m +14°, (Rasalgethi, 'kneeler's head'), about 500 l.y. distant, is a red giant star some 600 times the Sun's diameter, making it one of the largest stars known. Like most red giants it is erratically variable, in this case fluctuating from about mag. 3 to mag. 4. It is actually a double star, with a mag. 5.4 blue-green companion visible in small telescopes. The estimated orbital period of the pair is over 3000 years.

β (beta) Her, 16h 30m +21°, mag. 2.8, is a yellow giant 120 l.y. away.

γ (gamma) Her, 16h 22m +19°, mag. 3.8, is a giant white star 160 l.y. away, with a wide, unrelated mag. 9.6 companion visible in small telescopes.

δ (delta) Her, 17h 15m +25°, mag. 3.1, is a blue-white star 110 l.y. away. Small telescopes show a nearby mag. 8.2 star which is physically unrelated.

ζ (zeta) Her, 16h 41m +32°, 32 l.y. away, is a mag. 2.8 yellow star with a close mag. 5.5 orange companion that orbits it every 34 years. They are closing together during the 1990s, requiring an aperture of at least 150 mm to split, and at their closest in 2002 are too tight for most amateur instruments. Thereafter they open out again, coming into the range of 150 mm apertures after 2005.

κ (kappa) Her, 16h 08m +17°, is a mag. 5.0 yellow giant 285 l.y. away with an unrelated mag. 6.3 yellow giant companion, 520 l.y. away, easily seen in small telescopes.

ρ (rho) Her, 17h 24m +37°, 320 l.y. away, is a pair of blue-white stars of mags. 4.6 and 5.4, visible in small telescopes.

95 Her, 18h 01m +22°, 300 l.y. away, is a double star for small telescopes, consisting of two giant stars of mags. 5.0 and 5.2, appearing silver and gold.

100 Her, 18h 08m +26°, 210 l.y. away, is an easy binary for small telescopes consisting of identical white stars of mag. 5.9, like a pair of celestial cat's eyes.

M13 (NGC 6205), 16h 42m +36°, is a 6th-mag. globular cluster of 300,000 stars, the brightest of its kind in northern skies. It can be seen with the naked eye and is unmistakable in binoculars, spanning about half the apparent width of the full Moon. The cluster lies 23,500 l.y. away and is at least 100 l.y. in diameter. A small telescope resolves individual stars throughout the cluster, giving a mottled, sparkling effect. ▶

40
HERCULES
Her · Herculis

HOROLOGIUM The Pendulum Clock

One of the constellations representing mechanical instruments introduced in the 1750s by the Frenchman Nicolas Louis de Lacaille. As with so many of his constellations, Horologium is faint and obscure.

α (alpha) Horologii, 4h 14m –42°, mag. 3.9, is a yellow giant star 180 l.y. away.

β (beta) Hor, 2h 59m –64°, mag. 5.0, is a giant white star 280 l.y. away.

NGC 1261, 3h 12m –55°, is an 8th-mag. globular cluster 44,000 l.y. away.

NGC 1512 is an 11th-mag. barred spiral galaxy 50 million l.y. away in Horologium with a smaller, 13th-mag. elliptical companion, NGC 1510. Anglo-Australian Telescope Board.

◀ M92 (NGC 6341), 17h 17m +43°, is a globular cluster only slightly inferior to its more famous neighbour, M13, which overshadows it. M92 is easily seen in binoculars. It is smaller and more condensed at the centre than M13, and needs a larger telescope to resolve its stars. It lies 25,500 l.y. away.

NGC 6210, 16h 45m +24°, is a 9th-mag. planetary nebula which a telescope of 75 mm or larger shows as a blue-green ellipse. It is 4000 l.y. away.

41
HOROLOGIUM
Hor · Horologii

Magnitudes: −1 0 1 2 3 4 5 <5

Double or multiple
Variable
Open cluster
Globular cluster
Galaxy
Diffuse neb.
Planetary neb.

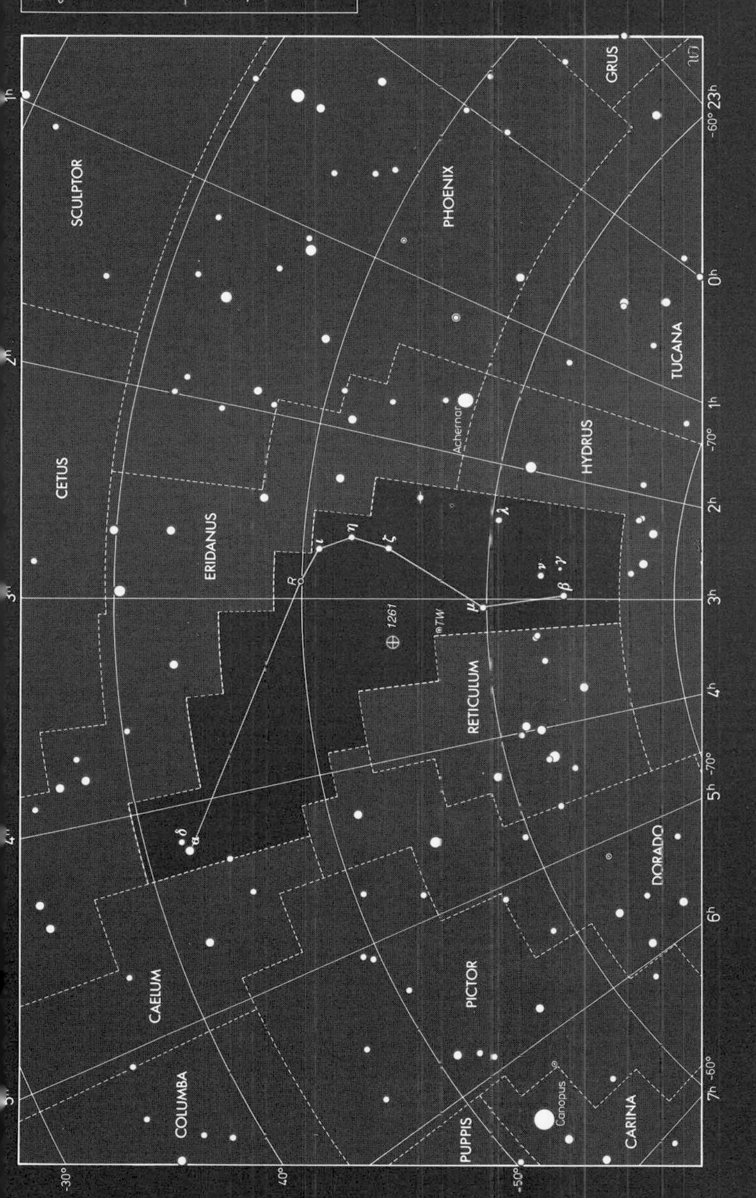

HYDRA The Water Snake

The largest constellation in the sky, but by no means easy to identify because of its faintness. Apart from its brightest star, Alphard, which marks the heart of the Water Snake, Hydra's only readily recognizable feature is its head, an attractive group of six stars. Hydra winds its way from the head in the northern celestial hemisphere, on the borders of Cancer, to the tip of its tail, south of the celestial equator adjacent to Libra and Centaurus. Its total length is over 100°. In mythology, Hydra is usually identified with the multi-headed monster slain by Hercules. Another legend links it with Corvus the Crow and Crater the Cup, which are found on its back, the bird having returned to the god Apollo with Hydra in its claws as an excuse for its aborted mission to fetch water in the cup.

α (alpha) Hydrae, 9h 28m –9°, (Alphard, 'the solitary one'), mag. 2.0, is an orange giant 65 l.y. away.

β (beta) Hya, 11h 53m –34°, mag. 4.3, is a blue-white star 330 l.y. away.

γ (gamma) Hya, 13h 19m –23°, mag. 3.0, is a yellow giant about 100 l.y. away.

δ (delta) Hya, 8h 38m +6°, mag. 4.2, is a blue-white star 150 l.y. away.

ε (epsilon) Hya, 8h 47m +6°, 250 l.y. away, is a beautiful but difficult double star of contrasting colours, needing high power on a telescope of at least 75 mm aperture. The stars are yellow and blue, of mags. 3.4 and 6.8, and orbit each other in about 900 years. The brighter star is also a very close binary with a period of 15 years, but is too close to be split by amateur telescopes.

54 Hya, 14h 46m –25°, 70 l.y. away, is an easy double for small telescopes, consisting of yellow and purple stars of mags. 4.9 and 7.2.

R Hya, 13h 30m –23°, is a red giant variable star similar to Mira in Cetus. It fluctuates between mags. 3 and 11 every 390 days. Its distance is unknown.

U Hya, 10h 38m –13°, is a deep-red variable star that fluctuates irregularly between mags. 4.3 and 6.5. Its distance is unknown.

M48 (NGC 2548), 8h 14m –6°, is a large cluster of about 80 stars, 2000 l.y. away, just visible to the naked eye under clear skies and a fine sight in binoculars. It is somewhat triangular in shape, wider than the apparent size of the full Moon, and is well shown by amateur telescopes under low power.

M83 (NGC 5236), 13h 37m –30°, is a large, face-on spiral galaxy of 8th mag., visible in a small telescope. It has a bright nucleus and its spiral arms can be traced with an aperture of 150 mm. (See the photograph on page 156.)

NGC 3242, 10h 25m –19°, is a 9th-mag. planetary nebula of similar apparent size to the disk of Jupiter; it is popularly termed the 'ghost of Jupiter'. This often-overlooked object 2600 l.y. away is prominent enough to be picked up in small telescopes at low magnification as a blue-green disk, while larger instruments show a bright inner disk surrounded by a fainter halo.

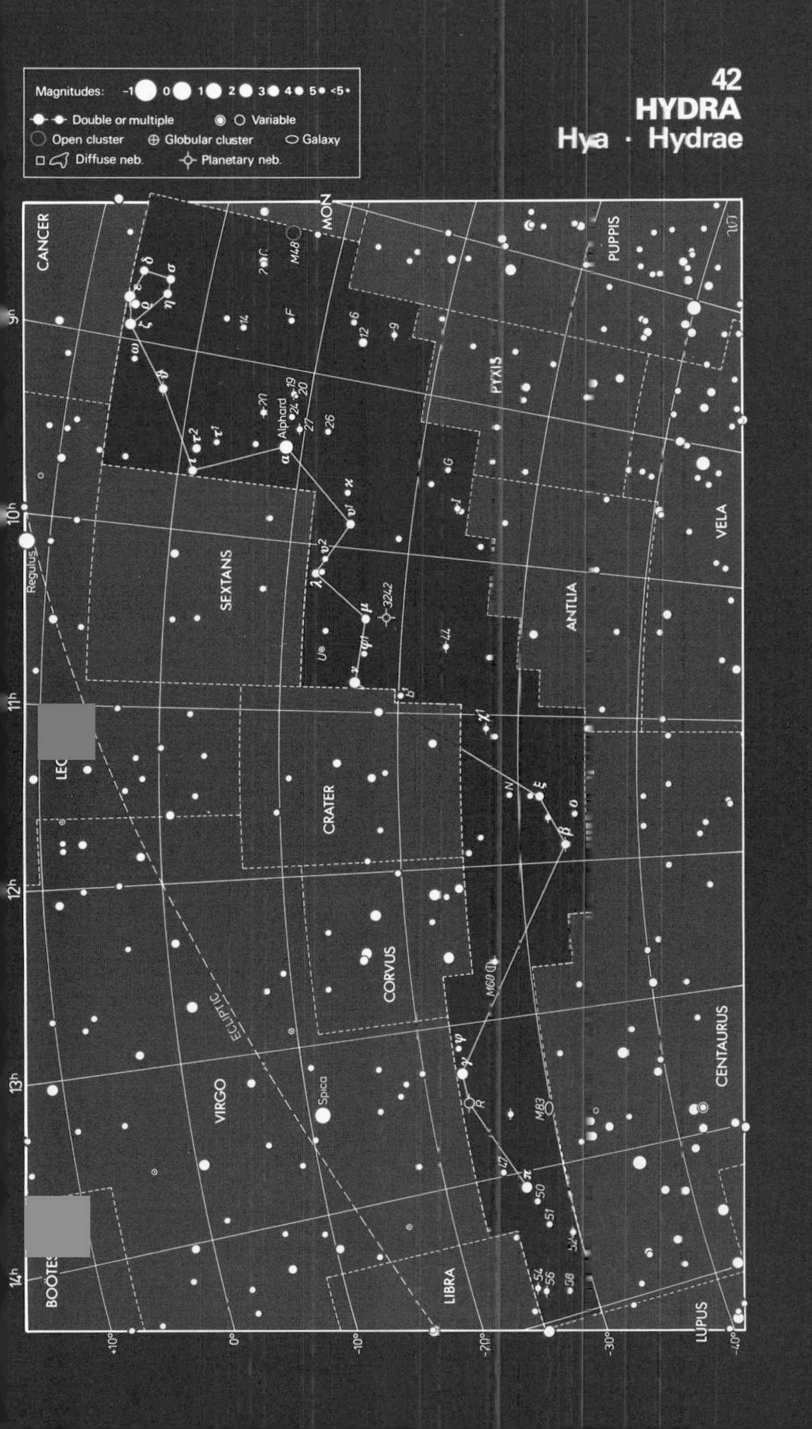

42
HYDRA
Hya · Hydrae
Magnitudes: -1 0 1 2 3 4 5 <5
Double or multiple
Variable
Open cluster
Globular cluster
Galaxy
Diffuse neb.
Planetary neb.
CANCER
MON
PUPPIS
PYXIS
VELA
SEXTANS
ANTLIA
CRATER
CORVUS
CENTAURUS
VIRGO
LIBRA
LUPUS
BOÖTES
Alphard
Regulus
Spica
ECLIPTIC
M48
M68
M83
3242
9h
10h
11h
12h
13h
14h
+10°
0°
-10°
-20°
-30°
-40°

HYDRUS The Lesser Water Snake

The Dutch navigators Pieter Dirkszoon Keyser and Frederick de Houtman introduced this constellation at the end of the 16th century as a smaller southern counterpart to the great Water Snake, Hydra. Keyser and de Houtman's Hydrus, lying between the two Magellanic Clouds, almost bridges the gap between Eridanus and the south celestial pole. There is little to interest the casual observer.

α (alpha) Hydri, 1h 59m –62°, mag. 2.9, is a white star 78 l.y. away.

β (beta) Hyi, 0h 26m –77°, mag. 2.8, is a yellow star 21 l.y. away.

γ (gamma) Hyi, 3h 47m –74°, mag. 3.2, is a red giant star 230 l.y. away.

π (pi) Hyi, 2h 14m –68°, is a binocular pair of red and orange giants, unrelated: π^1 is mag. 5.6 and lies 620 l.y. away; π^2, mag. 5.7, is 360 l.y. distant.

The face-on spiral galaxy M83 in Hydra, a turbulent swirl of stars and gas. For a description see page 154. Anglo-Australian Telescope Board.

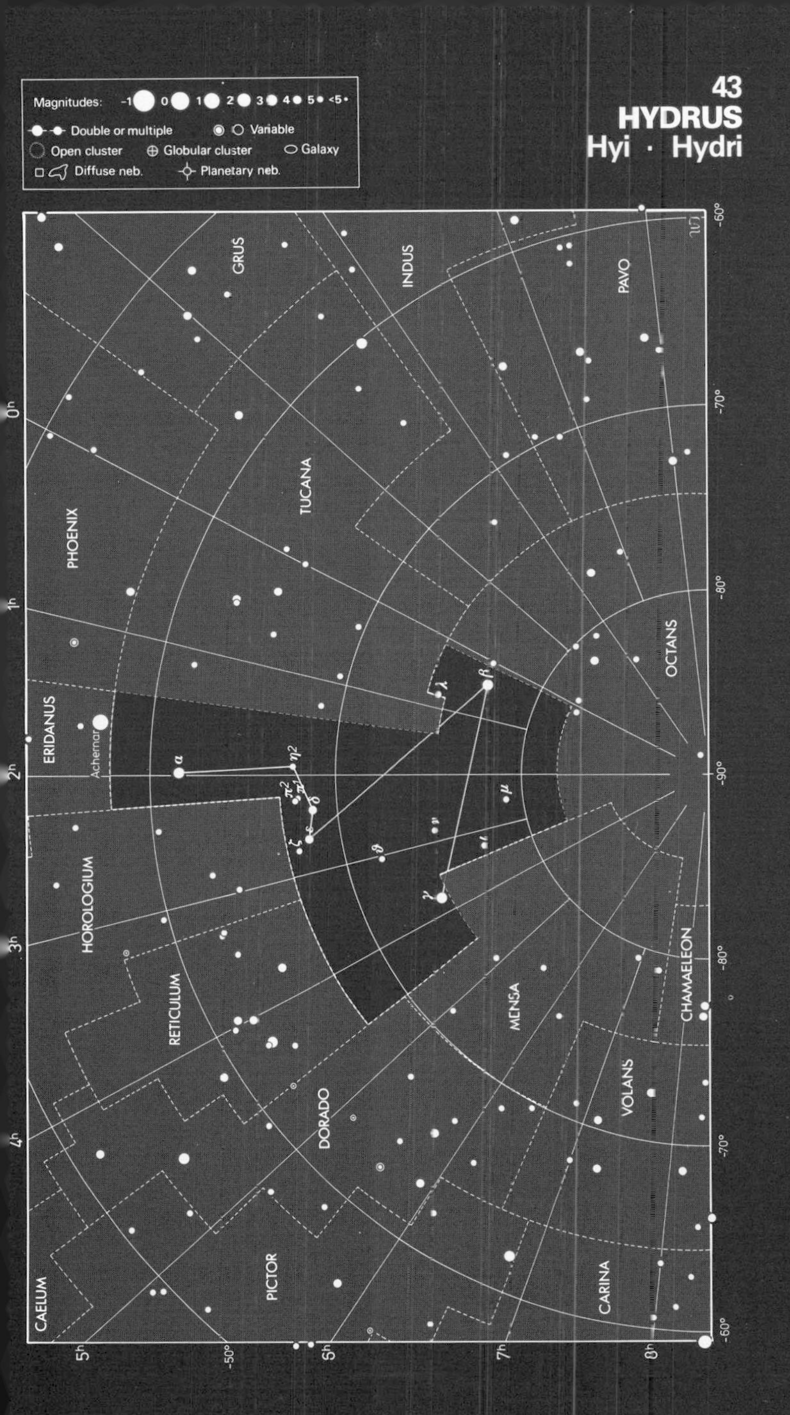

43
HYDRUS
Hyi · Hydri
Magnitudes: -1 0 1 2 3 4 5 <5
Double or multiple
Variable
Open cluster
Globular cluster
Galaxy
Diffuse neb.
Planetary neb.
GRUS
INDUS
PAVO
PHOENIX
TUCANA
OCTANS
ERIDANUS
Achernar
HOROLOGIUM
RETICULUM
MENSA
CHAMAELEON
DORADO
VOLANS
PICTOR
CARINA
CAELUM
0h
1h
2h
3h
4h
5h
6h
7h
8h
-50°
-60°
-70°
-80°
-90°

INDUS The Indian

A constellation representing an American native Indian, introduced at the end of the 16th century by the Dutch navigators Pieter Dirkszoon Keyser and Frederick de Houtman. It has no star brighter than 3rd magnitude.

α (alpha) Indi, 20h 38m –47°, mag. 3.1, is an orange giant star 125 l.y. away.

β (beta) Ind, 20h 55m –58°, mag. 3.7, is an orange giant 110 l.y. away.

δ (delta) Ind, 21h 58m –55°, mag. 4.4, is a white star 115 l.y. away.

ε (epsilon) Ind, 22h 03m –57°, mag. 4.7, is a yellow dwarf similar to the Sun, but slightly smaller and cooler. At 11.2 l.y. away, it is one of the Sun's closest neighbours.

ϑ (theta) Ind, 21h 20m –53°, 91 l.y. away, is a pair of white stars of mags. 4.4 and 7.2, divisible in a small telescope.

Between Indus and Hydrus lies the Small Magellanic Cloud, an irregularly shaped splash of stars. To the right of it in this photograph is 47 Tucanae, a globular cluster which lies in our own Galaxy. For descriptions see page 240. Royal Observatory Edinburgh.

44
INDUS
Ind · Indi

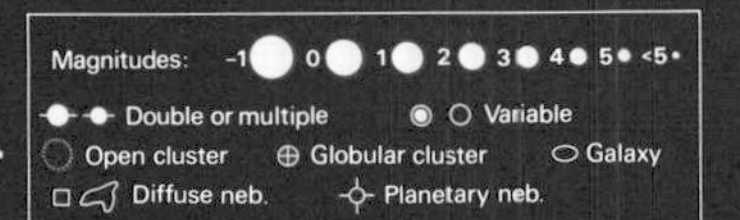

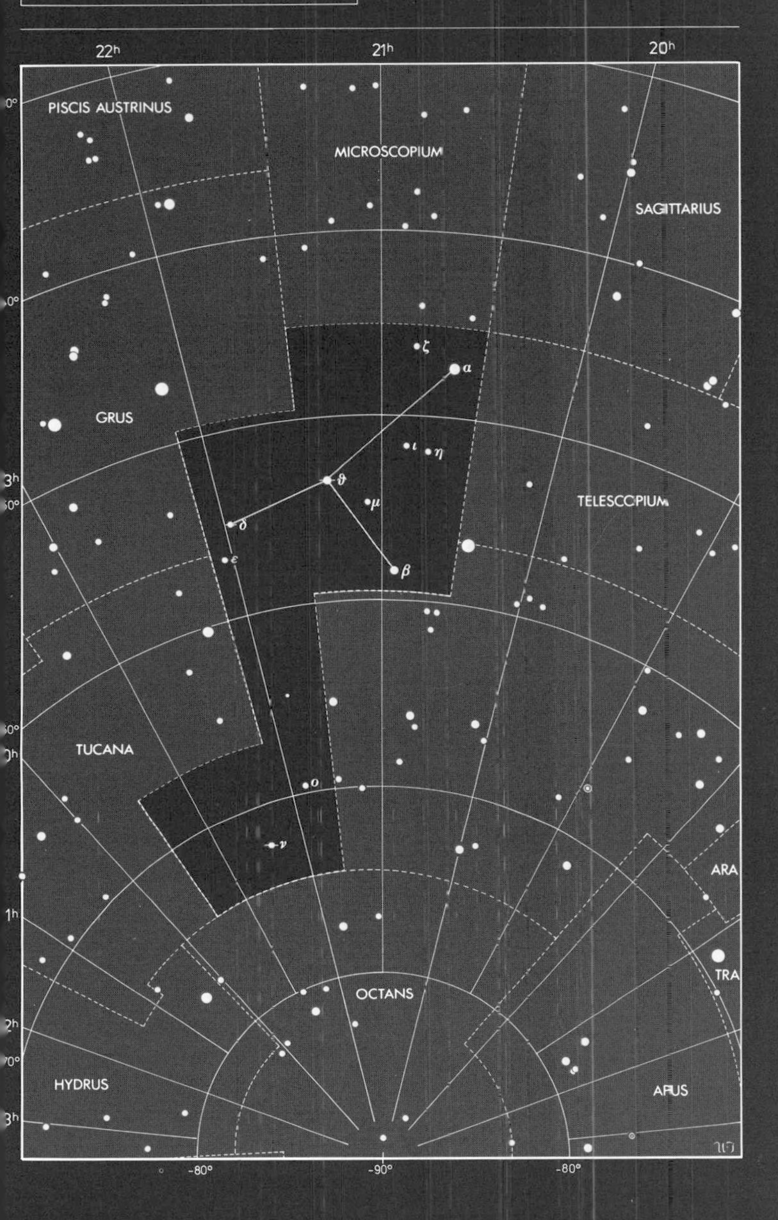

LACERTA The Lizard

An inconspicuous constellation sandwiched between Cygnus and Andromeda, introduced in 1687 by the Polish astronomer Johannes Hevelius. An alternative constellation in this area was Sceptrum, the Sceptre and Hand of Justice, created in 1679 by the Frenchman Augustin Royer to commemorate King Louis XIV. In 1787 the German Johann Elert Bode called this region Frederick's Glory in honour of King Frederick the Great of Prussia. Both these alternatives have been discarded. The constellation's most celebrated object is BL Lacertae, location 22h 02.7m +42° 17′, originally thought to be a peculiar 14th-magnitude variable star. It is now known to be the prototype of a group of objects believed to be giant elliptical galaxies with variable centres, lying far off in the Universe and evidently related to quasars. Three bright novae have appeared within the boundaries of Lacerta this century.

α (alpha) Lacertae, 22h 31m +50°, mag. 3.8, is a blue-white star 110 l.y. away.

β (beta) Lac, 22h 24m +52°, mag. 4.4, is a yellow giant 205 l.y. away.

Johannes Hevelius (1611–1687)

Johannes Hevelius of Danzig was one of the finest observers of his day. His masterwork was a catalogue of 1564 stars, published posthumously in 1690. The catalogue was accompanied by a set of sky maps, beautifully engraved by Hevelius himself, on which he introduced seven constellations still in use today: Canes Venatici, Lacerta, Leo Minor, Lynx, Scutum, Sextans and Vulpecula. Another important cartographic product of Hevelius was his map of the Moon, published in 1647 in his book *Selenographia*. It was the first major Moon map to be made, and introduced the first system of lunar nomenclature. He named lunar formations after features on Earth: for instance, the crater Copernicus he called Etna, while Tycho was Mount Sinai and Mare Imbrium the Mediterranean. Only a few of the names given by Hevelius remain, such as the lunar Alps and Apennines; his names have mostly been discarded in favour of the system of Giovanni Battista Riccioli (1598–1671), who named the craters after famous philosophers and astronomers. Fittingly enough, the craters commemorating Hevelius and Riccioli are to be found close together on the Moon.

LACERTA
Lac · Lacertae

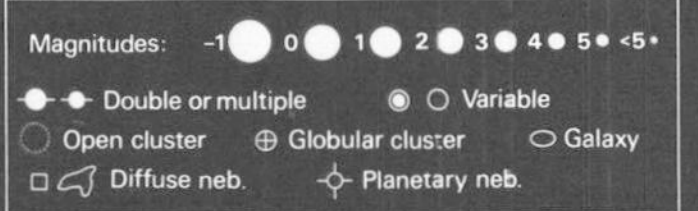

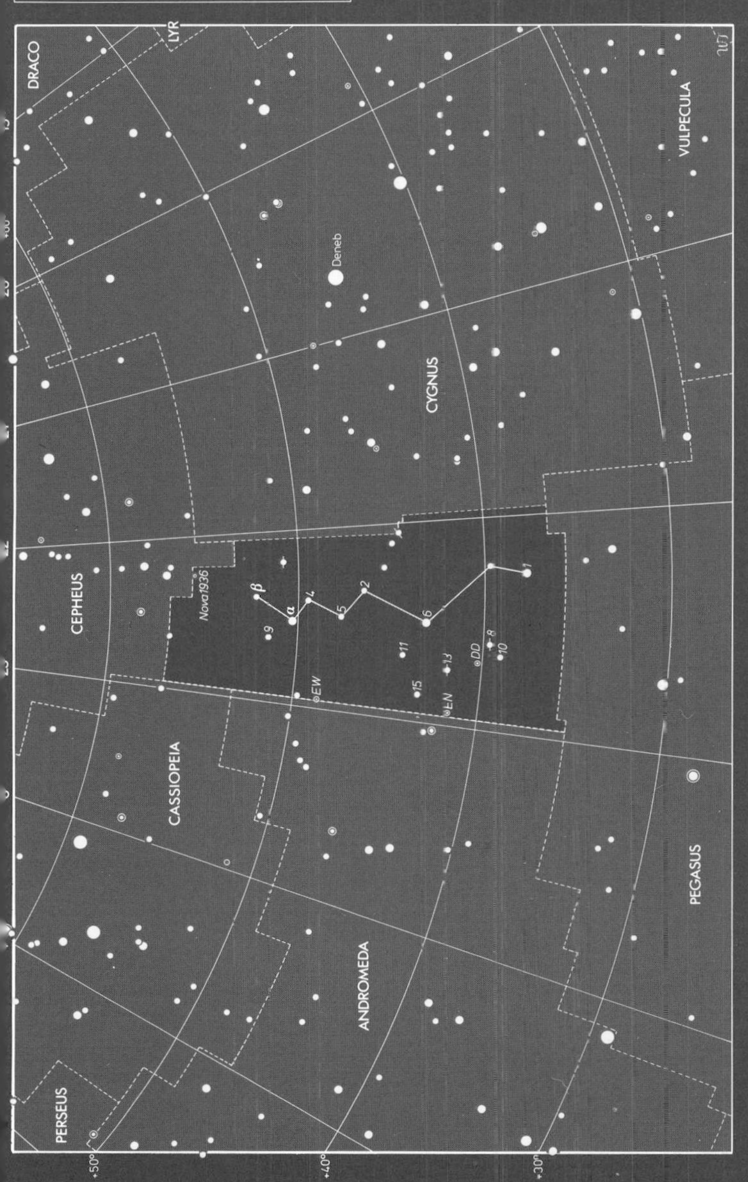

LEO The Lion

One of the few constellations that looks like the figure it is supposed to represent – in this case, a crouching lion. The Lion's head is outlined by a sickle shape of six stars from ε (epsilon) to α (alpha) Leonis, with the Lion's body stretching out behind, its tail marked by β (beta) Leonis. This is the Lion reputedly slain by the hero Hercules as the first of his 12 labours. Leo is a large and prominent constellation, containing many stars and galaxies of interest. The Sun passes through the constellation from mid-August to mid-September. Every November, the Leonid meteors radiate from a point near γ (gamma) Leonis. Usually the numbers seen are low, peaking at about 10 per hour on November 17–18, but occasionally spectacular storms of meteors have been recorded with as many as 100,000 in an hour, like celestial snowflakes. The last such Leonid meteor storm was seen from the United States in 1966 and another may occur in 1999. Leo contains a little-known star, Wolf 359, which is the third-nearest star to the Sun. It is a red dwarf 7.7 l.y. away, appearing of magnitude 13.5; it is located at 10h 56.5m +7° 01′, near the border with Sextans. Leo contains numerous distant galaxies, the brightest of which are mentioned on page 164, plus two faint dwarf members of our Local Group, beyond the reach of amateur telescopes.

α (alpha) Leonis, 10h 08m +12°, (Regulus, 'the little king'), mag. 1.4, is a blue-white star 91 l.y. away. It has a wide companion of mag. 7.7, visible in binoculars or small telescopes.

β (beta) Leo, 11h 49m +15°, (Denebola, 'lion's tail'), mag. 2.1, is a white star 42 l.y. away.

γ (gamma) Leo, 10h 20m +20°, (Algieba, 'the forehead'), 170 l.y. away, consists of a pair of golden-yellow giant stars of mags. 2.2 and 3.5, orbiting in about ▶

Trails of over 20 Leonid meteors swept across the stars of the Big Dipper in this 43-second exposure, made during the great Leonid meteor storm of 1966. Photograph by Dave McLean, Kitt Peak.

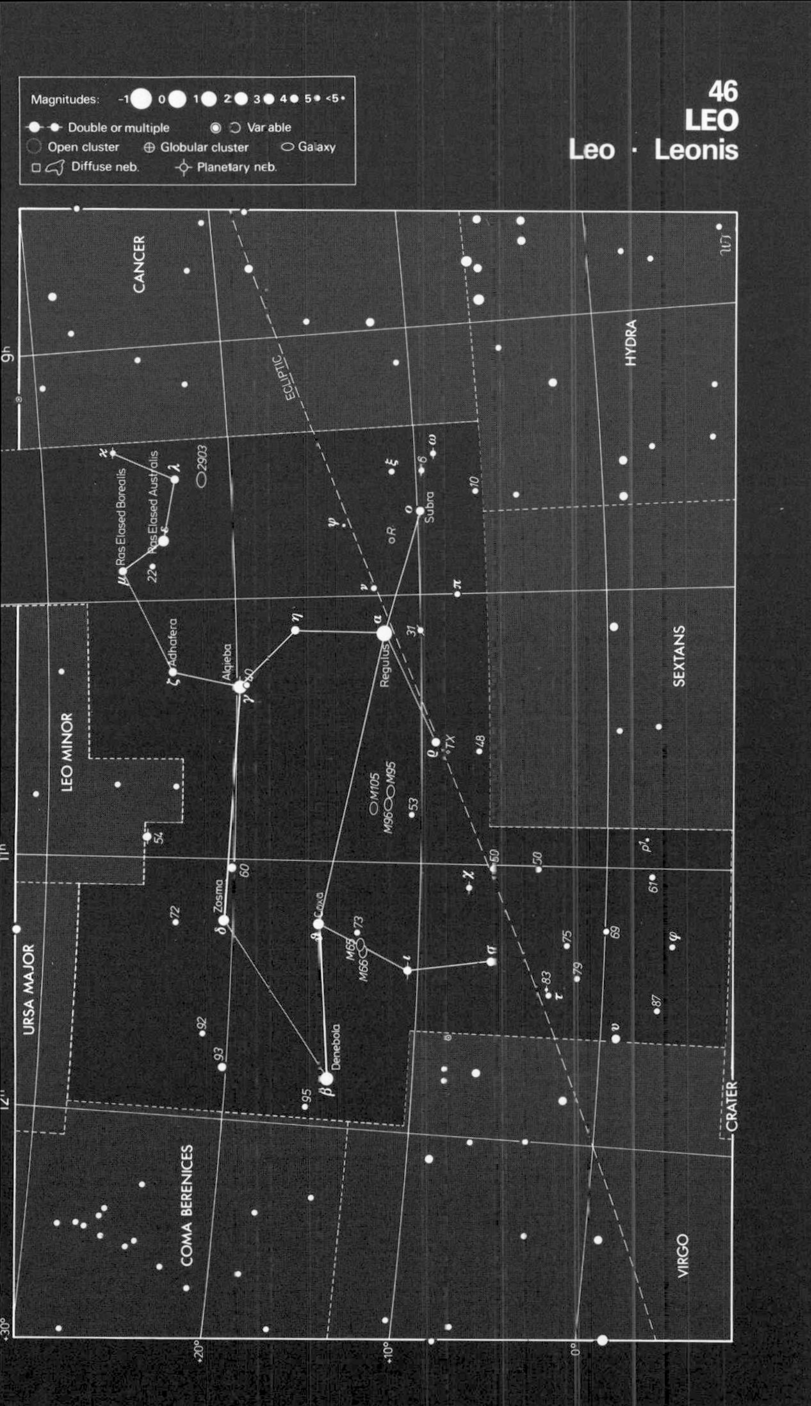
46
LEO
Leo · Leonis
Magnitudes: -1 0 1 2 3 4 5 <5
Double or multiple
Variable
Open cluster
Globular cluster
Galaxy
Diffuse neb.
Planetary neb.
CANCER
HYDRA
LEO MINOR
SEXTANS
URSA MAJOR
COMA BERENICES
CRATER
VIRGO
ECLIPTIC
Ras Elased Borealis
Ras Elased Australis
Adhafera
Algieba
Regulus
Subra
Zosma
Coxa
Denebola
2903
M105
M96
M95
M65
M66
9h
11h
12h
+30°
+20°
+10°
0°

LEO MINOR The Lesser Lion

Leo Minor lies between the larger and brighter constellations of Leo and Ursa Major. Johannes Hevelius, the Polish astronomer, introduced Leo Minor in 1687. There is little of interest in it, and the labelling of its stars is fragmentary, testimony to cavalier treatment by successive generations of celestial cartographers.

β (beta) Leonis Minoris, 10h 28m +37°, mag. 4.2, is a yellow giant star 190 l.y. away.

46 LMi, 10h 53m +34°, mag. 3.8, is an orange giant 78 l.y. away.

◀ 600 years. They form an exceptionally handsome double in small telescopes, one of the finest in the sky. In binoculars, an unrelated mag. 4.8 yellowish star, 40 Leonis, is visible nearby.

δ (delta) Leo, 11h 14m +21°, (Zosma), mag. 2.6, is a blue-white star 46 l.y. away.

ε (epsilon) Leo, 9h 46m +24°, mag. 3.0, is a yellow giant 310 l.y. away.

ζ (zeta) Leo, 10h 17m +23°, mag. 3.4, is a giant white star 130 l.y. away. Binoculars show an unrelated orange background star nearby, of mag. 5.9. A wider third star, mag. 5.8 and also unrelated, can be seen in binoculars, making this an optical triple.

ι (iota) Leo, 11h 24m +11°, 69 l.y. away, is a close and difficult double star. It shines to the naked eye as a yellow-white star of mag. 3.9, but actually consists of components of mags. 4.0 and 6.7 orbiting every 190 years. The stars are currently moving apart, and will continue to do so until the year 2070; apertures of 150 mm should be sufficient to separate them by the year 2000, and thereafter they will get progressively easier.

54 Leo, 10h 56m +25°, 150 l.y. away, is a double for small telescopes, consisting of blue-white components of mags. 4.5 and 6.3.

R Leo, 9h 48m +11°, is a red giant variable of the Mira type, lying more than 3000 l.y. away. It appears strongly red when at maximum. R Leonis normally varies between 6th and 10th mags. every 312 days on average, but on occasion can become as bright as mag. 4.4.

M65, M66 (NGC 3623, NGC 3627), 11h 19m +13°, a pair of spiral galaxies about 20 million l.y. away. At 9th mag. they can be detected in large binoculars under clear conditions, but at least 100 mm aperture at low power is required for their elongated shape and condensed centres to be clearly seen.

M95, M96 (NGC 3351, NGC 3368), 10h 44m +12°, 10h 47m +12°, a pair of spiral galaxies of 10th and 9th mag. respectively, 22 and 25 million l.y. away, visible as circular nebulosities in small telescopes. About 1° away lies the smaller M105 (NGC 3379), 10h 48m +13°, a 9th-mag. elliptical galaxy 25 million l.y. distant.

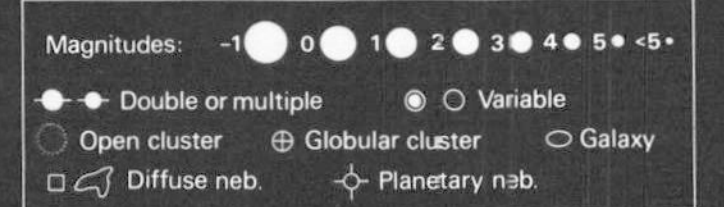

47
LEO MINOR
LMi · Leonis Minoris

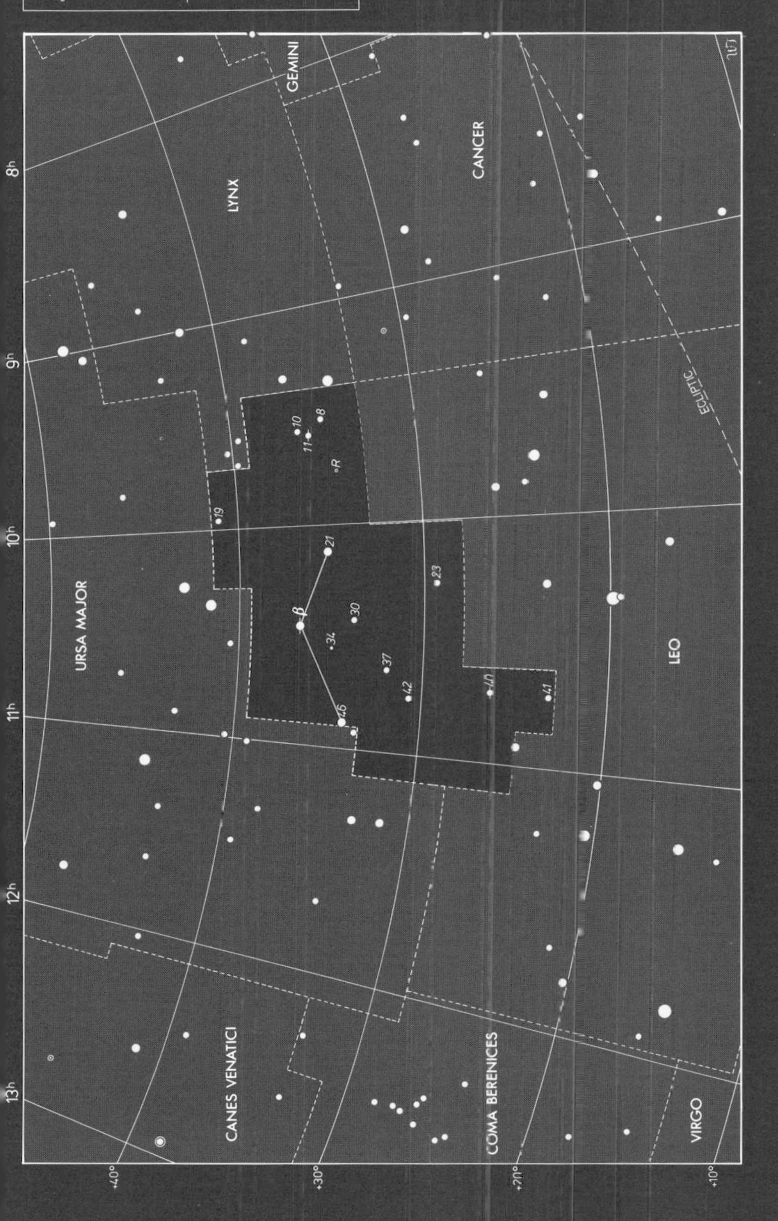

LEPUS The Hare

Lepus is a constellation known since Ancient Greek times. It represents a hare, cunningly located at the feet of its hunter, Orion, and pursued endlessly across the sky by Canis Major, the hunter's dog. The Hare is also associated in many legends with the Moon. For instance, the familiar figure of the man in the Moon is sometimes interpreted as a hare or rabbit, so perhaps Lepus is another incarnation of the lunar hare. Lepus is overshadowed by Orion's brilliance, but is not without interest for amateur observers.

α (alpha) Leporis, 5h 33m –18°, (Arneb, 'hare'), mag. 2.6, is a white supergiant star, 950 l.y. away.

β (beta) Lep, 5h 28m –21°, mag. 2.8, is a yellow giant star 290 l.y. away.

γ (gamma) Lep, 5h 44m –22°, 27 l.y. away, is an attractive binocular duo consisting of a yellow star of mag. 3.6 with an orange companion of mag. 6.2.

δ (delta) Lep, 5h 51m –21°, mag. 3.8, is an orange giant 125 l.y. away.

ε (epsilon) Lep, 5h 05m –22°, mag. 3.2, is an orange giant 160 l.y. away.

$\varkappa$ (kappa) Lep, 5h 13m –13°, 220 l.y. away, is a mag. 4.4 blue-white star with a close mag. 7.4 companion, difficult to see in the smallest telescopes because of the magnitude contrast.

R Lep, 5h 00m –15°, is an intensely red star known as Hind's Crimson Star after the English observer John Russell Hind, who described it in 1845 as 'like a drop of blood on a black field'. R Leporis is a Mira-type variable that ranges from mag. 5.5 at its brightest to as faint as 12th mag. in a period of around 430 days.

M79 (NGC 1904), 5h 25m –25°, is a small but rich globular cluster 43,000 l.y. away, visible as a fuzzy 8th-mag. star in small telescopes. Nearby in the same low-power field is the multiple star Herschel 3752, consisting of a mag. 5.4 primary with two companions, a wide one of mag. 9.1 and a close one of mag. 6.6, all visible in small telescopes.

NGC 2017, 5h 39m –18°, is a small but remarkable star cluster, also known as the multiple star Herschel 3780. Modest amateur telescopes reveal a group of five well-spaced stars ranging from 6th to 10th mag. In addition, the brightest star has a mag. 7.9 companion that requires a telescope of at least 200 mm aperture to split, while an aperture of at least 100 mm shows that one of the other stars is a close double. There is also a 12th-mag. component which should be visible with 100 mm or upwards, so this is actually a family of at least eight related stars.

48
LEPUS
Lep · Leporis

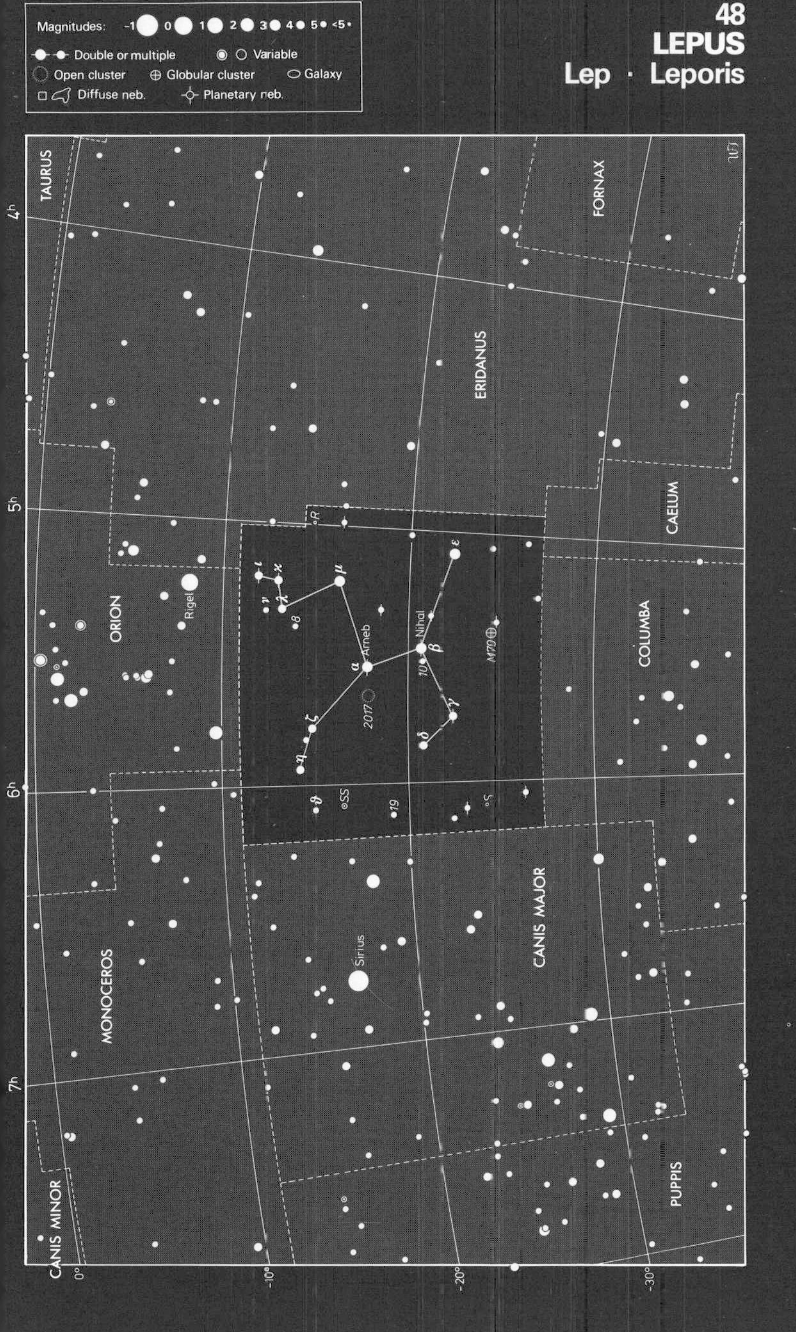

LIBRA The Scales

A small, faint constellation of the zodiac, through which the Sun passes during November. The Ancient Greeks knew it as the Claws of the Scorpion, an extension of neighbouring Scorpius, an identification that lives on in the names of its stars. But the Romans made it into a separate constellation in the time of Julius Caesar in the first century BC. Since then the Scales have come to be regarded as the symbol of justice, held aloft by the goddess of justice, Astraeia, represented by the neighbouring figure of Virgo. Although faint, Libra contains several interesting stars.

α (alpha) Librae, 14h 50m –16°, (Zubenelgenubi, 'the southern claw'), 72 l.y. away, is a wide binocular double consisting of a blue-white star of mag. 2.8 with a white companion of mag. 5.2.

β (beta) Lib, 15h 17m –9°, (Zubeneschamali, 'the northern claw'), mag. 2.6, is celebrated as one of the few bright stars to show a distinct greenish tinge. It lies 140 l.y. away.

γ (gamma) Lib, 15h 36m –15°, (Zubenelakrab, 'the scorpion's claw'), mag. 3.9, is a yellow giant 75 l.y. away.

δ (delta) Lib, 15h 01m –9°, 290 l.y. away, is an eclipsing variable of the Algol type. It varies between mags. 4.9 and 5.9 in 2 days 8 hours.

ι (iota) Lib, 15h 12m –20°, is a multiple star 142 l.y. away. Its main blue-white component, of mag. 4.5, has a wide mag. 9.4 companion that is difficult to see in the smallest telescopes because of the brightness difference. An aperture of 75 mm or above with high magnification will split this fainter companion into two stars of 10th and 11th mags. The brighter star is itself a binary with a 22-year period, but too close for amateur telescopes. Binoculars show a mag. 6.1 star nearby called 25 Librae, which may also be associated with the system.

μ (mu) Lib, 14h 49m –14°, is a close double star consisting of components of mags. 5.6 and 6.7, divisible in a telescope of 75 mm aperture. It lies about 500 l.y. away.

48 Lib, 15h 58m –14°, mag. 4.9, is a shell star similar to γ (gamma) Cassiopeiae and Pleione in Taurus. It is a blue giant with an abnormally high speed of rotation that causes it to throw off rings of gas from its equator, varying irregularly by about 0.1 mag. as it does so. It also bears the variable-star designation FX Lib.

NGC 5897, 15h 17m –21°, is a large but loosely scattered 9th-mag. globular cluster 40,000 l.y. away, unspectacular in small instruments.

LIBRA
Lib · Librae

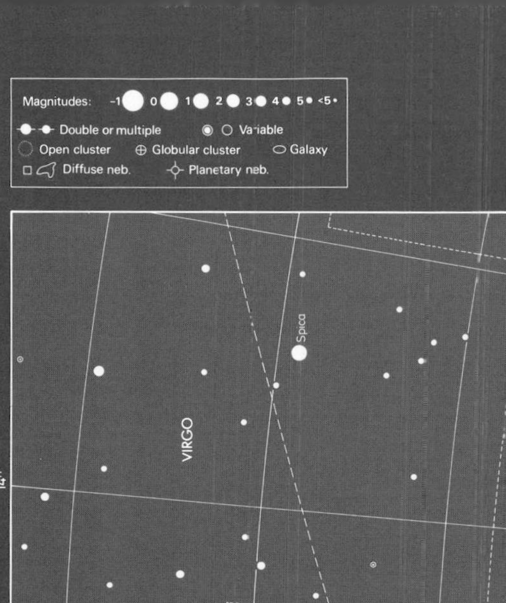

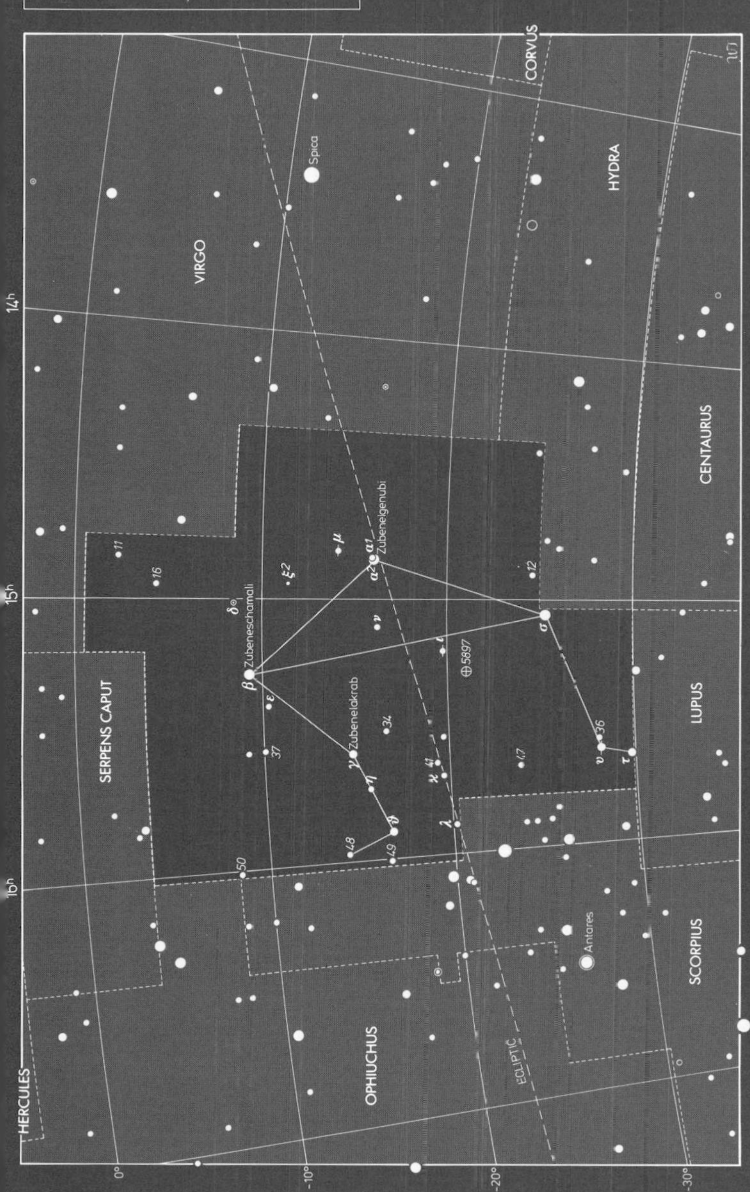

LUPUS The Wolf

Lupus is stocked with numerous interesting objects, although it is often overlooked in favour of its more spectacular neighbours Scorpius and Centaurus. The constellation was regarded by the Greeks and Romans as an unspecified wild animal, impaled on a pole by Centaurus the Centaur. Its identification as a wolf seems to have become common in Renaissance times. Lupus lies in the Milky Way and is rich in double stars.

α (alpha) Lupi, 14h 42m –47°, mag. 2.3, is a blue giant star 620 l.y. away.

β (beta) Lup, 14h 59m –43°, mag. 2.7, is a blue-white giant 680 l.y. away.

γ (gamma) Lup, 15h 35m –41°, mag. 2.8, is a blue-white star 460 l.y. away. It is a close binary with a 150-year orbital period, splittable only in apertures of 200 mm and above.

ε (epsilon) Lup, 15h 23m –45°, 620 l.y. away, is a blue-white star of mag. 3.4 with a wide mag. 8.8 companion visible in a small telescope. The primary is itself a close double, splittable only in large apertures.

η (eta) Lup, 16h 00m –38°, 520 l.y. away, is a double star consisting of a mag. 3.4 blue-white primary with a mag. 7.9 companion, not easy to see in a small telescope because of the magnitude contrast.

κ (kappa) Lup, 15h 12m –49°, is an easy double star for small telescopes, consisting of blue-white components of mags. 3.9 and 5.7, distances 170 and 275 l.y.

μ (mu) Lup, 15h 19m –48°, 250 l.y. away, is a multiple star. Small telescopes reveal a mag. 4.3 blue-white primary with a wide mag. 7.2 companion. But in telescopes of at least 100 mm, with high magnification, the primary itself is seen to be double, consisting of two near-identical stars of mags. 5.1 and 5.2.

ξ (xi) Lup, 15h 57m –34°, 130 l.y. away, is a neat pair of mag. 5.1 and 5.6 blue-white stars, well seen in a small telescope.

π (pi) Lup, 15h 05m –47°, appears to the naked eye as mag. 3.9, but telescopes above 75 mm aperture show that it consists of two close blue-white stars of mags. 4.7 and 4.8 about 400 l.y. away.

NGC 5822, 15h 05m –54°, is a large, loose open cluster of about 150 stars, 1800 l.y. away, visible in binoculars or small telescopes.

NGC 5986, 15h 46m –38°, is a 7th-mag. globular cluster, 33,000 l.y. distant, visible as a rounded patch in a small telescope.

50
LUPUS
Lup · Lupi

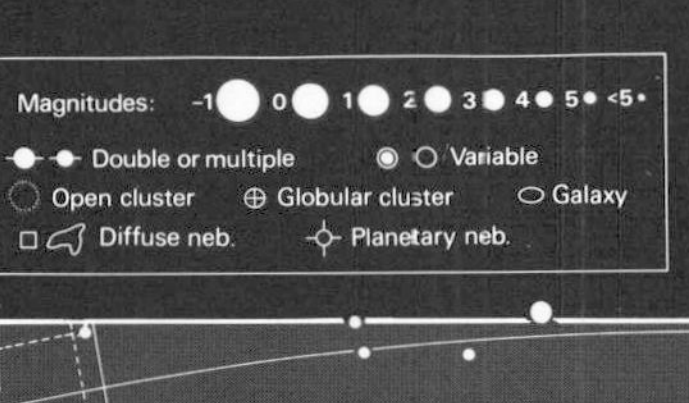

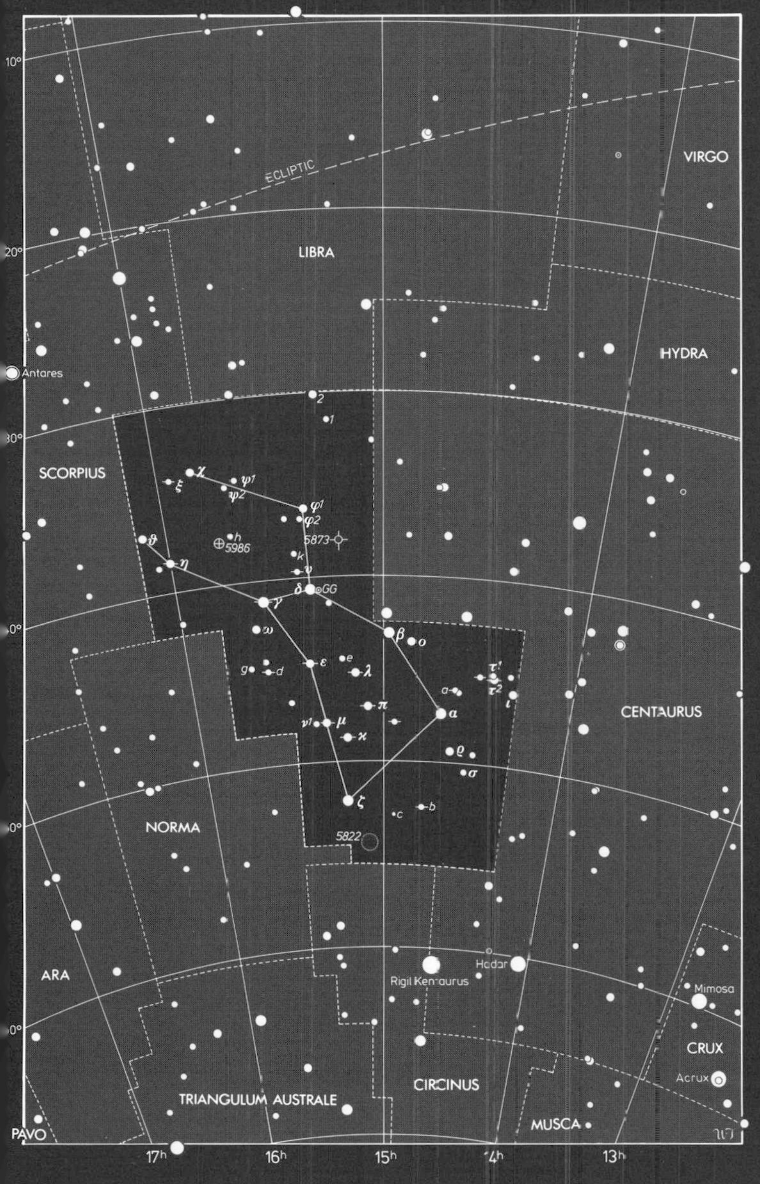

LYNX The Lynx

A decidedly obscure constellation, despite its considerable size (larger than Gemini, for instance). It was introduced in 1687 by the Polish astronomer Johannes Hevelius to fill the gap between Ursa Major and Auriga. He named it Lynx, it is said, because only the lynx-eyed would be able to see its stars – a reference to the fact that his own eyesight was exceptionally keen. Despite its faintness, owners of small telescopes will find many exquisite double stars within it.

α (alpha) Lyncis, 9h 21m +34°, mag. 3.2, is a red giant about 150 l.y. away.

5 Lyn, 6h 27m +58°, 360 l.y. away, is an orange giant star of mag. 5.2 with a wide, unrelated mag. 7.9 companion, visible in a small telescope.

12 Lyn, 6h 46m +59°, 140 l.y. away, is a fascinating triple star. A small telescope will show a mag. 4.9 blue-white star with a fainter mag. 7.3 companion. Telescopes of 75 mm aperture and over reveal that the brighter component is itself a binary, of mags. 5.4 and 6.0, which orbit every 700 years.

15 Lyn, 6h 57m +58°, 100 l.y. away, is a close double star for telescopes of 150 mm aperture and above. The components are of mags. 4.8 and 5.9, the brighter star appearing a deep yellow colour.

19 Lyn, 7h 23m +55°, is an attractive triple star for small telescopes, with blue-white components of mags. 5.5 and 6.5. The very wide third star is of mag. 7.7. The trio lies about 400 l.y. away.

38 Lyn, 9h 19m +37°, 115 l.y. away, is a pair of mag. 3.9 and 6.6 blue-white stars, difficult in the smallest telescopes because of their closeness.

41 Lyn, 9h 29m +46°, 180 l.y. away, is actually over the border in Ursa Major. It is a yellow giant of mag. 5.4, and small telescopes reveal a wide companion of mag. 8.0. A 10th-mag. star nearby forms a triangle, making this an apparent triple.

Flamsteed Numbers

Apart from α Lyncis, all the main stars in Lynx are referred to not by Greek letters but by so-called Flamsteed numbers. These numbers originate in a catalogue of 2935 stars, *Historia Coelestis Britannica*, compiled by the first Astronomer Royal of England, John Flamsteed (1646–1719). The catalogue was published posthumously in 1725. Flamsteed listed the stars in each constellation in order of right ascension. The numbers that are now known as Flamsteed numbers were added to these stars by later astronomers.

51 LYNX
Lyn · Lyncis

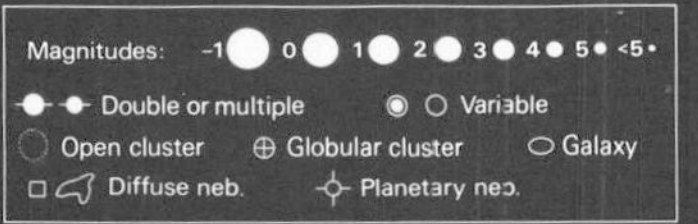

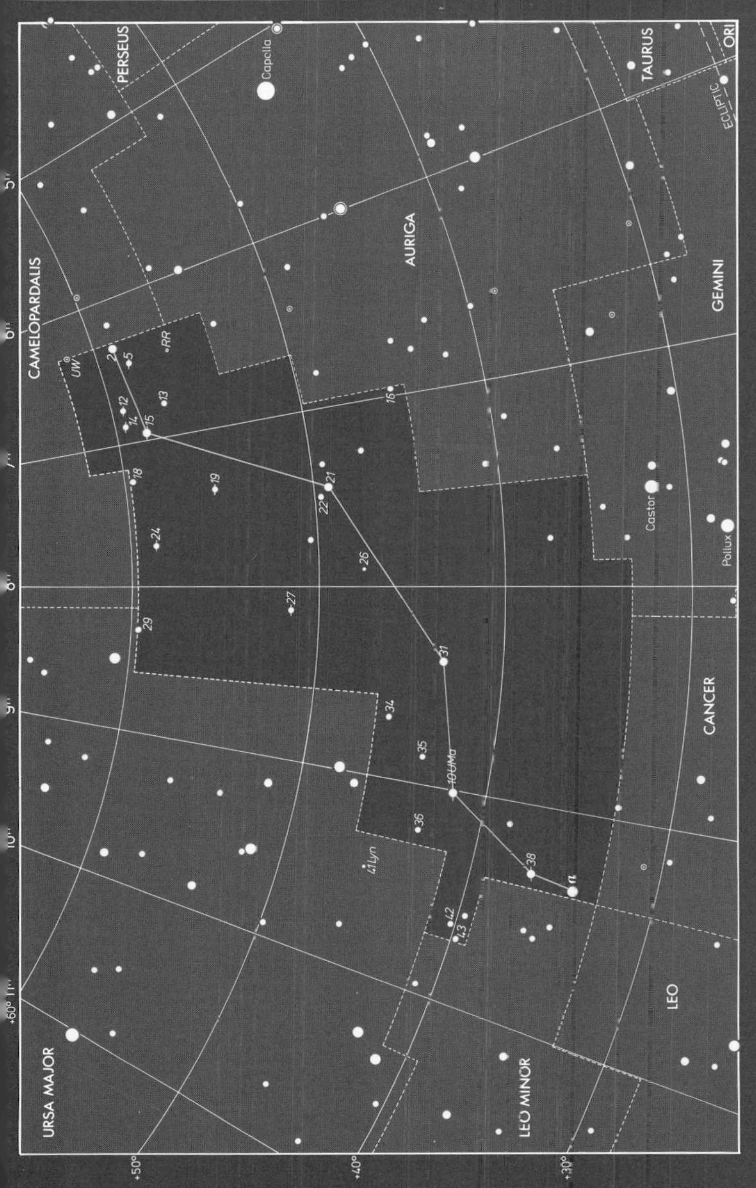

LYRA The Lyre

A constellation dating from ancient times, representing the stringed instrument invented by Hermes and subsequently given by his half-brother Apollo to the great musician Orpheus. This constellation has also been visualized as an eagle or vulture. Although small, Lyra is bright and prominent. It contains the fifth-brightest star in the sky, Vega, which forms one corner of the Summer Triangle (Deneb in Cygnus and Altair in Aquila mark the other two corners). Our Sun's motion around the Galaxy is carrying us in the direction of Vega at a velocity of 20 km per second relative to nearby stars. Because of precession, Vega will become the Pole Star around AD 14,000. The Lyrid meteor shower emanates from the constellation each year, reaching a peak of about 10 per hour on April 21–22.

α (alpha) Lyrae, 18h 37m +39°, (Vega, 'the swooping eagle'), mag. 0.03, is a brilliant blue-white star 26 l.y. away. It is the fifth-brightest star in the sky.

β (beta) Lyr, 18h 50m +33°, (Sheliak, 'the harp'), 1500 l.y. away, is a remarkable multiple star. Small telescopes easily resolve it as a double star of cream and blue components. The fainter, blue star is of mag. 7.2, while the brighter star is an eclipsing binary that varies between mags. 3.3 and 4.3 in 12.9 days. These eclipsing stars are so close together that gravity distorts them into egg-shapes, and hot gas spirals off them into space.

γ (gamma) Lyr, 18h 59m +33°, (Sulafat, 'the tortoise'), mag. 3.2, is a blue-white giant 205 l.y. away.

δ (delta) Lyr, 18h 54m +37°, is a wide naked-eye or binocular double consisting of two unrelated stars: δ^1, blue-white, mag. 5.6, 1150 l.y. away; and δ^2, a red giant 720 l.y. away, which varies erratically from mag. 4.2 to 4.3. ▶

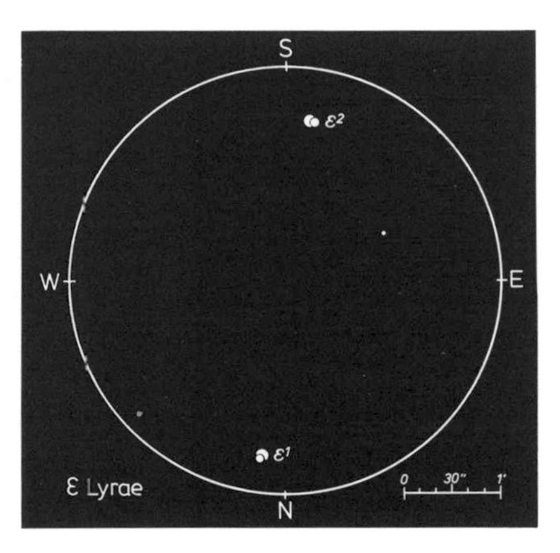

The Double Double, ε (epsilon) Lyrae, as seen through a telescope.
Wil Tirion.

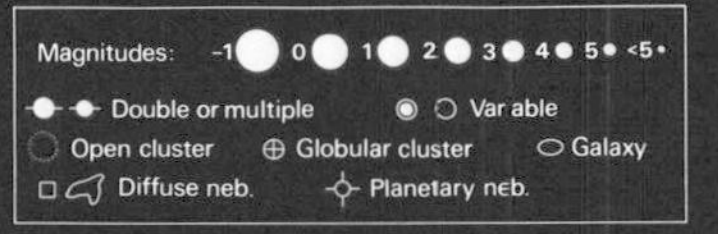
Magnitudes: -1 0 1 2 3 4 5 <5
Double or multiple
Variable
Open cluster
Globular cluster
Galaxy
Diffuse neb.
Planetary neb.

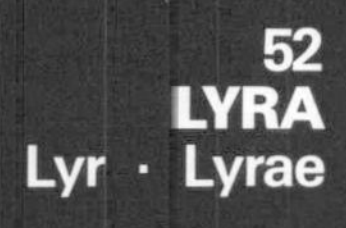
52
LYRA
Lyr · Lyrae

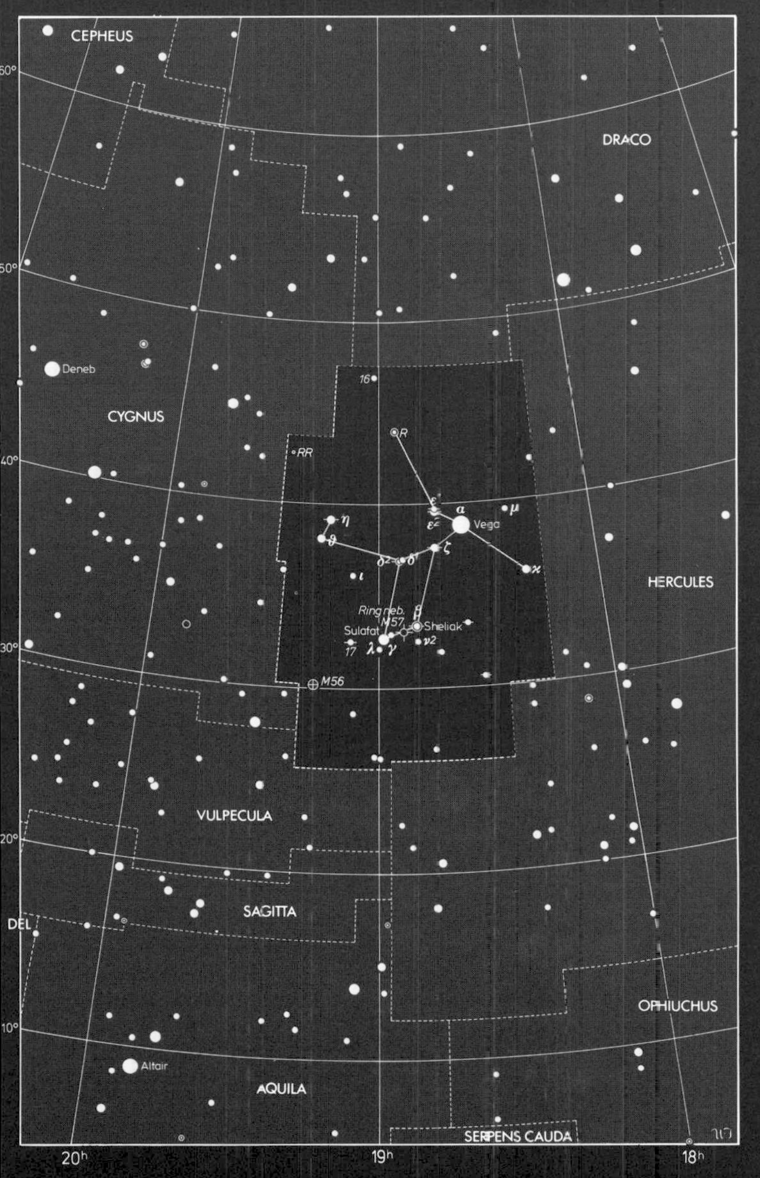
CEPHEUS
DRACO
Deneb
CYGNUS
16
R
RR
α
Vega
μ
ε
η
ϑ
ζ
δ²
δ¹
ι
κ
HERCULES
Ring neb.
M57
β
Sheliak
Sulafat
ν2
17
λ
γ
M56
VULPECULA
SAGITTA
DEL
OPHIUCHUS
Altair
AQUILA
SERPENS CAUDA
60°
50°
40°
30°
20°
10°
20h
19h
18h

MENSA The Table Mountain

A constellation introduced by the Frenchman Nicolas Louis de Lacaille to commemorate Table Mountain at the Cape of Good Hope from where he surveyed the southern skies in 1751–52. Part of the Large Magellanic Cloud strays from neighbouring Dorado over the border into Mensa, possibly reminding Lacaille of the cloud that frequently caps the real Table Mountain. Unfortunately, the constellation itself is faint and unimportant.

α (alpha) Mensae, 6h 10m –75°, mag. 5.1, is a yellow star similar to the Sun, 30 l.y. away.

β (beta) Men, 5h 03m –71°, mag. 5.3, is a yellow giant 305 l.y. away.

γ (gamma) Men, 5h 32m –76°, mag. 5.2, is an orange giant 390 l.y. away.

η (eta) Men, 4h 55m –75°, mag. 5.5, is an orange giant 460 l.y. away.

◀ ε (epsilon) Lyr, 18h 44m +40°, 135 l.y. away, is a celebrated quadruple star commonly called the Double Double. It is easily separated into two stars of mags. 4.7 and 4.6 by binoculars or even keen eyesight. But a telescope of 60 to 75 mm aperture and high magnification reveals that each star is itself double, with the orientations of the two pairs almost at right angles to each other (see diagram on page 174). The ε^1 pair has mags. of 5.0 and 6.1 and an orbital period of over 1000 years; the ε^2 pair has mags. of 5.2 and 5.5, and period about 600 years. Quadruple stars are rare, and this is the finest of them.

ζ (zeta) Lyr, 18h 45m +38°, 210 l.y. away, is a double star, easily split in small telescopes or binoculars into components of mags. 4.4 and 5.7.

η (eta) Lyr, 19h 14m +39°, 910 l.y. away, is a blue-white star of mag. 4.4 with a wide mag. 9.1 companion visible in a small telescope.

RR Lyr, 19h 25m +43°, 200 l.y. away, is the prototype of an important class of variable stars used as 'standard candles' for indicating distances in space. RR Lyrae variables are often found in globular clusters and are thus known as cluster-type variables. They are giant stars, related to Cepheid variables, that pulsate in size, varying by about one magnitude usually in less than a day. RR Lyrae itself varies from mag. 7.1 to 8.1 in 13.6 hours.

M57 (NGC 6720), 18h 54m +33°, the Ring Nebula, is a famous 9th-mag. planetary nebula 2000 l.y. away, conveniently placed between β (beta) and γ (gamma) Lyrae. On photographs taken through large telescopes it looks like a celestial smoke ring. A small telescope shows it as a noticeably elliptical misty disk, but a larger telescope is needed to see the central hole. It is one of the brightest planetary nebulae and appears larger in the sky than the planet Jupiter. (For a photograph see page 266.)

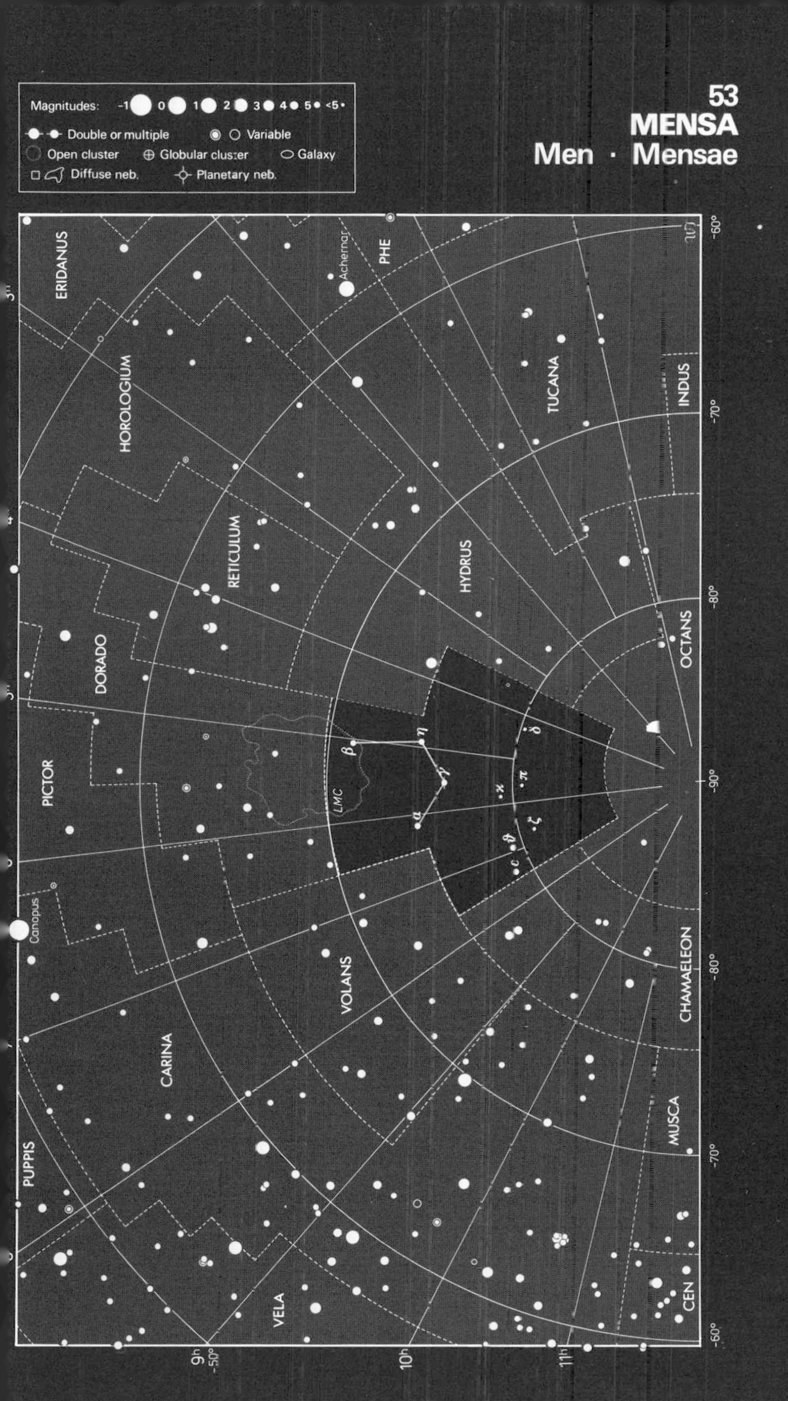

53
MENSA
Men · Mensae
Magnitudes: -1 0 1 2 3 4 5 <5
Double or multiple
Variable
Open cluster
Globular cluster
Galaxy
Diffuse neb.
Planetary neb.
ERIDANUS
HOROLOGIUM
RETICULUM
DORADO
PICTOR
Canopus
CARINA
PUPPIS
VELA
VOLANS
LMC
HYDRUS
TUCANA
Achernar
PHE
INDUS
OCTANS
CHAMAELEON
MUSCA
CEN
9h
10h
11h
-50°
-60°
-70°
-80°
-90°

MICROSCOPIUM The Microscope

Another of the southern hemisphere constellations representing scientific instruments that were introduced in the 1750s by the Frenchman Nicolas Louis de Lacaille. As with so many of his constellations, Microscopium is little more than a filler, encompassing a few faint stars between better-known constellations.

α (alpha) Microscopii, 20h 50m –34°, mag. 4.9, is a yellow giant star 250 l.y. away. It has a 10th-mag. companion, visible in small telescopes.

γ (gamma) Mic, 21h 01m –32°, mag. 4.7, is a yellow giant 245 l.y. away.

ε (epsilon) Mic, 21h 18m –32°, mag. 4.7, is a blue-white star 190 l.y. away.

The Rosette Nebula NGC 2237 in Monoceros, perhaps the most strikingly beautiful nebula in the heavens, surrounds the star cluster NGC 2244. For descriptions see page 180. Hale Observatories photograph.

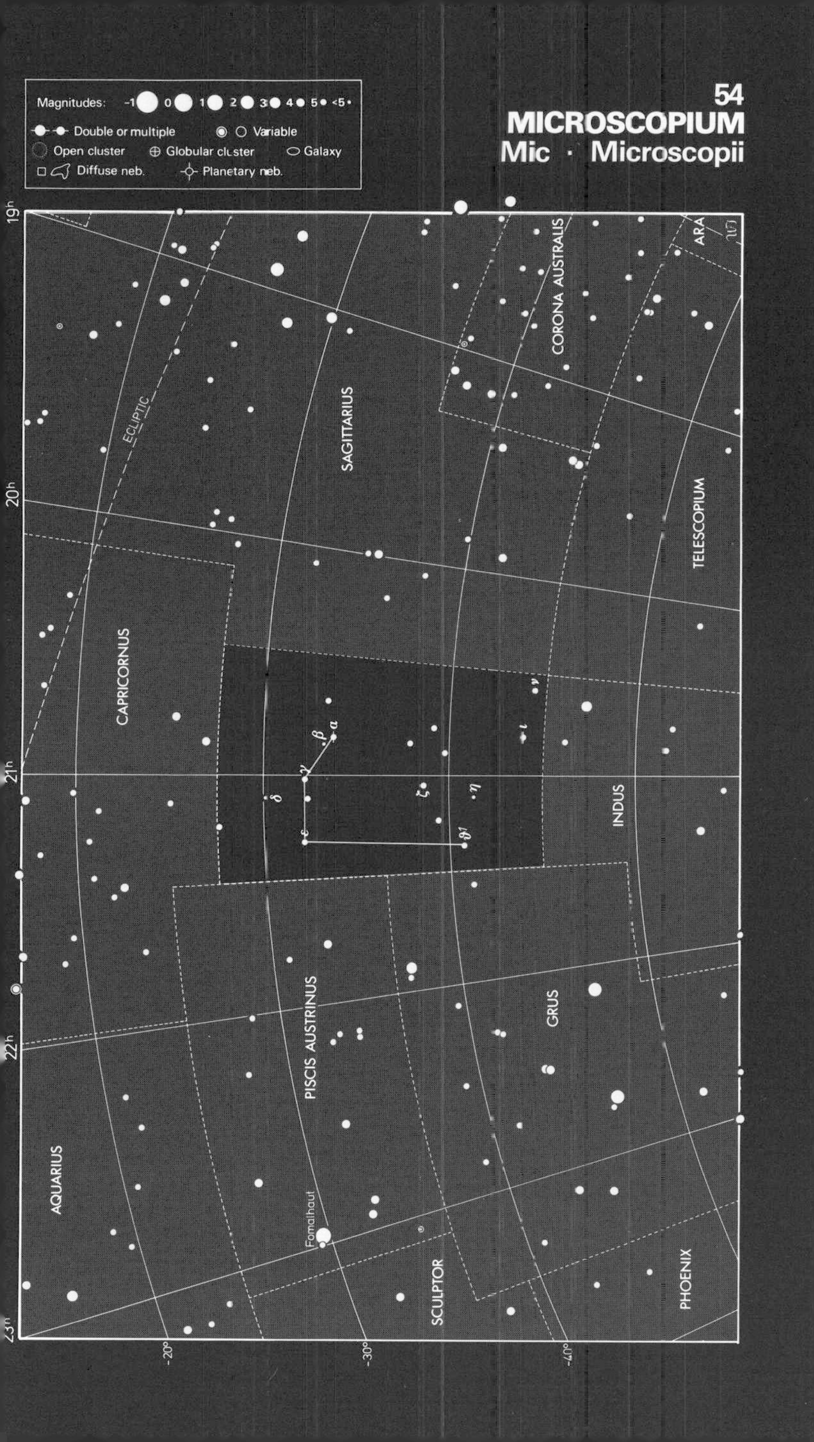
54
MICROSCOPIUM
Mic · Microscopii
Magnitudes: -1 0 1 2 3 4 5 <5
Double or multiple
Variable
Open cluster
Globular cluster
Galaxy
Diffuse neb.
Planetary neb.
19h
20h
21h
22h
23h
-20°
-30°
-40°
ECLIPTIC
SAGITTARIUS
CORONA AUSTRALIS
ARA
TELESCOPIUM
CAPRICORNUS
INDUS
PISCIS AUSTRINUS
GRUS
AQUARIUS
Fomalhaut
SCULPTOR
PHOENIX

MONOCEROS The Unicorn

A faint but fascinating constellation between Orion and Canis Minor, introduced by the Dutch theologian and astronomer Petrus Plancius in 1613, apparently because of references to a unicorn in the Old Testament of the Bible. Its location in the Milky Way ensures that it is well stocked with nebulae and clusters. Among the most celebrated stars of Monoceros is Plaskett's Star, a mag. 6.1 spectroscopic binary named after the Canadian astronomer John S. Plaskett, who found in 1922 that it is the most massive pair of stars known; each star is presently estimated to be at least 55 times the mass of the Sun. Plaskett's Star lies at 6h 37.4m, +6° 08′, near the cluster NGC 2244, of which it may be an outlying member.

α (alpha) Monocerotis, 7h 41m –10°, mag. 3.9, is an orange giant star 175 l.y. away.

β (beta) Mon, 6h 29m –7°, 590 l.y. away, is rated as perhaps the finest triple star in the heavens. The smallest of telescopes should separate the three components, of mags. 4.6, 5.4 and 5.6, on a steady night. They form a curving arc of blue-white stars, the faintest two being the closest together.

δ (delta) Mon, 7h 12m 0°, mag. 4.2, is a blue-white star 115 l.y. away. It has a wide, unrelated naked-eye companion, 21 Mon, of mag. 5.5.

ε (epsilon) Mon, 6h 24m +5°, also known as 8 Mon, 155 l.y. away, is an easy double star for small telescopes, consisting of yellow and blue components of mags. 4.4 and 6.7 in an attractive low-power field.

S Mon, 6h 41m +10°, is an intensely luminous blue-white star of mag. 4.7, 2500 l.y. away, situated in the star cluster NGC 2264 (see page 182). It is a double star, with a close companion of mag. 7.6 visible in a small telescope. S Mon is slightly variable, fluctuating erratically by about 0.1 mag.

M50 (NGC 2323), 7h 03m –8°, is an open cluster of about 80 stars, half the size of the full Moon, visible in binoculars and small telescopes. Apertures of 100 mm or so resolve it into stars of 8th mag. and fainter, with a red star at the centre. M50 is 3000 l.y. away.

NGC 2232, 6h 27m –5°, is a scattered cluster of 20 stars for binoculars, containing the mag. 5.1 blue-white star 10 Mon. The cluster, which covers the same area of sky as the full Moon, lies 1300 l.y. away.

NGC 2237, NGC 2244, 6h 32m +5°, is a complex combination of a faint diffuse nebula, known as the Rosette Nebula, and a cluster of stars, all about 5500 l.y. away. Long-exposure photographs show the nebula as a pink loop, twice the apparent diameter of the full Moon. Visual observations with large amateur telescopes reveal only the brightest parts of the nebula, each of which is given a separate NGC number. The associated cluster, NGC 2244, consists of stars that have been born from the Rosette Nebula's gas; it is just visible to the naked eye and is an easy binocular object. The six most prominent stars of the cluster form a rectangular shape, although the brightest of them, the mag. 5.8 ▶

MONOCEROS
Mon · Monocerotis

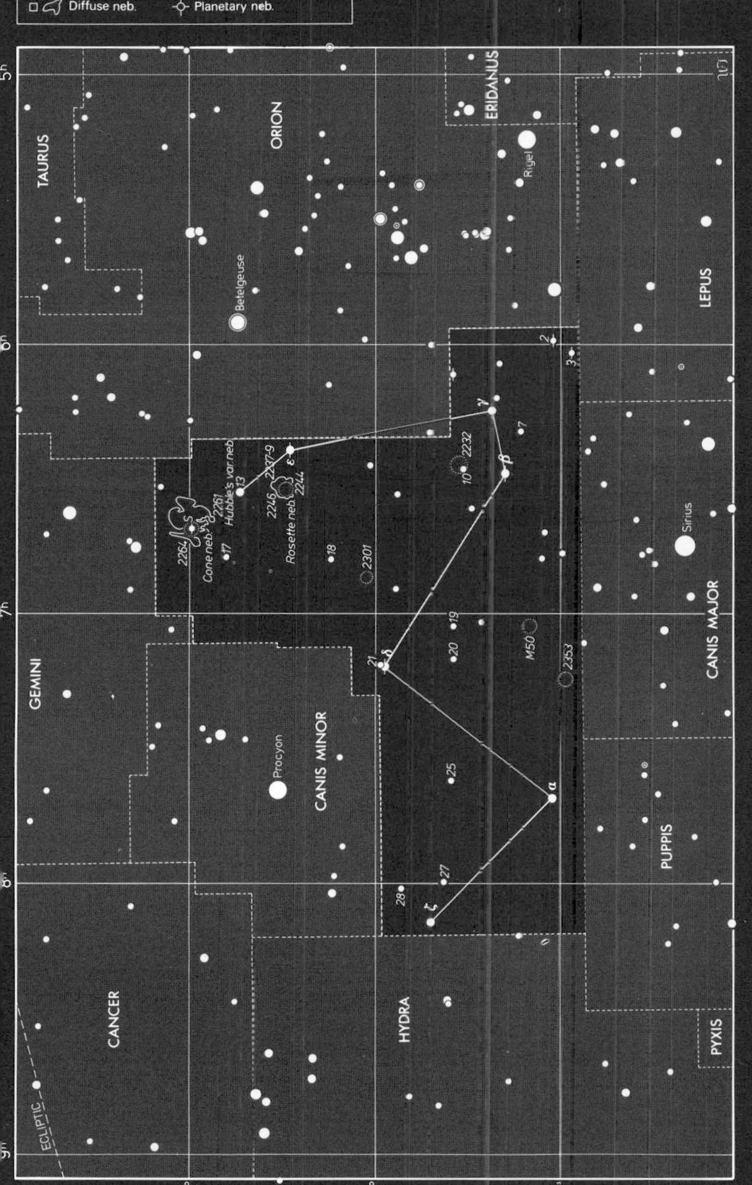

MUSCA The Fly

A small southern constellation lying at the foot of the Southern Cross. It is one of the 12 constellations introduced at the end of the 16th century by the Dutch navigators Pieter Dirkszoon Keyser and Frederick de Houtman, originally under the name of Apis, the Bee. There is little of note in the constellation other than part of the Coalsack Nebula, which spills into Musca from the neighbouring Southern Cross.

α (alpha) Muscae, 12h 37m –69°, mag. 2.7, is a blue-white star 390 l.y. away.

β (beta) Mus, 12h 47m –68°, 300 l.y. away, is a close pair of stars of mags. 3.7 and 4.0, requiring 75 mm aperture and high magnification to split. The orbital period of the pair is 400 years or so.

δ (delta) Mus, 13h 02m –72°, mag. 3.6, is an orange giant 185 l.y. away.

ϑ (theta) Mus, 13h 08m –65°, is a double star of mags. 5.7 and 7.3 for small telescopes. The brighter star is a blue supergiant, while its companion is a Wolf–Rayet star, a rare type of very hot star; it is the second-brightest such star in the sky, γ (gamma) Velorum being the brightest of all.

NGC 4833, 13h 00m –71°, is a fairly large, 7th-mag. globular cluster 18,000 l.y. away, visible in binoculars and small telescopes and resolvable into stars with an aperture of 100 mm.

◀ yellow dwarf 12 Mon, seems to be not a true member but an unrelated foreground star. The cluster is likely to be the only part of this celebrated object visible in a small telescope, but the pale outline of the nebula can just be made out in binoculars under clear, dark skies. (See photograph on page 178.)

NGC 2261, 6h 39m +9°, Hubble's Variable Nebula, is a small, faint nebula containing the remarkable variable star R Mon. At its brightest, this star is of mag. 9.5. Its erratic brightness fluctuations, down to about 12th mag., may be caused by the pangs of its birth from the surrounding nebula. This star and its nebula are open to study only by larger amateur telescopes; the star's distance is uncertain but it may be associated with the nearby NGC 2264 complex (see below), which would place it at about 2500 l.y.

NGC 2264, 6h 41m +10°, is another combination of star cluster and nebula. The cluster, visible in binoculars, has about 40 members, including the 5th-mag. star S Mon (see page 180). The nebula, known as the Cone Nebula because of its tapered shape, shows up well only on long-exposure photographs and is beyond the reach of amateur telescopes. The distance of NGC 2264 is 2500 l.y.

NGC 2301, 6h 52m 0°, is a binocular cluster of 80 or so stars of 8th mag. and fainter, 2500 l.y. away.

MUSCA
Mus · Muscae

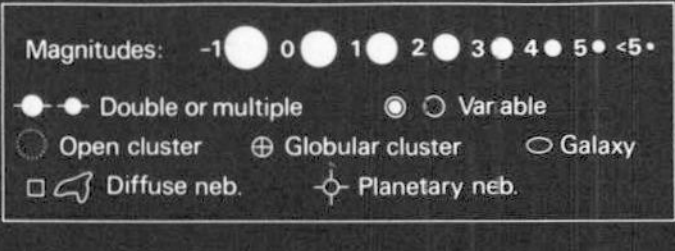

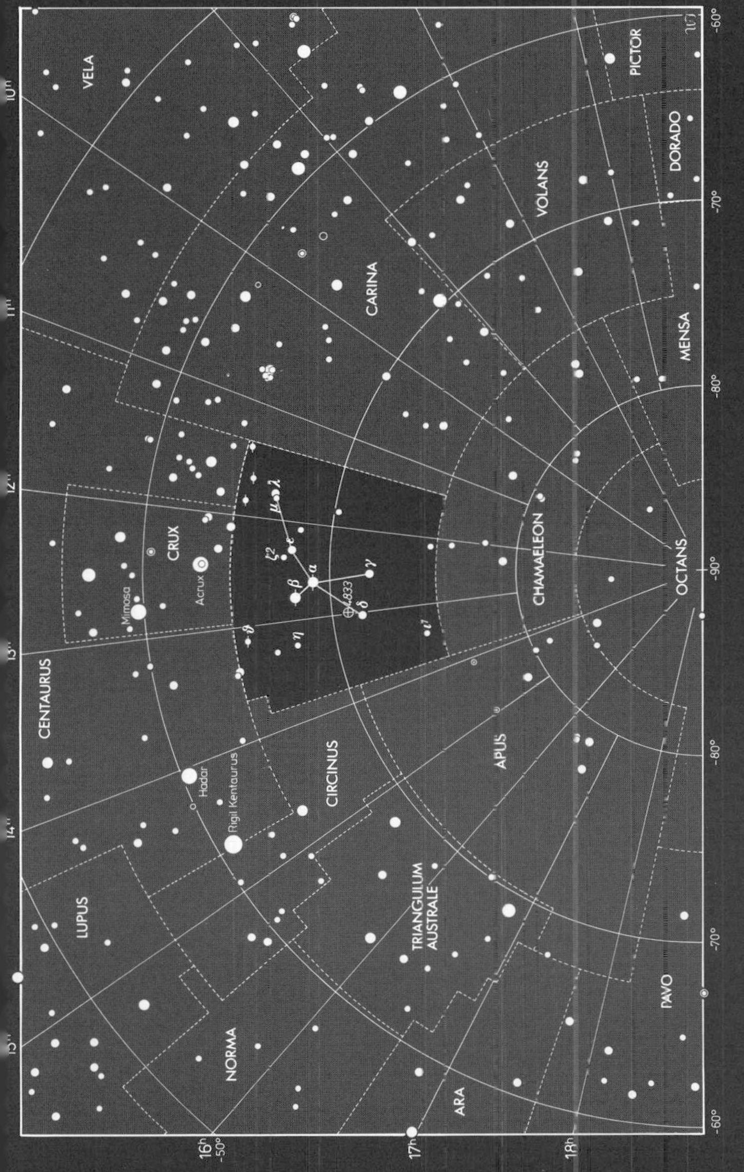

NORMA The Level

A superfluous constellation invented in the 1750s by Nicolas Louis de Lacaille, who populated the southern skies with several constellations representing scientific instruments, in this case a surveyor's level. Originally its constituent stars were part of Ara and Lupus. Since Lacaille's time the boundaries of Norma have been altered, so that the stars which were formerly α (alpha) and β (beta) Normae have been swallowed up by neighbouring constellations. Norma lies in a rich region of the Milky Way.

γ^2 (gamma^2) Normae, 16h 20m –50°, mag. 4.0, is a yellow giant star 145 l.y. away. Next to it lies the far more distant yellow supergiant γ^1 (gamma^1) Normae, mag. 5.0, estimated to be an astounding 10,000 l.y. away.

δ (delta) Nor, 16h 06m –45°, mag. 4.7, is a white star about 150 l.y. away.

ε (epsilon) Nor, 16h 27m –48°, 490 l.y. away, is a double star with components of mags. 4.5 and 7.5 visible in small telescopes. The brighter star is also a spectroscopic binary.

ι^1 (iota^1) Nor, 16h 04m –58°, 220 l.y. away, appears in small telescopes as a double star of mags. 4.6 and 8.1. The brighter star is itself a very close binary, divisible only in very large telescopes, with an orbital period of 27 years.

NGC 6087, 16h 19m –58°, is a binocular cluster of about 40 stars, 3000 l.y. away. At the centre of the cluster lies the Cepheid variable S Nor, which varies from mag. 6.1 to 6.8 in 9.8 days.

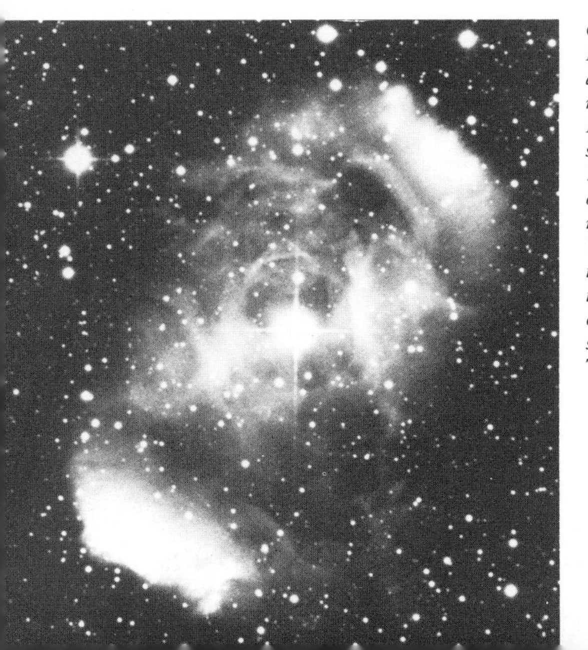

On the border of Norma with Ara lies an unusual patch of nebulosity called NGC 6164–5. It surrounds a young, hot 7th-mag. star known only by its catalogue number HD 148937. The gas that makes up the nebula has been thrown off in a series of outbursts by the central star. Anglo-Australian Telescope Board.

NORMA
Nor · Normae

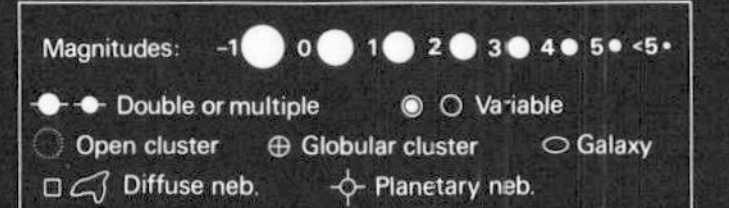

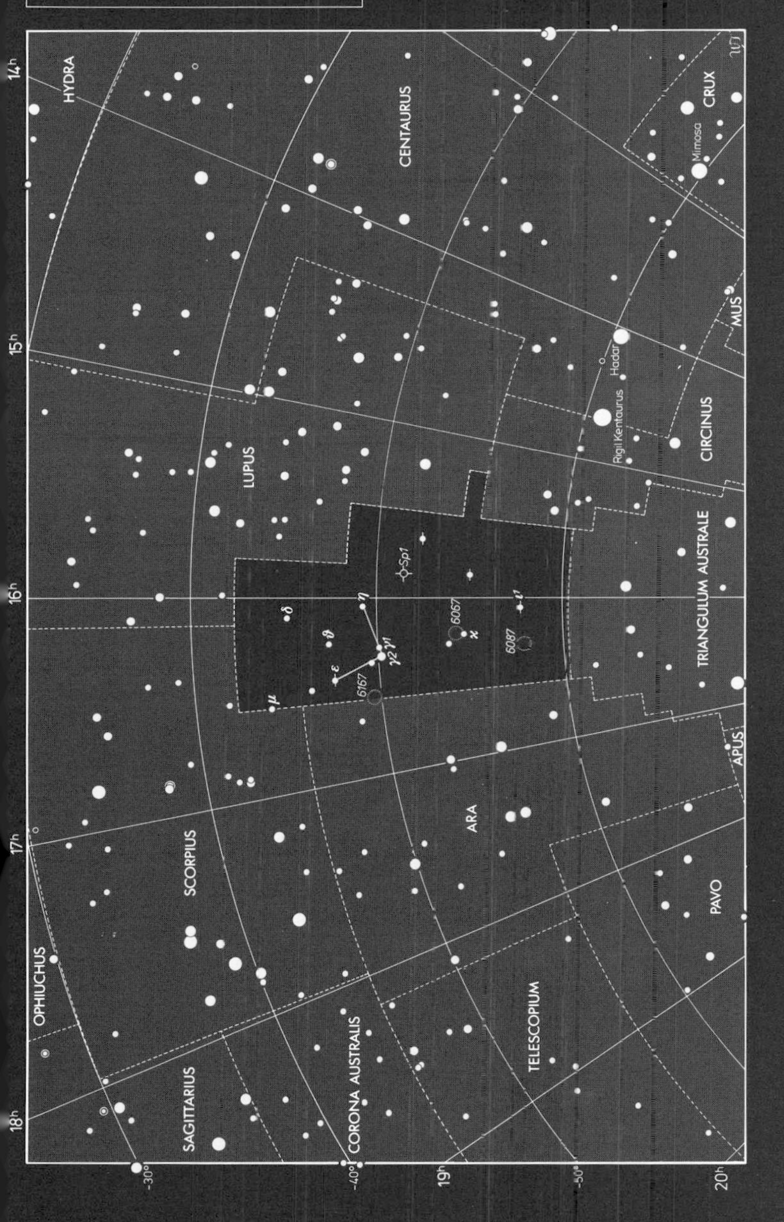

OCTANS The Octant

The constellation that contains the south pole of the sky. Despite this privileged position, Octans is faint and unremarkable. There is no southern equivalent of Polaris, the North Pole Star; the nearest moderately bright star to the south celestial pole is σ (sigma) Octantis, which lies about 1° from the pole. The constellation of Octans commemorates the instrument known as the octant, a forerunner of the sextant, invented by the Englishman John Hadley and used by him for measuring star positions. The constellation itself was introduced in the 1750s by Nicolas Louis de Lacaille during his stay at the Cape of Good Hope, and its dullness is a memorial to his dreadful lack of imagination.

α (alpha) Octantis, 21h 05m –77°, mag. 5.2, is a spectroscopic binary consisting of white and yellow giants 390 l.y. away.

β (beta) Oct, 22h 46m –81°, mag. 4.2, is a white star 110 l.y. away.

δ (delta) Oct, 14h 27m –84°, mag. 4.3, is an orange giant 185 l.y. away.

λ (lambda) Oct, 21h 51m –83°, 300 l.y. away, is a double star with yellow and white components of mags. 5.4 and 7.7, individually visible in small telescopes.

ν (nu) Oct, 21h 41m –77°, mag. 3.8, is a yellow giant 72 l.y. away.

σ (sigma) Oct, 21h 09m –89°, mag. 5.4, is a giant white star 300 l.y. away.

Detail chart showing the position of the south celestial pole and its movement over a period of a century as a result of precession. Wil Tirion.

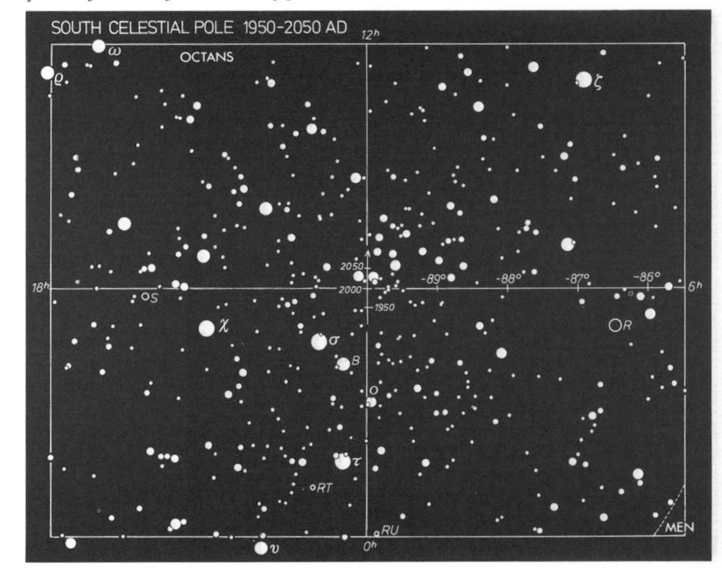

OCTANS
Oct · Octantis

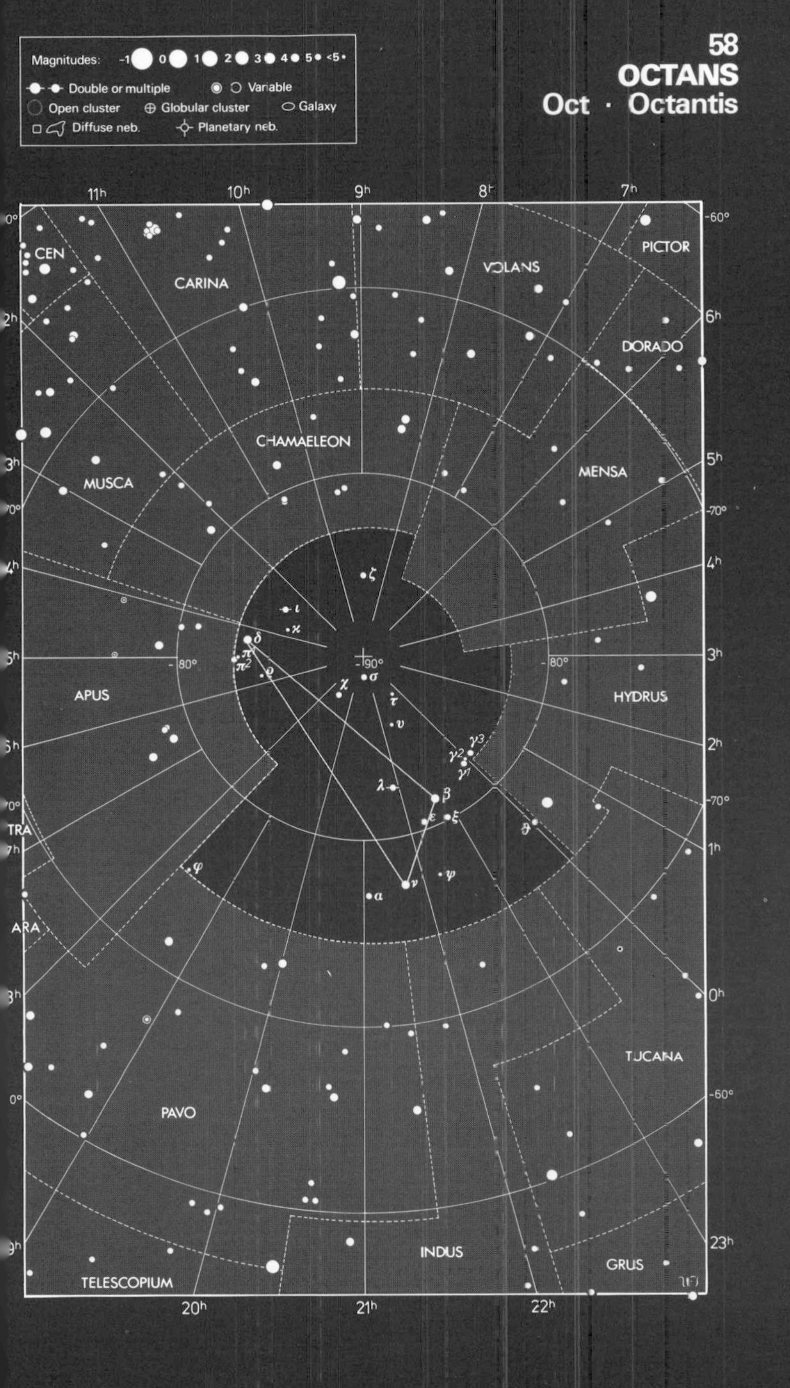

OPHIUCHUS The Serpent Holder

An ancient constellation, representing a man encoiled by a serpent (the constellation Serpens). Ophiuchus is usually identified as Aesculapius, a mythical healer and forerunner of Hippocrates, whose reputed powers included the ability to raise the dead. The serpent he holds is a symbol of this power, since snakes are seemingly reborn every year when they shed their skin. Perhaps the most celebrated star in Ophiuchus is the mag. 9.5 red dwarf Barnard's Star, 6 l.y. away, the second-closest star to the Sun. It lies at 17h 57.8m, +4° 34′, and is named after the American astronomer E. E. Barnard who found in 1916 that it has the greatest proper motion of any star, covering the apparent diameter of the Moon every 180 years. Barnard's Star is of particular interest because it may have planets. The southernmost regions of Ophiuchus extend into rich starfields of the Milky Way, looking towards the centre of the Galaxy; consequently the constellation is noted for its star clusters. Ophiuchus was the site of the last supernova seen to erupt in our Galaxy, which appeared at 17h 30.6m, –21° 29′, in 1604.

α (alpha) Ophiuchi, 17h 35m +13°, (Rasalhague, 'head of the serpent charmer'), mag. 2.1, is a white star 59 l.y. away.

β (beta) Oph, 17h 43m +5°, (Cebalrai), mag. 2.8, is a yellow giant 110 l.y. away.

γ (gamma) Oph, 17h 48m +3°, mag. 3.8, is a blue-white star 115 l.y. away.

δ (delta) Oph, 16h 14m –4°, (Yed Prior, 'the preceding star of the hand'), mag. 2.7, is an orange giant 160 l.y. away.

ε (epsilon) Oph, 16h 18m –5°, (Yed Posterior, 'the following star of the hand'), mag. 3.2, is an orange giant 125 l.y. away.

ζ (zeta) Oph, 16h 37m –11°, mag. 2.6, is a blue-white star 550 l.y. away.

η (eta) Oph, 17h 10m –16°, mag. 2.4, is a blue-white star 68 l.y. away.

ρ (rho) Oph, 16h 26m –23°, 680 l.y. away, is a striking multiple star for small instruments. The brightest star is mag. 5.0. This has a close companion of mag. 5.9 visible in a small telescope under high magnification; either side of this pair are wide binocular companions of mags. 6.7 and 7.3.

τ (tau) Oph, 18h 03m –8°, 62 l.y. away, is a close pair of cream-coloured stars, mags. 5.2 and 5.9, orbiting every 280 years and currently gradually closing, so that they require at least 75 mm aperture by the year 2000 and 100 mm by 2030.

36 Oph, 17h 15m –27°, 18 l.y. away, is a pair of mag. 5.3 orange dwarf stars split by small apertures. Their calculated orbital period is 500 years or more.

70 Oph, 18h 05m +2°, 16 l.y. away, is a celebrated double star, consisting of yellow and orange components of mags. 4.2 and 6.0 which orbit each other every 88 years. They are now rapidly widening so that an aperture of 75 mm should separate them by the year 1995, and they will be easy in the smallest apertures after 2000. ▶

59
OPHIUCHUS
Oph · Ophiuchi

Magnitudes: -1 0 1 2 3 4 5 <5
Double or multiple — Variable
Open cluster — Globular cluster — Galaxy
Diffuse neb. — Planetary neb.

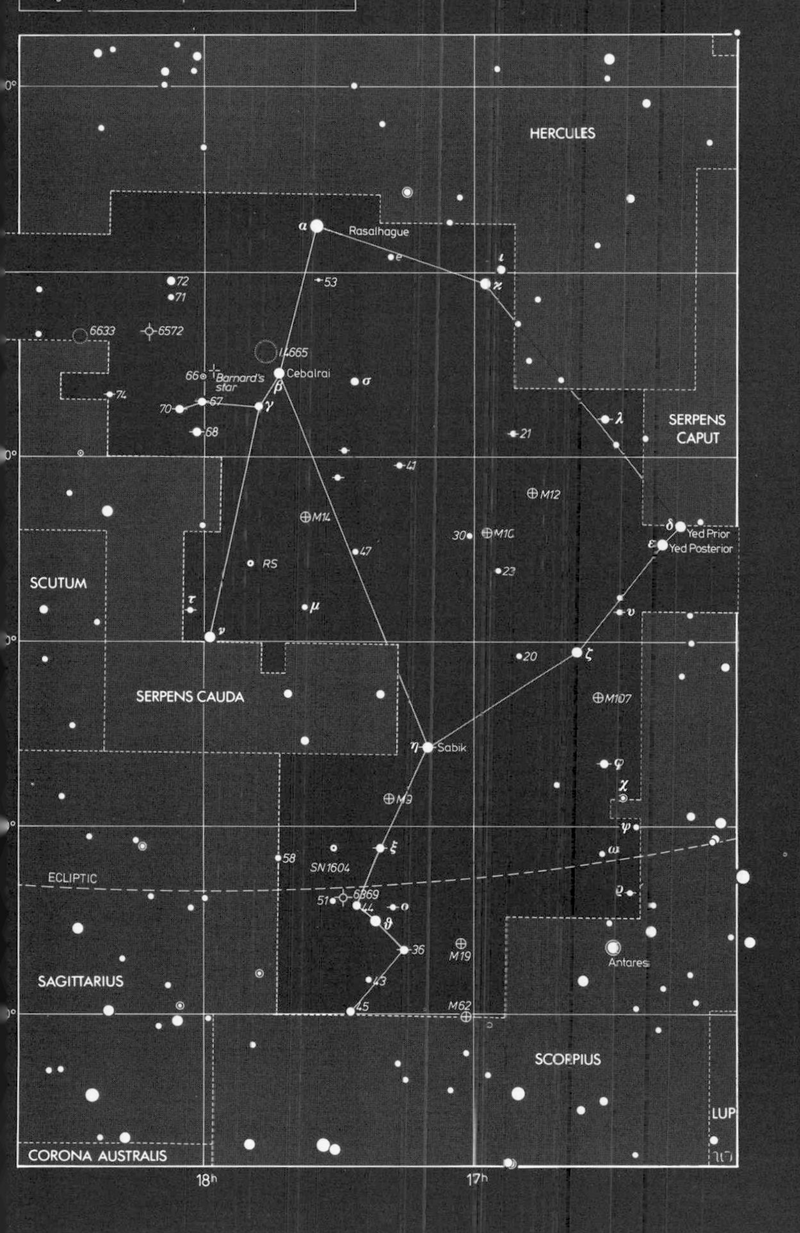

ORION The Hunter

Without doubt the brightest and grandest constellation in the sky, crammed with objects of interest for all sizes of instrument. Orion's impressiveness stems in large measure from the fact that it is an area of star formation in a nearby arm of the Galaxy, centred on the famous Orion Nebula. The Orion Nebula, M42, marks the Hunter's sword, hanging from his belt. The belt itself is formed by a line of three bright stars. Orion is depicted as brandishing a club and shield at the snorting Taurus, the Bull. In one story the boastful Orion was stung to death by a scorpion and is now placed in the sky so that he sets as the constellation Scorpius, the Scorpion, rises. Each year the Orionid meteors radiate from a point near the border with Gemini. As many as 25 Orionid meteors per hour may be seen around October 22.

α (alpha) Orionis, 5h 55m +7°, (Betelgeuse), 425 l.y. away, is a red supergiant star so large that it is unstable. It fluctuates erratically in size between about 300 and 400 times the Sun's diameter, changing in brightness as it does so from mag. 0.0 to 1.3.

β (beta) Ori, 5h 15m –8°, (Rigel, 'giant's leg'), at mag. 0.1 the brightest star in Orion, is a blue-white supergiant 1050 l.y. away; note its colour contrast with Betelgeuse. Rigel has a mag. 6.8 companion, difficult to see in small telescopes, particularly in poor seeing, because of glare from the primary.

γ (gamma) Ori, 5h 25m +6°, (Bellatrix, 'the warrior'), mag. 1.6, is a blue giant star 360 l.y. away. ▶

◀ RS Oph, 17h 50m –7°, is a recurrent nova seen to have erupted five times, a record it shares with T Pyxidis and U Scorpii. Normally it is around 12th mag., but it flared up to naked-eye brightness in 1898, 1933, 1958, 1967 and 1985.

M10 (NGC 6254), 16h 57m –4°, is a 7th-mag. globular cluster visible in binoculars or a small telescope. It is 14,000 l.y. away, somewhat closer than its neighbour M12. Individual stars can be resolved with telescopes of 100 mm aperture.

M12 (NGC 6218), 16h 47m –2°, is a 7th-mag. globular cluster 18,000 l.y. away, visible in binoculars or a small telescope. In small apertures it appears slightly larger than its neighbour M10, and its stars are more loosely scattered. There are several other globular clusters in Ophiuchus worthy of attention, but M10 and M12 are the finest.

NGC 6572, 18h 12m +7°, is a 9th-mag. planetary nebula 2000 l.y. away, visible in at least a 75 mm aperture as a tiny blue-green ellipse.

NGC 6633, 18h 28m +7°, is a scattered binocular cluster of about 30 stars, 1000 l.y. away.

IC 4665, 17h 46m +6°, is a loose and irregular cluster of two dozen or so stars of 7th mag. and fainter, 1400 l.y. away, larger than the apparent size of the Moon and best seen in binoculars.

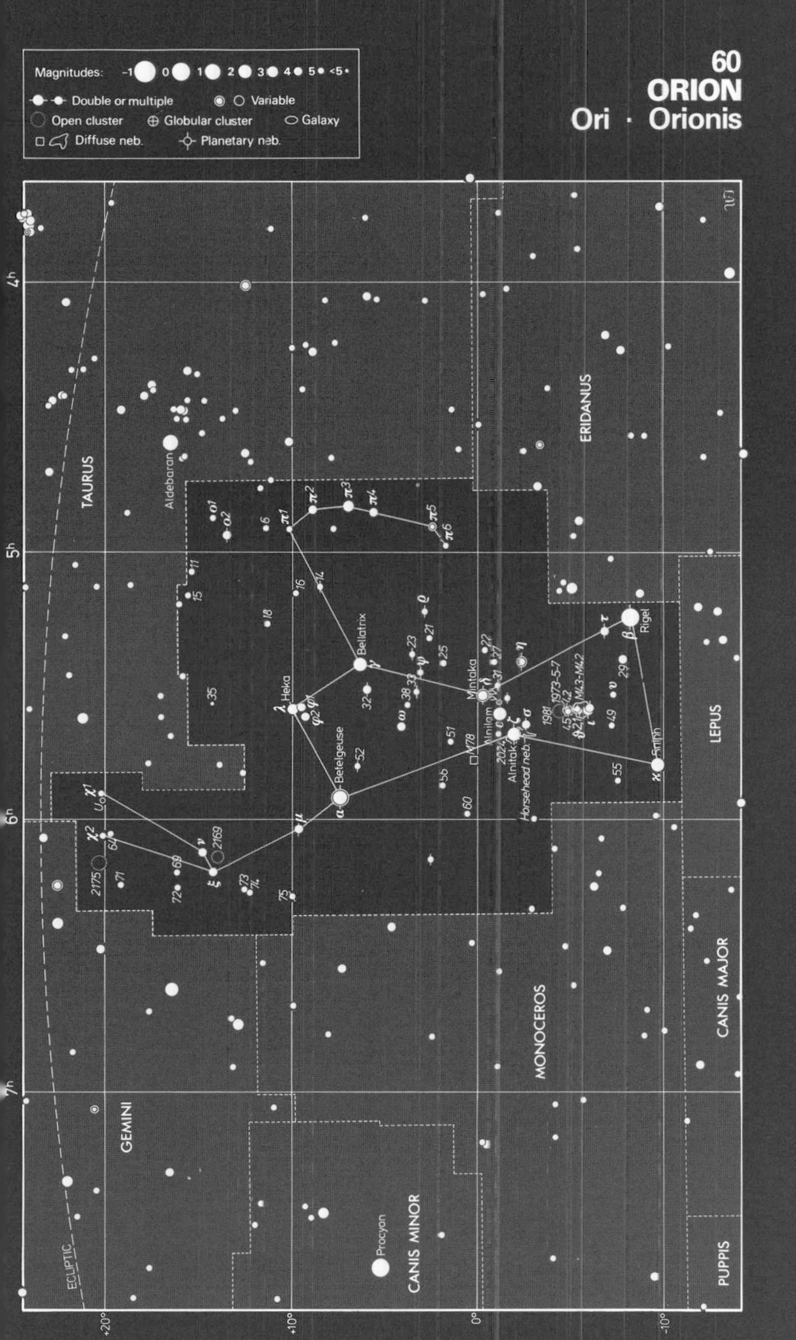
60
ORION
Ori · Orionis
Magnitudes: -1 0 1 2 3 4 5 <5
Double or multiple
Variable
Open cluster
Globular cluster
Galaxy
Diffuse neb.
Planetary neb.
TAURUS
GEMINI
ERIDANUS
LEPUS
MONOCEROS
CANIS MAJOR
CANIS MINOR
PUPPIS
ECLIPTIC
Aldebaran
Betelgeuse
Bellatrix
Heka
Mintaka
Alnilam
Alnitak
Rigel
Saiph
Procyon
Horsehead neb.
M78
M43-M42
4h
5h
6h
7h
+20°
+10°
0°
-10°

◀ δ (delta) Ori, 5h 32m 0°, (Mintaka, 'the belt'), 1500 l.y. away, is a complex multiple star. To the naked eye it appears as a blue-white star of mag. 2.2. Binoculars or small telescopes reveal a wide companion of mag. 6.9 that is unrelated, lying 2150 l.y. away. The brighter star is also an eclipsing binary that varies by about 0.2 mag. every 5.7 days.

ε (epsilon) Ori, 5h 36m –1°, (Alnilam, 'string of pearls'), mag. 1.7, is a blue supergiant 1400 l.y. away.

ζ (zeta) Ori, 5h 41m –2°, (Alnitak, 'the girdle'), 1100 l.y. away, appears to the naked eye as a blue-white star of mag. 1.8, but telescopes of 75 mm aperture and above reveal a close companion of mag. 4.0 that is estimated to orbit it every 1500 years. There is also a much wider 10th-mag. star.

η (eta) Ori, 5h 24m –2°, 1500 l.y. away, is a complex multiple–variable. A telescope of at least 100 mm aperture at high magnification is needed to show that it consists of two close stars, of mags. 3.8 and 4.8. The brighter star is also an eclipsing binary, varying by about 0.2 mag. every 8 days.

ϑ^1 (theta^1) Ori, 5h 35m –5°, 1500 l.y. away, is a multiple star at the heart of the Orion Nebula, from which it has recently formed and which it illuminates. This star is commonly known as the Trapezium, because a small telescope shows four stars here; but a telescope of 100 mm aperture also reveals two others, of 11th mag. The four main stars of the Trapezium are of mags. 5.1, 6.7, 6.7 and 8.0. Nearby lies ϑ^2 (theta^2) Orionis, a wide double of mags. 5.1 and 6.4.

ι (iota) Ori, 5h 35m –6°, 1900 l.y. away, is a double star on the southern edge of the Orion Nebula, divisible in a small telescope. Its components are of mags. 2.8 and 6.9. Also visible in the same field of view is a wider double of blue-white stars, Struve 747, of mags. 4.8 and 5.7.

$\varkappa$ (kappa) Ori, 5h 48m –10°, (Saiph, 'sword'), mag. 2.1, is a blue supergiant 1300 l.y. away.

σ (sigma) Ori, 5h 39m –3°, 1500 l.y. away, is perhaps the most impressive of all Orion's stellar treasures. To the naked eye it appears as a blue-white star of mag. 3.8, but small telescopes reveal much more. On one side of the star are blue-white companions of mags. 6.7 and 6.6, the wider of which can be glimpsed in binoculars; it is an eclipsing binary with a range of about 0.1 mag. On the opposite side is a closer 9th-mag. companion, which is more difficult to see because of glare from the primary. The effect is like a planet with moons. To complete the picture, in the same telescopic field of view is a faint triple star called Struve 761, consisting of a narrow triangle of 8th- and 9th-mag. stars. An extraordinarily rich and unexpected sight, to be returned to again and again.

M42, M43 (NGC 1976, NGC 1982), 5h 35m –5°, is the most celebrated of Orion's deep-sky wonders – a gigantic nebula of gas and dust, 1500 l.y. away and 15 l.y. in diameter, from which a star cluster is being born. Behind the visible part of the nebula, illuminated by the stars of the Trapezium (see ϑ Ori), radio and infrared astronomers have detected an even larger dark cloud in which more stars are forming. M42 covers an area greater than 1° × 1°, and is indisputably the finest diffuse nebula in the sky, clearly visible to the naked eye as a hazy cloud. Binoculars and small telescopes reveal some of the more prominent wreaths and swirls of gas, which become more complex and breathtaking with increasing aperture. Although colour photographs depict the nebula as red and blue, to the eye it appears distinctly greenish because of the different colour sensitivities of photographic film and the human eye. A dark lane of dust known as the Fish Mouth separates M42 from M43, a smaller and rounder patch that is really part of the same cloud; M43 is centred on a 7th-mag. star.

M78 (NGC 2068), 5h 47m 0°, is a small, elongated nebula centred on a 10th-mag. double star. In large amateur telescopes it shows wispy structure.

NGC 1977, 5h 36m –5°, 1500 l.y. away, is an elongated nebulosity just north of the Orion Nebula, centred on the mag. 4.6 blue giant star 42 Orionis, also known as c Orionis. This object would be more celebrated were it not so overshadowed by M42.

NGC 1981, 5h 35m –4°, is a scattered cluster of about 20 stars of 6th mag. and fainter, 1300 l.y. away, to the north of the nebulosity NGC 1977. Included in this cluster is the double star Struve 750, a neat pair of 6th- and 8th-mag. stars.

NGC 2024, 5h 41m –2°, is a mushroom-shaped cloud of gas about ½° wide, surrounding the star ζ (zeta) Orionis. Running south from ζ Ori is a strip of nebulosity, IC 434, into which is indented the celebrated Horsehead Nebula, a dark cloud of obscuring dust shaped like a horse's head. Although long-exposure photographs show NGC 2024 and the Horsehead Nebula well (see page 267), they are notoriously difficult to see with amateur telescopes.

Detailed chart of the Orion Nebula region. Wil Tirion.

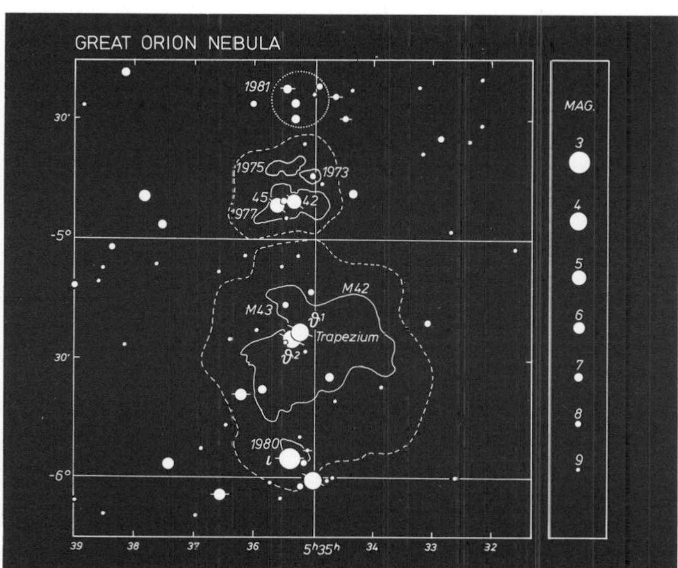

PAVO The Peacock

A constellation introduced at the end of the 16th century by the Dutch navigators Pieter Dirkszoon Keyser and Frederick de Houtman. It is one of several celestial birds in the region, including Apus, Tucana, Grus and Phoenix. In mythology the peacock was sacred to Juno, goddess of the heavens, from whose breast the Milky Way sprang. According to legend, Juno set a creature with a hundred eyes called Argus to watch over a white heifer, into which she guessed her husband Jupiter had turned one of his illicit lovers, the nymph Io. At Jupiter's request, Mercury decapitated the watchful Argus and released the heifer. Juno placed the hundred eyes of Argus on the peacock's tail.

α (alpha) Pavonis, 20h 26m –57°, (Peacock), mag. 1.9, is a blue-white star 360 l.y. away.

β (beta) Pav, 20h 45m –66°, mag. 3.4, is a white star 88 l.y. away.

δ (delta) Pav, 20h 09m –66°, mag. 3.6, is a yellow star 19 l.y. away.

η (eta) Pav, 17h 46m –65°, mag. 3.6, is a yellow giant 145 l.y. away.

$\varkappa$ (kappa) Pav, 18h 57m –67°, is one of the brightest Cepheid variables in the sky. It is a yellow-white supergiant, varying between mags. 3.9 and 4.8 in 9.1 days. Its distance is uncertain, but is several hundred light years.

ξ (xi) Pav, 18h 23m –61°, 460 l.y. away, is a red giant of mag. 4.4 with a close companion of mag. 8.6, lost in the primary's glare in small telescopes.

NGC 6752, 19h 11m –60°, is a large 5th-mag. globular cluster, visible in binoculars, covering half the apparent diameter of the Moon. It lies 14,000 l.y. away.

NGC 6744 in Pavo, a stately 9th-mag. galaxy with a short central bar and extensive spiral arms. Anglo-Australian Telescope Board.

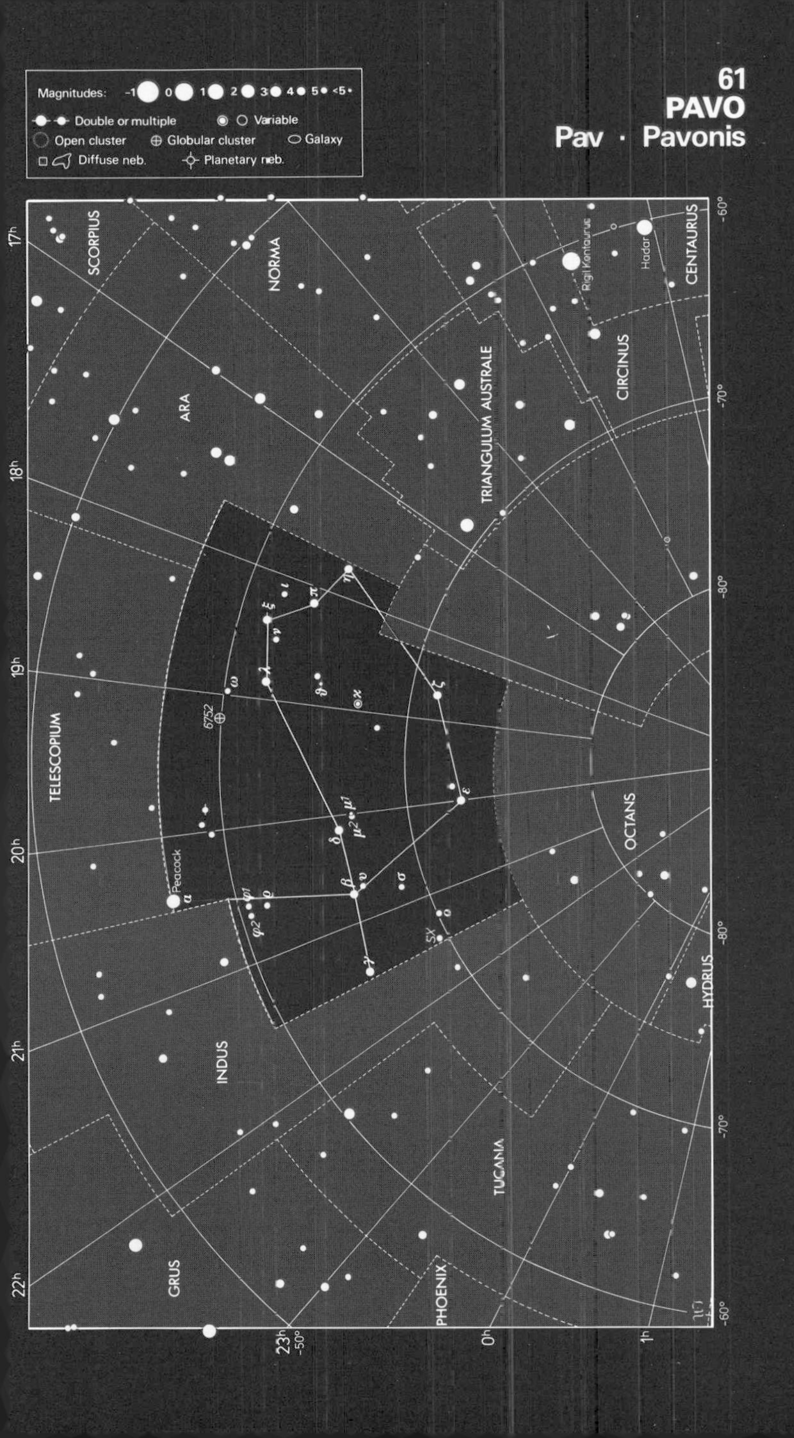

61
PAVO
Pav · Pavonis
Magnitudes: -1 0 1 2 3 4 5 <5
Double or multiple
Variable
Open cluster
Globular cluster
Galaxy
Diffuse neb.
Planetary neb.
SCORPIUS
NORMA
ARA
TRIANGULUM AUSTRALE
CIRCINUS
CENTAURUS
Rigil Kentaurus
Hadar
TELESCOPIUM
Peacock
6752
OCTANS
INDUS
TUCANA
HYDRUS
GRUS
PHOENIX
SX
17h
18h
19h
20h
21h
22h
23h
0h
1h
-50°
-60°
-70°
-80°

PEGASUS

The winged horse of Greek mythology, born from the blood of Medusa after she was slain by Perseus, who lies nearby in the sky. The most famous feature of Pegasus is the Great Square, outlined by four stars. One of these stars, which astronomers of old knew as δ (delta) Pegasi, is now assigned to Andromeda. The Great Square of Pegasus contains surprisingly few naked-eye stars for such a large area of sky. In fact, the entire constellation is poor in objects of interest, apart from a major globular cluster, M15.

α (alpha) Pegasi, 23h 05m +15°, (Markab, 'shoulder'), mag. 2.5, is a blue-white giant star 165 l.y. away.

β (beta) Peg, 23h 04m +28°, (Scheat, 'shin'), 180 l.y. away, is a red giant that varies from mag. 2.3 to 2.7 with no definite period.

γ (gamma) Peg, 0h 13m +15°, (Algenib, 'the side'), mag. 2.8, is a blue-white star 590 l.y. away. It is a variable of the β Cephei type, but its variations every 3 hours 45 minutes are less than 0.1 mag., too slight to be discernible to the naked eye.

ε (epsilon) Peg, 21h 44m +10°, (Enif, 'nose'), 520 l.y. away, is a yellow supergiant of mag. 2.4. A small telescope, or even good binoculars, reveals a wide bluish mag. 8.4 companion star. Larger telescopes also show an 11th-mag. companion closer to ε Peg, making this an apparent triple system.

ζ (zeta) Peg, 22h 41m +11°, (Homam), mag. 3.4, is a white star 155 l.y. away.

η (eta) Peg, 22h 43m +30°, (Matar), mag. 3.0, is a yellow giant 310 l.y. away.

π (pi) Peg, 22h 10m +33°, is a very wide binocular duo of white and yellow stars, both giants. They are of mags. 4.3 and 5.6, and lie 170 and 310 l.y. away respectively.

1 Peg, 21h 22m +20°, mag. 4.1, is a yellow giant 225 l.y. away with a mag. 8.2 companion visible in a small telescope.

M15 (NGC 7078), 21h 30m +12°, is an outstanding 6th-mag. globular cluster, 30,000 l.y. distant, at the limit of naked-eye visibility but easily seen in binoculars, with a 6th-mag. star nearby as a sure guide to its location. A telescope shows it as a glorious misty sight in an attractive field. With an aperture of 150 mm or so, its outer regions can be resolved into a mottled ground of sparkling stars, and larger apertures show stars all the way to the bright and condensed core.

NGC 7331, 22h 37m +34°, is a 10th-mag. spiral galaxy, visible under good conditions in apertures of 100 mm or so as an elongated smudge. It lies about 40 million l.y. away.

62 PEGASUS
Peg · Pegasi

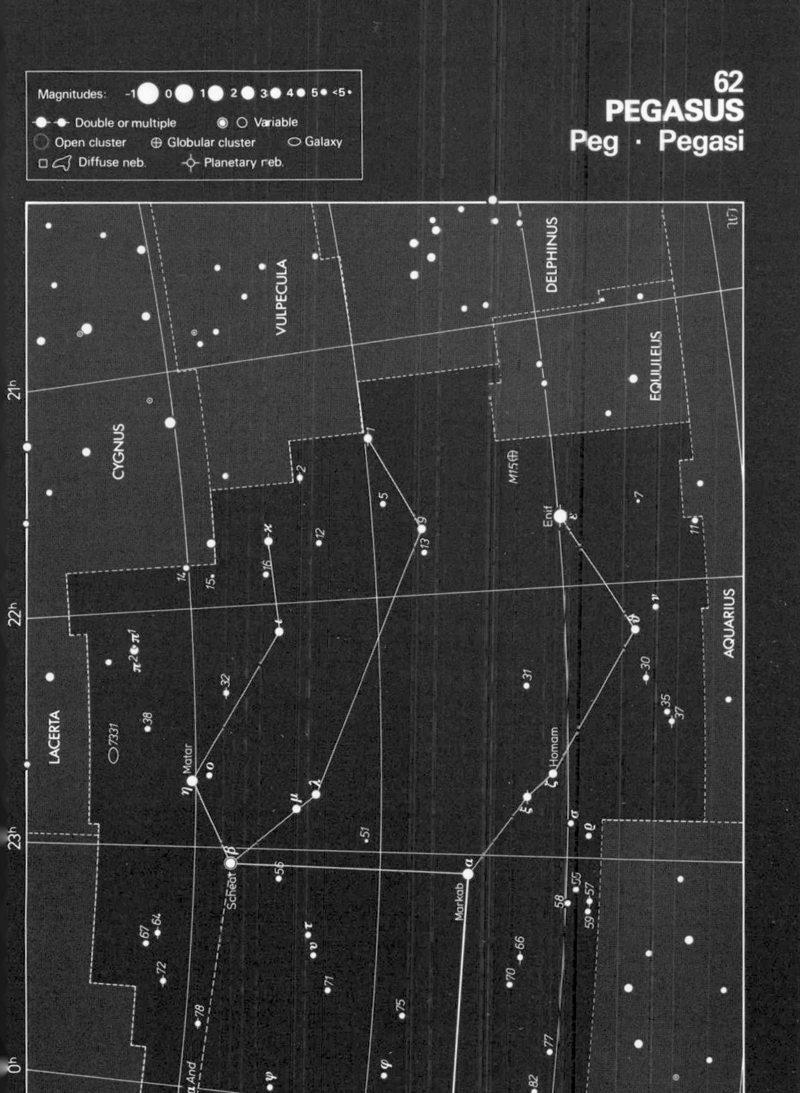

PERSEUS

Perseus was the hero of Greek mythology who rescued the chained maiden Andromeda from the clutches of the sea monster Cetus. Previously, Perseus had slain Medusa the Gorgon, whose head he is pictured holding in one hand. The Gorgon's head is marked by the winking star Algol, sometimes imagined as Medusa's evil eye. Perseus lies in a rich part of the Milky Way and is well worth sweeping with binoculars; note, in particular, the Double Cluster. In 1901 Nova Persei flared up to magnitude 0.2 at 3h 31.2m, +43° 54′; it threw off a shell of gas that is now visible in large telescopes. NGC 1499, the California Nebula, so named because its shape resembles California, spans the apparent width of five full Moons north of ξ (xi) Persei, which illuminates it. Despite its considerable size the California Nebula is elusive visually, but shows up well on long-exposure photographs. At 3h 19.8m, +41° 31′, lies the radio source Perseus A, associated with the 12th-magnitude peculiar galaxy NGC 1275, which is at the centre of the Perseus cluster of galaxies, 250 million l.y. away. Near γ (gamma) Persei lies the radiant of the Perseid meteors, the most glorious meteor shower of the year: around August 12–13 as many as 75 bright meteors can be seen flashing from Perseus each hour.

α (alpha) Persei, 3h 24m +50°, (Algenib, 'the side', or Mirfak, 'elbow'), mag. 1.8, is a yellow supergiant 550 l.y. away. Binoculars reveal a brilliant scattering of stars in this region, forming a loose cluster known as Melotte 20.

β (beta) Per, 3h 08m +41°, (Algol, 'the demon'), 100 l.y. away, is one of the most celebrated variable stars in the sky. It is the prototype of the eclipsing binary class of variables, in which two close stars periodically eclipse one another as they orbit their common centre of gravity. Algol's eclipses occur every 2.87 days, when the star's apparent brightness sinks from mag. 2.1 to 3.4 before returning to normal about 10 hours later.

γ (gamma) Per, 3h 05m +54°, mag. 2.9, is a yellow giant 95 l.y. away.

δ (delta) Per, 3h 43m +48°, mag. 3.0, is a blue giant 390 l.y. away.

ε (epsilon) Per, 3h 58m +40°, 680 l.y. away, is a blue-white star of mag. 2.9 with an unrelated mag. 7.6 companion, difficult to see through the smallest telescopes because of the magnitude contrast.

ζ (zeta) Per, 3h 54m +32°, 1600 l.y. away, is a blue supergiant of mag. 2.9 with a mag. 9.5 companion visible in a small telescope.

η (eta) Per, 2h 51m +56°, 880 l.y. away, is an orange supergiant of mag. 3.8 with a mag. 8.5 blue companion that forms an attractive double for small telescopes. The field of view contains a sprinkling of background stars.

ρ (rho) Per, 3h 05m +39°, 200 l.y. away, is a red giant that varies between mags. 3.3 and 4.0 in semi-regular fashion every 7 weeks or so. ▶

PERSEUS
Per · Persei

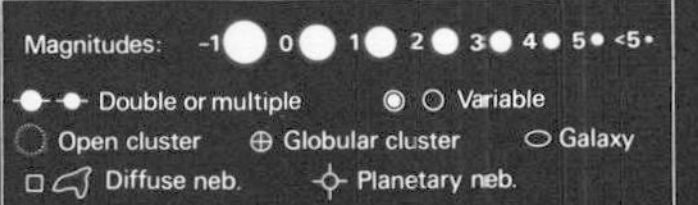

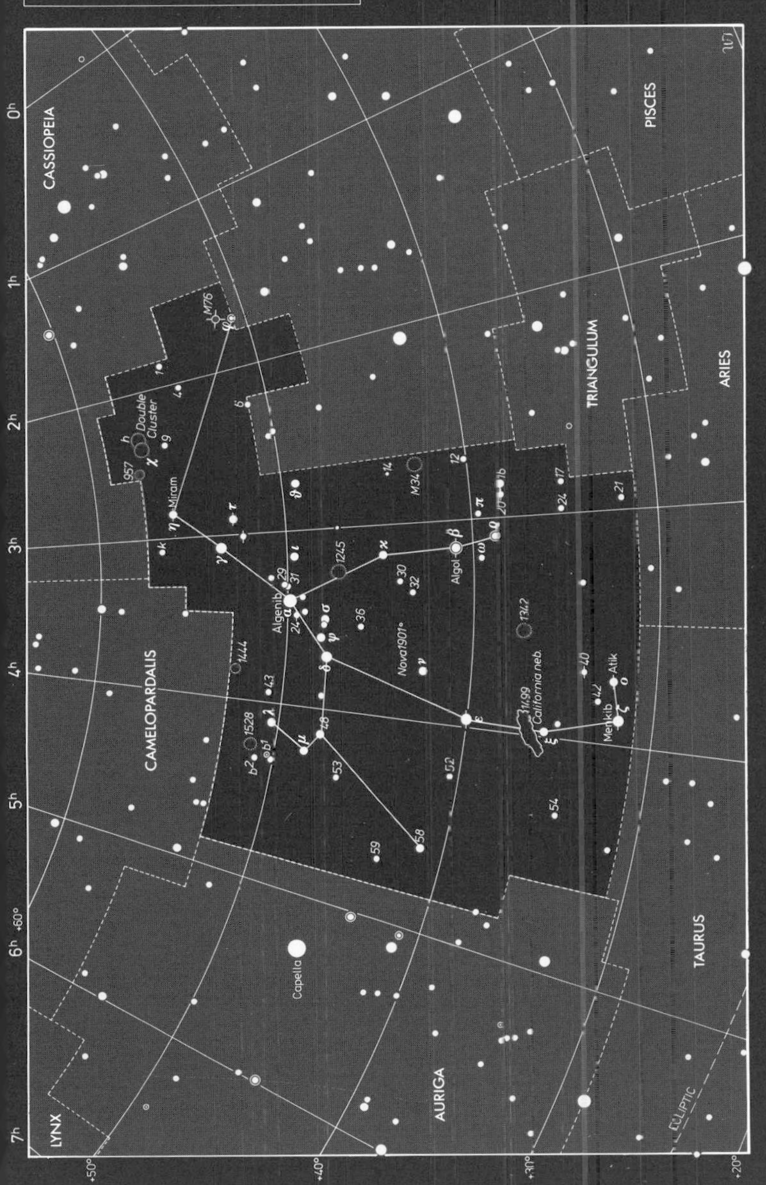

PHOENIX The Phoenix

An inconspicuous constellation near the southern end of Eridanus, representing the mythical bird that was regularly reborn from its own ashes. It was introduced at the end of the 16th century by the Dutch navigators Pieter Dirkszoon Keyser and Frederick de Houtman in an area that had been known by the Arabs as the Boat, moored on the bank of the river Eridanus.

α (alpha) Phoenicis, 0h 26m –42°, mag. 2.4, is a yellow giant star 88 l.y. away.

β (beta) Phe, 1h 06m –47°, 130 l.y. away, appears to the naked eye as a yellow star of mag. 3.3. In fact, it is a close double with well-matched components of mags. 4.0 and 4.2, divisible in telescopes of 100 mm aperture.

γ (gamma) Phe, 1h 28m –43°, mag. 3.4, is an orange giant of uncertain distance.

ζ (zeta) Phe, 1h 08m –55°, 980 l.y. away, is a complex variable and multiple star. The main star is a blue-white eclipsing variable that fluctuates between mags. 3.9 and 4.4 every 1.67 days. It has a mag. 6.9 companion visible in a small telescope. There is also a much closer mag. 6.9 companion which requires a telescope above 200 mm aperture.

◀ M34 (NGC 1039), 2h 42m +43°, is a bright star cluster at the limit of naked-eye visibility. It is far less rich and condensed than the Double Cluster, containing about 60 stars splashed over an area larger than the apparent size of the full Moon. Binoculars resolve it into stars, and it is well seen in a small telescope. M34 lies about 1500 l.y. away.

M76 (NGC 650–1), 1h 42m +52°, the Little Dumbbell, is a planetary nebula, the faintest object on Messier's list with a magnitude of about 11.5 and hence difficult to see, although it can be picked up in 100 mm aperture on a dark night. It is relatively large for a planetary nebula, similar in size to the Ring Nebula in Lyra, but smaller than its namesake the Dumbbell Nebula in Vulpecula, which it resembles in shape. Each end of the Little Dumbbell has a separate NGC number. It lies 3600 l.y. away.

NGC 869, NGC 884, 2h 19m +57°, 2h 22m +57°, the famous Double Cluster in Perseus, also known as h and χ (chi) Persei. They are two open star clusters visible to the naked eye and superb in binoculars, each covering an area equal to the full Moon. NGC 869 is the brighter and richer of the pair, containing an estimated 200 stars, compared with its companion's 150. They both lie about 7400 l.y. away and are both relatively young, only a few million years old. Small telescopes have an advantage when observing these objects, for at low powers both clusters can be seen in the same field of view, which is not always the case with larger and more powerful telescopes. Most of the stars in the clusters are blue-white, but there are several red stars to be spotted among the masses of glittering points resolved by telescopes. A breathtaking sight in all apertures.

64
PHOENIX
Phe · Phoenicis

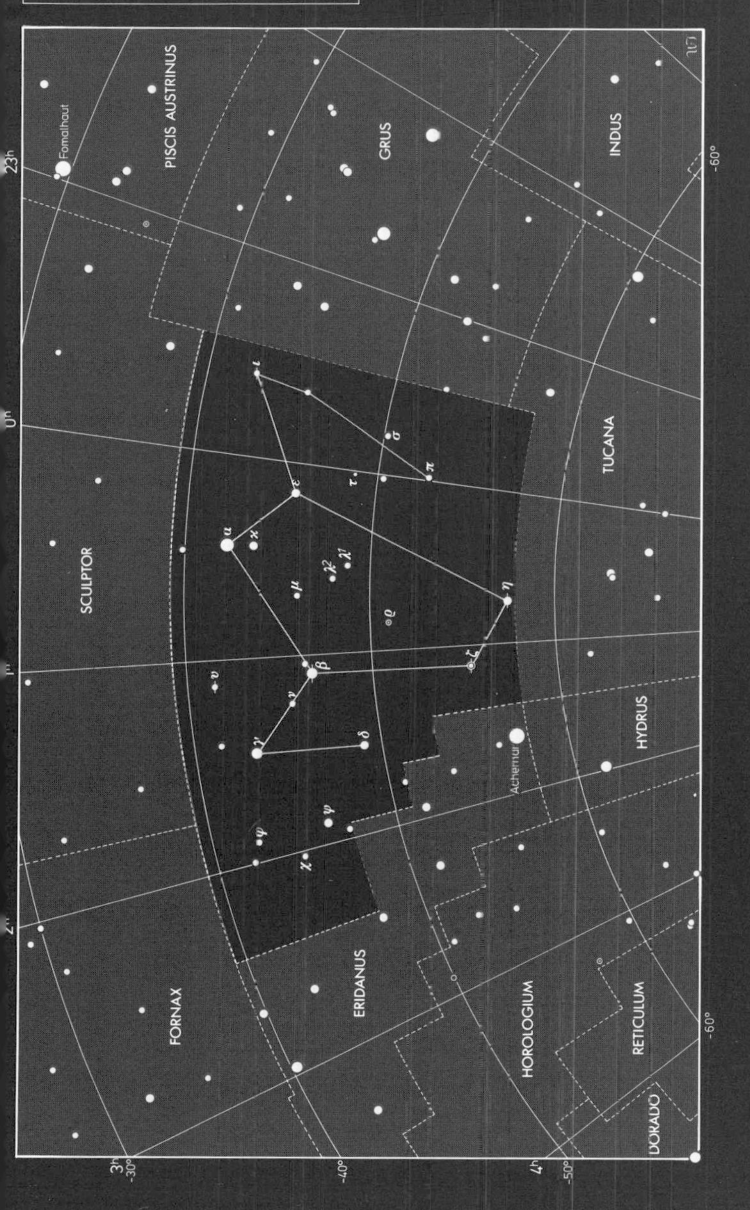

PICTOR The Painter's Easel

A faint constellation overshadowed by the neighbouring brilliant star Canopus in Carina on one side and the Large Magellanic Cloud in Dorado on the other. The constellation was invented in the 1750s by Nicolas Louis de Lacaille, who originally called it Equuleus Pictoris, which has since been shortened. At 5h 12m –45° lies the mag. 8.8 red dwarf known as Kapteyn's Star, 12.7 l.y. away, named after the Dutch astronomer who discovered in 1897 that it has the second-largest proper motion of any known star (the record is held by Barnard's Star in Ophiuchus).

α (alpha) Pictoris, 6h 48m –62°, mag. 3.3, is a white star 72 l.y. away.

β (beta) Pic, 5h 47m –51°, mag. 3.8, is a blue-white star 59 l.y. away. This star became famous in 1984 when astronomers photographed a disk of dust and gas around it, thought to be a planetary system in the process of formation.

γ (gamma) Pic, 5h 50m –56°, mag. 4.5, is a yellow giant 250 l.y. away.

δ (delta) Pic, 6h 10m –55°, is a blue-white star 2400 l.y. away. It is an eclipsing binary of the β Lyrae type, varying from mag. 4.7 to 4.9 every 1.67 days.

ι (iota) Pic, 4h 51m –53°, 160 l.y. away, is an easy double of mags. 5.6 and 6.4 for small telescopes.

M74 in Pisces is a spiral galaxy with loosely wound arms. For a description see page 204. Hale Observatories photograph.

65
PICTOR
Pic · Pictoris

Magnitudes: -1 0 1 2 3 4 5 <5

Double or multiple — Variable

Open cluster — Globular cluster — Galaxy

Diffuse neb. — Planetary neb.

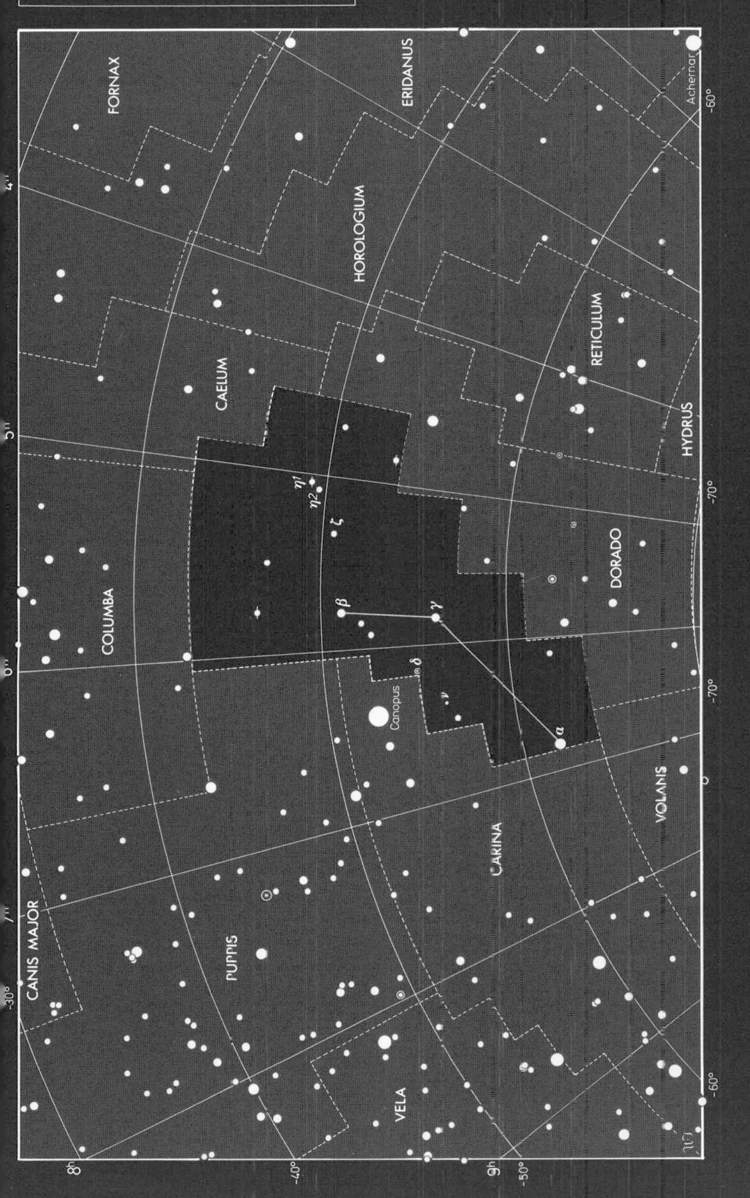

PISCES The Fishes

An ancient constellation representing two fishes tied by their tails, the knot being marked by the star α (alpha) Piscium. One legend identifies the constellation with Venus and her son Cupid, who swam away from the attack of the monster Typhon in the guise of fishes. The constellation contains the vernal equinox – the point at which the Sun crosses the celestial equator into the northern celestial hemisphere each year. This point originally lay in neighbouring Aries, but has now moved into Pisces because of precession; eventually it will move on into Aquarius. Although faint, Pisces contains numerous stars of interest.

α (alpha) Piscium, 2h 02m +3°, (Alrescha, 'the cord'), appears to the naked eye as a star of mag. 3.8, but in fact is a challenging double with an orbital period of over 900 years. Its components, of mags. 4.2 and 5.2, are gradually closing but will remain within range of 100 mm aperture for at least the first half of the 21st century. The brighter star is a spectroscopic binary, and the fainter one may be also. Their colour is bluish-white, although some observers see the brighter star as greenish. They lie about 100 l.y. away.

β (beta) Psc, 23h 04m +4°, mag. 4.5, is a blue-white star 320 l.y. away.

γ (gamma) Psc, 23h 17m +3°, mag. 3.7, is a yellow giant 155 l.y. away.

ζ (zeta) Psc, 1h 14m +8°, 150 l.y. away, is a wide double of mags. 5.2 and 6.3, divisible in the smallest telescopes.

η (eta) Psc, 1h 31m +15°, mag. 3.6, is the brightest star in the constellation. It is a yellow giant, 145 l.y. away.

κ (kappa) Psc, 23h 27m +1°, mag. 4.9, is a blue-white star about 100 l.y. away with a mag. 6.3 binocular companion, unrelated.

ρ (rho) Psc, 1h 26m +19°, mag. 5.4, is a white star 98 l.y. away that forms an easy binocular duo with the unrelated orange giant 94 Piscium, mag. 5.5, 390 l.y. away.

ψ^1 (psi^1) Psc, 1h 06m +21°, is a wide pair of blue-white stars of mags. 5.3 and 5.6 about 200 l.y. away, visible in small telescopes or even good binoculars.

TX Psc, 23h 46m +3°, also known as 19 Psc, is a deep-red irregular variable star, visible with the naked eye or binoculars. It fluctuates between about mags. 4.8 and 5.2. Its distance is unknown.

M74 (NGC 628), 1h 37m +16°, is a face-on spiral galaxy 25 million l.y. away. At 9th mag. it can be glimpsed in a small telescope under dark conditions, but needs an aperture of at least 150 mm to be well seen. (See photograph on page 202.)

66 PISCES
Psc · Piscium

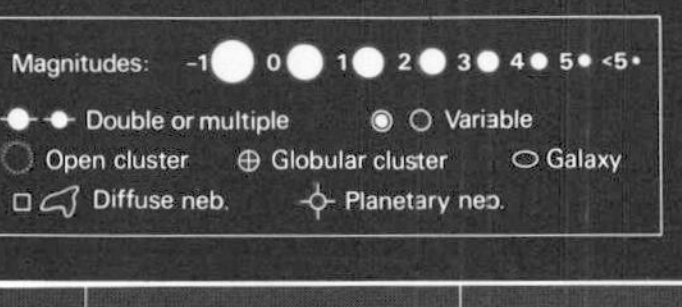

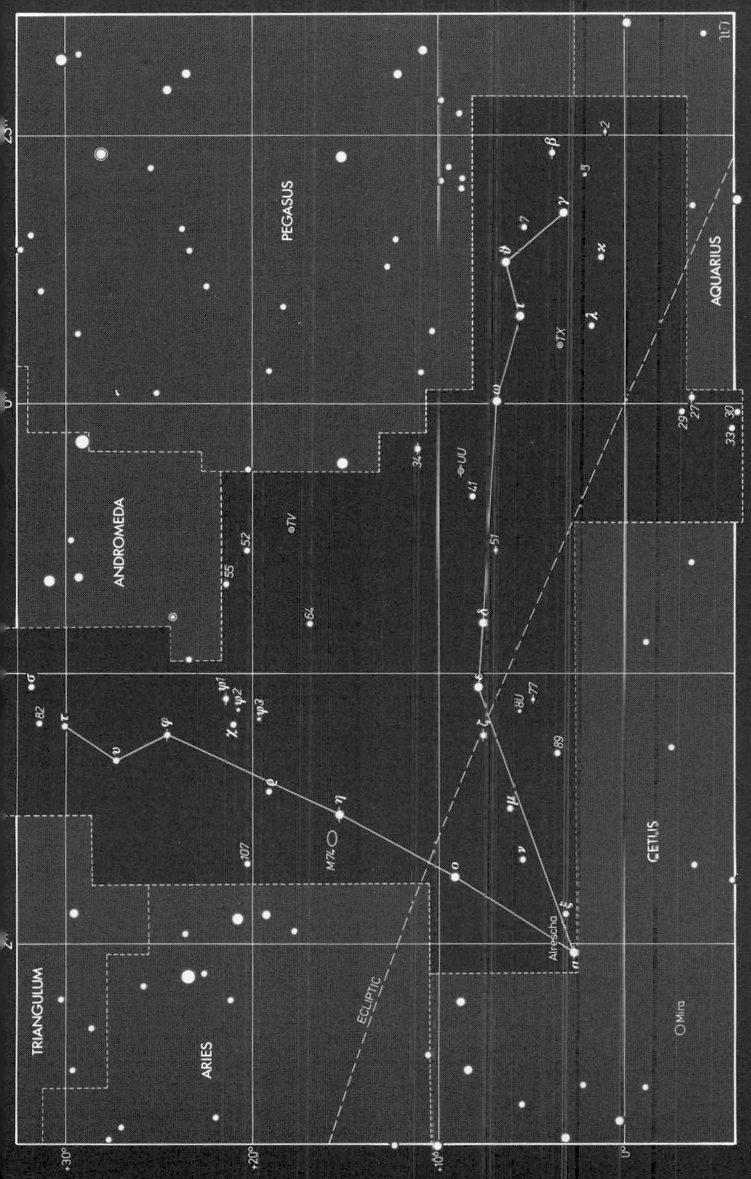

PISCIS AUSTRINUS The Southern Fish

Sometimes called Piscis Australis, this constellation has been known since ancient times and is often represented as a fish drinking the flow of water from the urn of neighbouring Aquarius. This fish was said to be the parent of the two zodiacal fishes, represented by Pisces.

α (alpha) Piscis Austrini, 22h 58m –30°, (Fomalhaut, 'the fish's mouth'), mag. 1.2, is a blue-white star 22 l.y. away.

β (beta) PsA, 22h 32m –32°, 135 l.y. away, is a wide double star consisting of a mag. 4.3 white primary and a mag. 7.7 companion visible in small telescopes.

γ (gamma) PsA, 22h 53m –33°, 325 l.y. away, is a double star of mags. 4.5 and 8.0, made difficult to split in small telescopes by the magnitude contrast.

η (eta) PsA, 22h 01m –28°, 420 l.y. away, is a close pair of blue-white stars of mags. 5.8 and 6.8, divisible with an aperture of 100 mm and high power.

Although Piscis Austrinus contains no notable deep-sky objects, its neighbour Sculptor has several galaxies of note, among them NGC 253, a spiral galaxy seen nearly edge-on. It has no central bulge, but its arms are a seething mass of stars and dust. For a description see page 224. Anglo-Australian Telescope Board.

PISCIS AUSTRINUS
PsA · Piscis Austrini

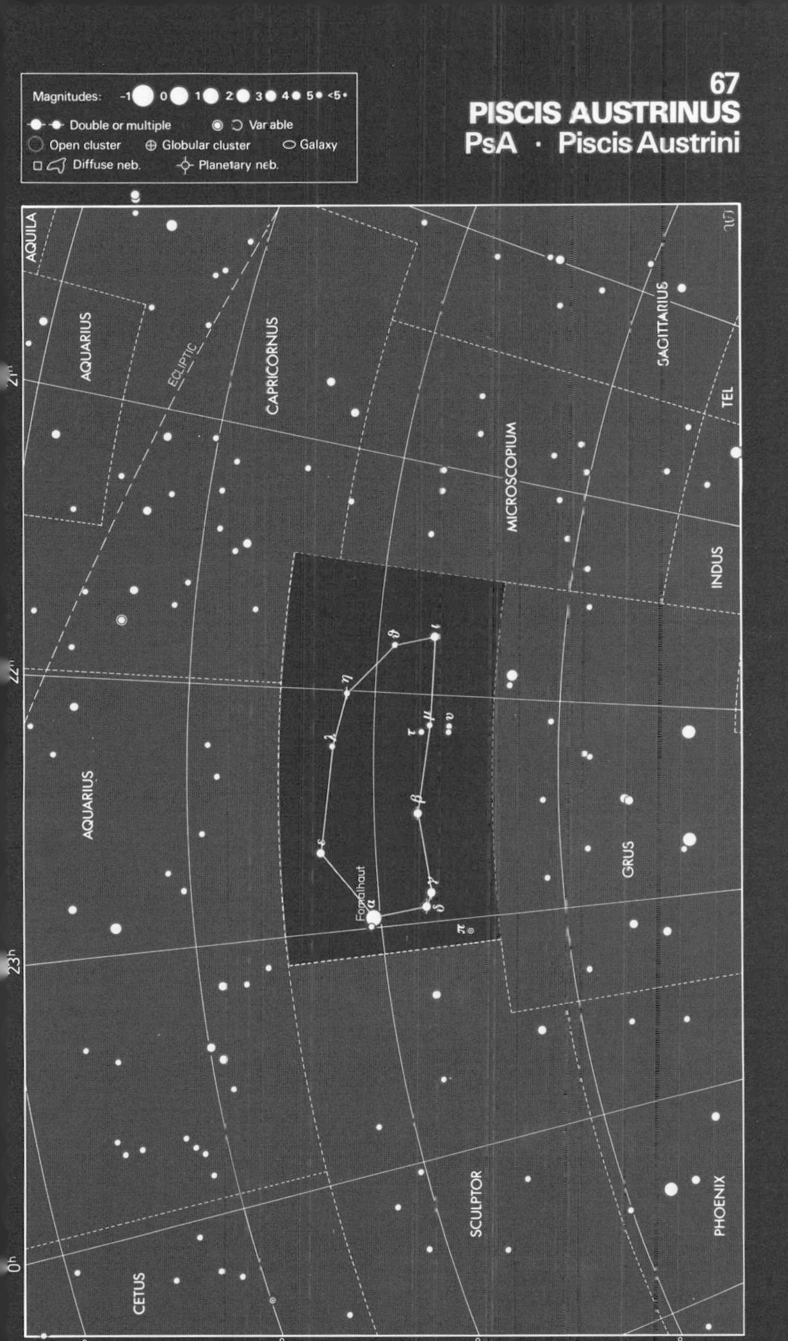

PUPPIS The Stern

This is the largest of the three sections into which the ancient constellation of Argo Navis, the ship of the Argonauts, was dismembered in 1763 by Nicolas Louis de Lacaille; the other sections are Carina and Vela. Puppis lies in the Milky Way and contains rich starfields for sweeping with binoculars.

ζ (zeta) Puppis, 8h 04m –40°, (Naos, 'ship'), mag. 2.2, is a brilliant blue-white supergiant 1500 l.y. away, one of the hottest stars known, with a surface temperature of about 35,000°C.

ξ (xi) Pup, 7h 49m –25°, mag. 3.4, is a yellow supergiant 650 l.y. away. Binoculars reveal a wide, unrelated mag. 5.3 orange giant companion, 280 l.y. away.

π (pi) Pup, 7h 17m –37°, mag. 2.7, is an orange giant 100 l.y. away.

ρ (rho) Pup, 8h 08m –24°, mag. 2.8, is a yellow giant 300 l.y. away. It is a variable of the δ Scuti type, changing by 0.1 mag. every 3 hours 23 minutes.

k Pup, 7h 39m –27°, about 350 l.y. away, is a striking double star with blue-white components of mags. 4.5 and 4.6 easily divisible in small telescopes.

L Pup, 7h 13m –45°, is an optical double consisting of two unrelated and contrasting stars. L^1 is mag. 4.9, a blue-white star about 500 l.y. away. L^2 is a red giant semi-regular variable, about 150 l.y. away, that fluctuates between mags. 2.6 and 6.2 every 140 days or so.

V Pup, 7h 58m –49°, is an eclipsing binary of the β Lyrae type. It varies from mag. 4.4 to 4.9 every 35 hours. Its distance is unknown.

M46 (NGC 2437), 7h 42m –15°, is a 6th-mag. cluster of about 100 faint stars of remarkably uniform brightness, most being around 10th mag. With its neighbour M47 it is visible to the naked eye as a brighter knot in the Milky Way. In binoculars it appears as a smudgy patch the size of the full Moon, while a small telescope shows it as a sprinkling of stardust. M46 lies 5400 l.y. away. On its northern edge lies the 10th-mag. planetary nebula NGC 2438. This is not associated with the cluster, but is a foreground object about 3000 l.y. away.

M47 (NGC 2422), 7h 37m –15°, is a scattered naked-eye cluster the same apparent size as the full Moon. It contains about 30 stars, the brightest being of mag. 5.7. M47 lies 1600 l.y. away, less than one-third of the distance of its richer neighbour M46.

M93 (NGC 2447), 7h 45m –24°, is a 6th-mag. binocular cluster, 3600 l.y. away, consisting of 80 stars of 8th mag. and fainter, arranged in a wedge-shape.

NGC 2451, 7h 45m –38°, is a large and bright cluster of 40 stars 850 l.y. away, centred on the mag. 3.6 orange giant star c Puppis.

NGC 2477, 7h 52m –39°, is a large 6th-mag. cluster of about 160 faint stars, looking in binoculars like a loose globular, seemingly with arms. An excellent object that would have featured on Messier's list had he lived farther south. It lies 4200 l.y. away.

68
PUPPIS
Pup · Puppis

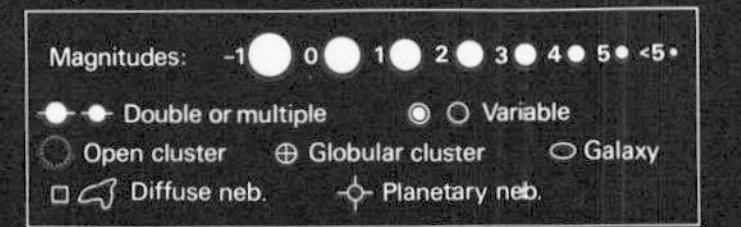

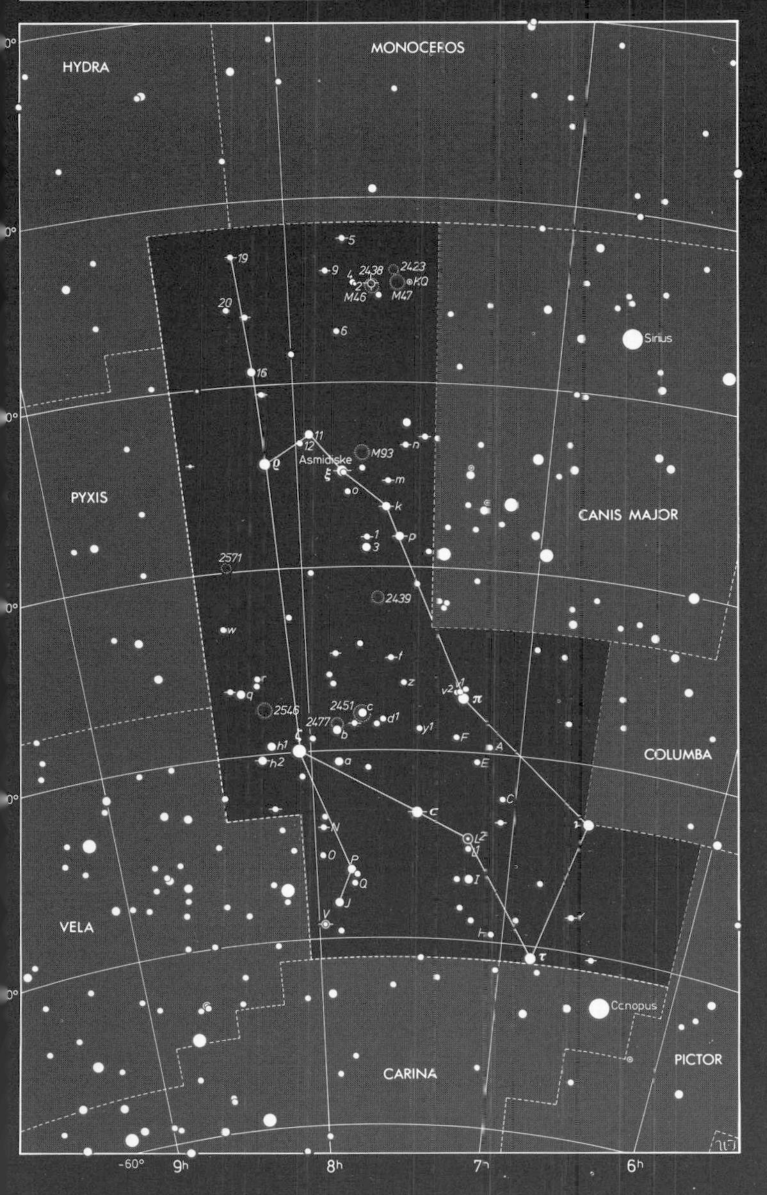

PYXIS The Compass

A small constellation invented by Nicolas Louis de Lacaille in the 1750s, representing a magnetic compass. It lies near Puppis, the stern of the Argonauts' ship. Pyxis contains no objects of particular interest to users of small telescopes, despite the fact that it lies in the Milky Way.

α (alpha) Pyxidis, 8h 44m –33°, mag. 3.7, is a blue-white giant star 1050 l.y. away.

β (beta) Pyx, 8h 40m –35°, mag. 4.0, is a yellow giant 320 l.y. away.

γ (gamma) Pyx, 8h 51m –28°, mag. 4.0, is an orange giant 215 l.y. away.

T Pyx, 9h 05m –32°, is a recurrent nova that has undergone five recorded eruptions, in 1890, 1902, 1920, 1944 and 1966. Normally it is of mag. 14, but brightens to 6th or 7th magnitude. Further outbursts may be expected.

Nicolas Louis de Lacaille (1713–1762)

Lacaille, a French astronomer, was the first person to map the southern skies comprehensively, as a result of which he became known as the Father of Southern Astronomy. He directed an expedition of the French Academy of Sciences to the Cape of Good Hope in 1750–54, where he systematically surveyed the southern celestial hemisphere, listing over 10,000 stars. The accurate positions of 2000 of these, along with a star map, were published posthumously in 1763 under the title *Coelum Australe Stelliferum*. Lacaille is usually best remembered for the 14 new constellations he introduced, representing instruments used in science and the fine arts: Antlia, Caelum, Circinus, Fornax, Horologium, Mensa, Microscopium, Norma, Octans, Pictor, Pyxis, Reticulum, Sculptor and Telescopium. Lacaille also dismantled one constellation, Robur Carolinum, Charles's Oak; this was formed in 1678 by Edmond Halley from some of the stars of Argo Navis to commemorate the oak tree in which his patron, King Charles II, hid after his defeat by Oliver Cromwell at the Battle of Worcester. It is said that this inspired piece of flattery earned Halley his master's degree from Oxford by the king's express command. Lacaille, less impressed, uprooted the oak and returned its stars to Argo Navis.

69
PYXIS
Pyx · Pyxidis

Magnitudes: -1 0 1 2 3 4 5 <5
Double or multiple — Variable
Open cluster — Globular cluster — Galaxy
Diffuse neb. — Planetary neb.

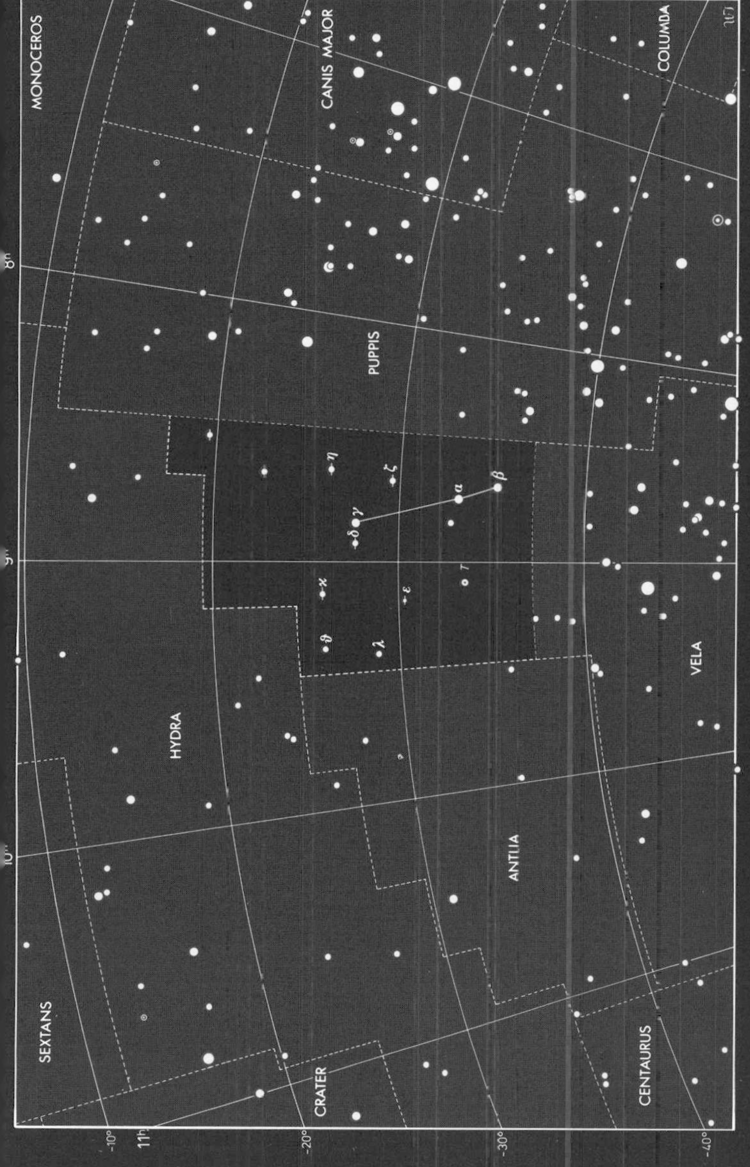

RETICULUM The Net

A constellation introduced in the 1750s by Lacaille to commemorate a grid-like device known as a reticle which he used for measuring star positions during his surveys of the southern sky. It is not a prominent constellation, but lies near the Large Magellanic Cloud.

α (alpha) Reticuli, 4h 14m –62°, mag. 3.4, is a yellow giant star 135 l.y. away.

β (beta) Ret, 3h 44m –65°, mag. 3.9, is an orange star 78 l.y. away.

ζ (zeta) Ret, 3h 18m –63°, 39 l.y. away, is a wide naked-eye or binocular double of near-identical yellow stars similar to the Sun, of mags. 5.2 and 5.5.

Across the border of Reticulum with neighbouring Dorado lies this attractive 9th-mag. spiral galaxy, NGC 1566. It is a type of galaxy known as a Seyfert galaxy, with a bright, variable centre. Anglo-Australian Telescope Board.

RETICULUM
Ret · Reticuli

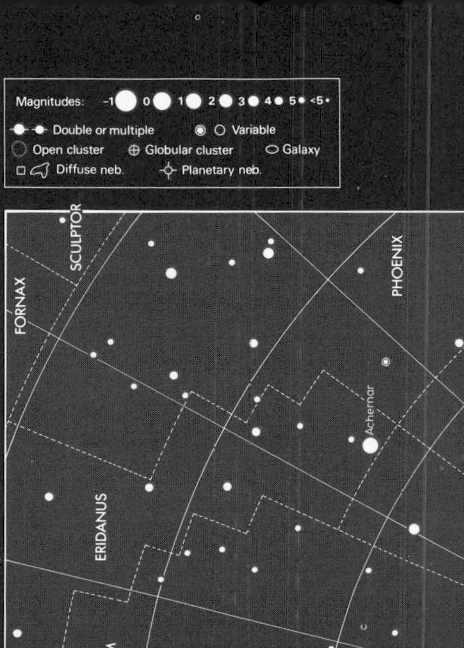

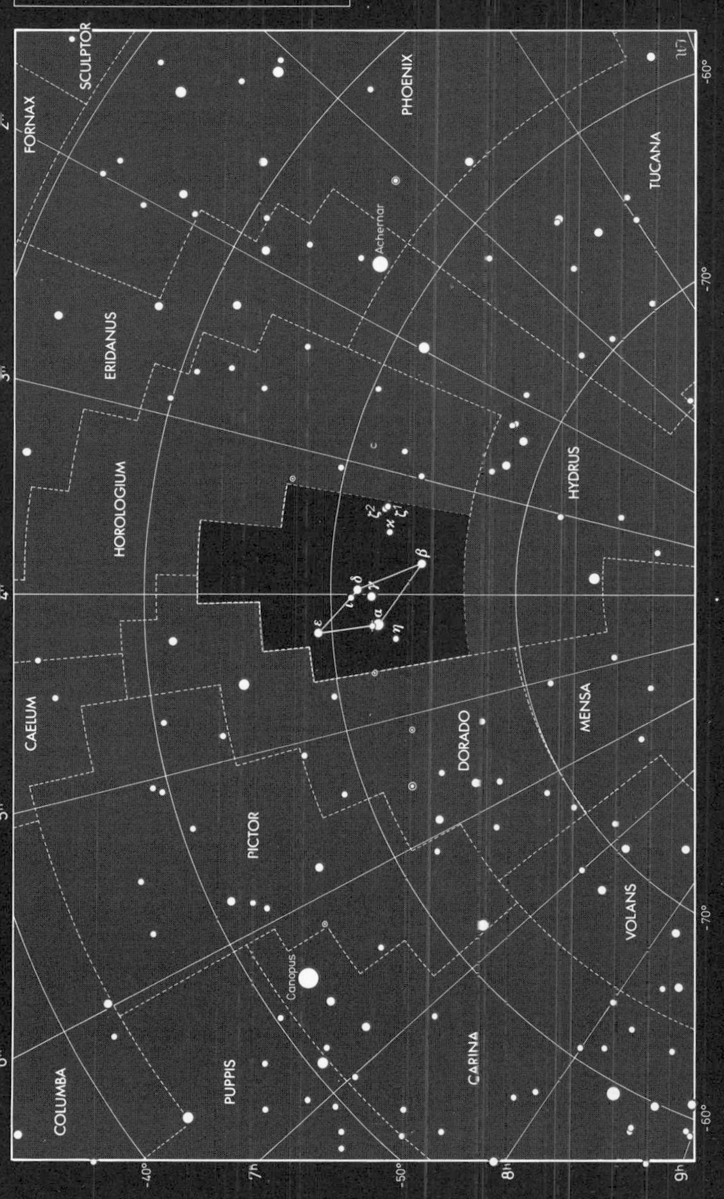

SAGITTA The Arrow

Despite its diminutive size – it is the third-smallest constellation in the sky – this distinctive arrow-shaped group was known to the Ancient Greeks. In the sky, the Arrow seems to be flying between Cygnus the Swan and Aquila the Eagle; in one legend, the Arrow was shot by Hercules. Like its neighbour Vulpecula, Sagitta lies in a rich part of the Milky Way.

α (alpha) Sagittae, 19h 40m +18°, mag. 4.4, is a yellow giant 590 l.y. away.

β (beta) Sge, 19h 41m +17°, mag. 4.4, is a yellow giant 590 l.y. away.

γ (gamma) Sge, 19h 59m +19°, mag. 3.5, the brightest star in the constellation, is an orange giant 175 l.y. away.

δ (delta) Sge, 19h 47m +19°, mag. 3.8, is a red giant 750 l.y. away.

ζ (zeta) Sge, 19h 49m +19°, 400 l.y. away, is a double star of mags. 5.0 and 8.7, divisible in a small telescope.

WZ Sge, 20h 08m +18°, is a recurrent nova that flared up from 15th mag. to 7th or 8th mag. in 1913, 1946 and 1978; its location is worth checking in case of another flare-up.

M71 (NGC 6838), 19h 54m +19°, is an 8th-mag. globular cluster 13,000 l.y. away, visible as a small, somewhat elongated misty patch in binoculars or a small telescope.

Scan northwards over the border from Sagitta into neighbouring Vulpecula with binoculars or telescope and you will encounter the Dumbbell Nebula, M27, one of the showpieces of the sky. For a description see page 254.
Hale Observatories photograph.

SAGITTA
Sge · Sagittae

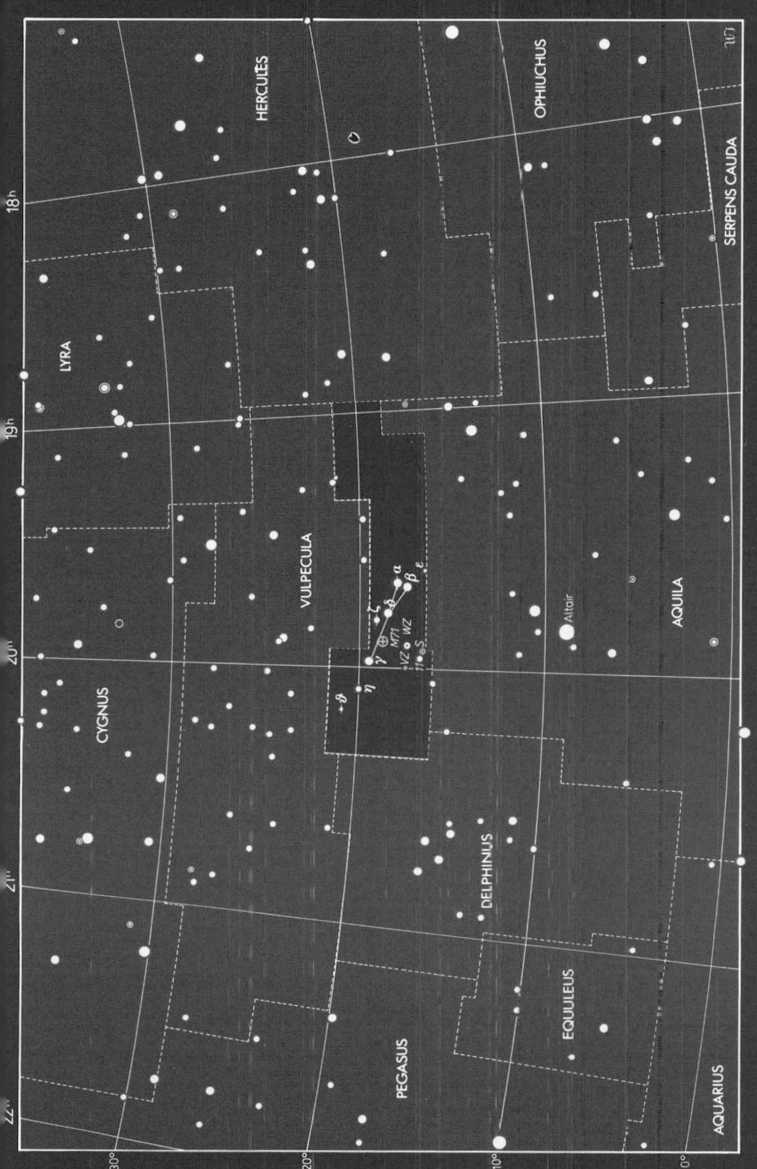

SAGITTARIUS The Archer

An ancient constellation depicting a centaur, half man, half beast, with a raised bow and arrow. It is an older constellation than the other celestial centaur, Centaurus, and is different in character. Whereas Centaurus is identified as a scholarly, beneficent creature, Sagittarius is depicted with a threatening look, aiming his arrow at the heart of Scorpius, the Scorpion. The bow is marked by the stars λ (lambda), δ (delta) and ε (epsilon) Sagittarii.

The main stars of Sagittarius are often popularly visualized as outlining the shape of a teapot, while a ladle shape known as the Milk Dipper is also seen here – a suitable implement to dip into this rich region of the Milky Way. The centre of our Galaxy lies in Sagittarius, so the Milky Way starfields are particularly rich here, as well as in neighbouring Scutum and Scorpius. The actual centre of the Galaxy is marked by a radio and infrared source known as Sagittarius A, lying at 17h 46.1m, –28° 51′. The main attraction of Sagittarius is its clusters and nebulae. Messier catalogued a total of 15 objects in Sagittarius, more than in any other constellation; only a selection of them can be mentioned here. The Sun passes through the constellation from mid-December to mid-January, and thus lies in Sagittarius at the winter solstice, its farthest point south of the equator.

α (alpha) Sagittarii, 19h 24m –41°, (Rukbat, 'knee', or Alrami, 'the archer'), mag. 4.0, is one of several examples in which the star labelled α in a constellation is not the brightest. It is a blue-white star 275 l.y. away.

β (beta) Sgr, 19h 23m –44°, consists of two unrelated naked-eye stars. β^1 Sgr, called Arkab Prior (Arkab comes from the Arabic for 'Achilles tendon'), is a blue-white star of mag. 4.0, 190 l.y. away; it has a mag. 7.2 companion visible in a small telescope. β^2 Sgr (Arkab Posterior), is a white star of mag. 4.3, 180 l.y. away. All three stars appear in the same line of sight by chance.

γ (gamma) Sgr, 18h 06m –30°, (Alnasl, 'the point of the arrow'), mag. 3.0, is a yellow giant 135 l.y. away.

δ (delta) Sgr, 18h 21m –30°, (Kaus Media, 'middle of the bow'), mag. 2.7, is an orange giant 100 l.y. away.

ε (epsilon) Sgr, 18h 24m –34°, (Kaus Australis, 'southern part of the bow'), mag. 1.8, is the brightest star in Sagittarius. It is a blue-white giant 95 l.y. away.

λ (lambda) Sgr, 18h 28m –25°, (Kaus Borealis, 'northern part of the bow'), mag. 2.8, is a yellow giant 72 l.y. away.

σ (sigma) Sgr, 18h 55m –26°, (Nunki), mag. 2.0, is a blue-white star 210 l.y. away.

RY Sgr, 19h 17m –34°, is the southern equivalent of R Coronae Borealis, a star like a reverse nova that normally shines at around 6th mag., but which can suddenly and unpredictably drop to 14th mag. ▶

SAGITTARIUS
Sgr · Sagittarii

Magnitudes: −1 0 1 2 3 4 5 <5

Double or multiple · Variable · Open cluster · Globular cluster · Galaxy · Diffuse neb. · Planetary neb.

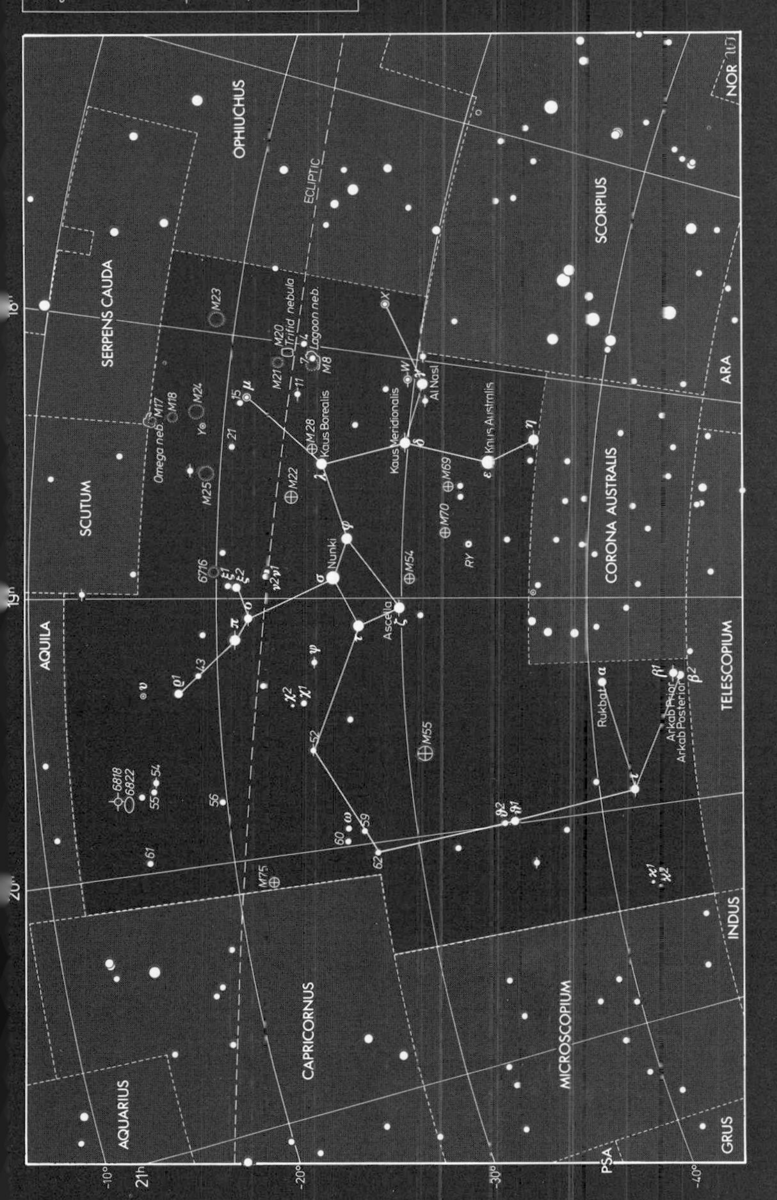

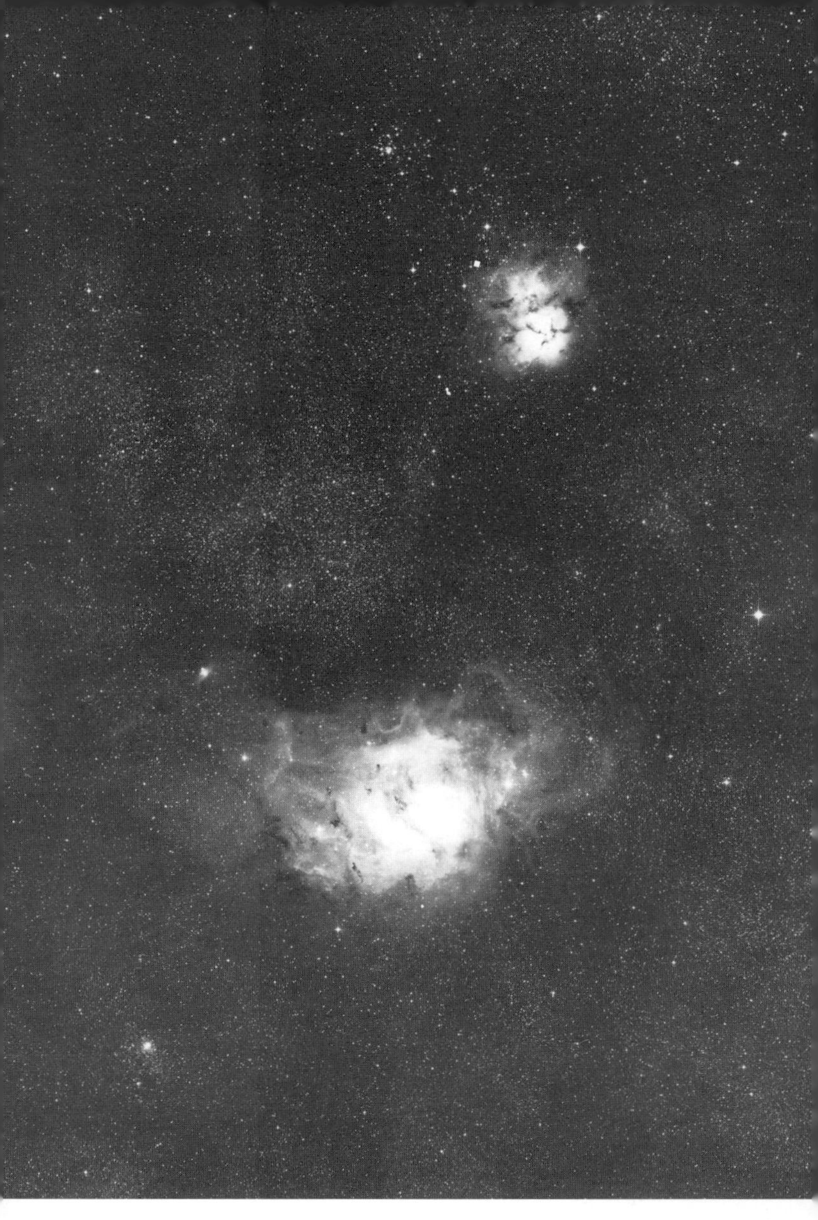

Among the Milky Way starfields of Sagittarius lie two outstanding bright nebulae: M20, the Trifid Nebula (top) and M8, the Lagoon Nebula (below). To the top left of the Trifid Nebula is the loose cluster M21. Royal Observatory Edinburgh.

◀ M8 (NGC 6523), 18h 04m –24°, the Lagoon Nebula, is a famous gaseous nebula, visible to the naked eye, elongated in shape and encompassing the star cluster NGC 6530. M8 is a fine object for binoculars or telescopes, covering the area of three full Moons, with a dark rift down its centre. In the eastern half of the nebula is NGC 6530, a cluster of about 25 stars of 7th mag. and fainter, formed recently from the surrounding gas. The other (western) side of the nebula is dominated by two main stars, the brighter of which is the blue-white supergiant 9 Sagittarii, mag. 6.0. Long-exposure photographs show the nebula as an intense red, but visually it appears milky-white. M8 is 5200 l.y. away.

M17 (NGC 6618), 18h 21m –16°, the Omega or Horseshoe Nebula, is another famous nebula. Binoculars show it as a wedge-shaped object about the size of the full Moon, while in larger instruments it appears arch-shaped, rather like a capital omega (Ω), which accounts for its popular name. M17 lies 5000 l.y. away. About 1° south of it is M18 (NGC 6613), a small, loose cluster of 20 stars of 9th mag. and fainter, unimpressive in binoculars.

M20 (NGC 6514), 18h 03m –23°, the Trifid Nebula, is a cloud of glowing gas far less impressive visually than photographically. Moderate-sized telescopes show it as only a diffuse patch of light centred on the double star HN 40, of 8th and 9th mags., which was evidently born from it and now illuminates it. The Trifid Nebula gets its name from three dark lanes of dust that trisect it, well shown on photographs but elusive in small apertures. It lies 5200 l.y. away, the same as M8. In the same low-power field of view lies the loose cluster M21 (NGC 6531), containing about 70 stars of 7th mag. and fainter, 4200 l.y. away.

M22 (NGC 6656), 18h 36m –24°, is a large, rich 5th-mag. globular cluster, one of the finest in the entire heavens and ranked third only to *ω* (omega) Centauri and 47 Tucanae. Visible to the naked eye, M22 is an excellent binocular object and a fine sight in small telescopes, which reveal its noticeably elliptical outline. A telescope of 75 mm aperture will begin to resolve its outer regions, and larger apertures show its brightest stars to be reddish. Its nucleus is not as condensed as that of many other globulars. M22 lies 10,000 l.y. away.

M23 (NGC 6494), 17h 57m –19°, is a widely spread cluster of 150 stars, 2200 l.y. away, just at the limit of resolution in binoculars. It is elongated in shape, consisting of a field of 9th-mag. stars of remarkably uniform appearance, some arranged in arcs.

M24, 18h 18m –18°, is a rich and extensive Milky Way starfield south of M17 and M18, grainy and shimmering in binoculars. Some observers restrict the name M24 to a fainter and far smaller cluster of about 100 faint stars, known also as NGC 6603. The whole Milky Way star cloud in this region measures about 2° by 1° and is one of the most prominent parts of the Milky Way to the naked eye.

M25 (IC 4725), 18h 32m –19°, 1900 l.y. away, is a scattered cluster of about 30 stars, well seen in binoculars. Its brightest star is the yellow supergiant Cepheid variable U Sgr, which varies from mag. 6.3 to 7.1 every 6 days 18 hours.

M55 (NGC 6809), 19h 40m –31°, is a 7th-mag. globular cluster, nebulous-looking in binoculars. Small telescopes resolve individual stars but show little central condensation of the cluster. M55 lies 17,000 l.y. away.

SCORPIUS The Scorpion

A resplendent constellation, lying in a rich area of the Milky Way and packed with exciting objects for users of small telescopes. In mythology, Scorpius was the scorpion whose sting killed Orion. In the sky Orion still flees from the scorpion, for Orion sets below the horizon as Scorpius rises. Originally, in Ancient Greek times and before, Scorpius was a much larger constellation, but the stars that once made up its claws now comprise the separate constellation of Libra. Scorpius clearly resembles the creature after which it is named, a distinctive curve of stars forming its stinging tail. Its heart is marked by Antares, which means 'like Mars', a reference to its strong red colour. North-east of β (beta) Scorpii, at 16h 19.9m, –15° 38′, lies the brightest X-ray source in the sky, Scorpius X-1. This has been identified with a 13th-magnitude spectroscopic binary 2300 l.y. away. The Sun passes briefly through Scorpius during the last week of November.

α (alpha) Scorpii, 16h 29m –26°, (Antares, 'like Mars'), 170 l.y. away, is a red supergiant 400 times the diameter of the Sun. It is a semi-regular variable, fluctuating between mags. 0.9 and 1.8 approximately every 5 years. Antares has a mag. 5.4 blue companion (given as 7th mag. in some sources) that requires at least 75 mm aperture and the steadiest atmospheric conditions to be visible against the primary's glare. The orbital period of the companion around Antares is estimated to be nearly 900 years.

β (beta) Sco, 16h 05m –20°, (Acrab, 'scorpion', or Graffias, 'claws') is a striking double star divisible in the smallest telescopes, consisting of blue-white stars of mags. 2.6 and 4.9, distances 520 and 720 l.y.

δ (delta) Sco, 16h 00m –23°, (Dschubba, 'forehead'), mag. 2.3, is a blue-white star 620 l.y. away.

ε (epsilon) Sco, 16h 50m –34°, mag. 2.3, is an orange giant 65 l.y. away.

ζ (zeta) Sco, 16h 54m –42°, is a naked-eye double of unrelated stars, ζ^2 being a mag. 3.6 orange giant 190 l.y. away, while ζ^1 is a blue-white supergiant of mag. 4.7; it is not certain whether ζ^1 is an outlying member of the open cluster NGC 6231 or a foreground object.

ϑ (theta) Sco, 17h 37m –43°, mag. 1.9, is a yellow-white giant 190 l.y. away.

λ (lambda) Sco, 17h 34m –37°, (Shaula, 'sting'), mag. 1.6, is a blue-white star 275 l.y. away.

μ (mu) Sco, 16h 52m –38°, 680 l.y away, is a naked-eye double consisting of an eclipsing binary that varies from mag. 2.8 to 3.1 every 34 hours 34 minutes, and a blue-white star of mag. 3.6.

ν (nu) Sco, 16h 12m –19°, 420 l.y. away, is a quadruple star similar to the famous Double Double in Lyra. A small telescope, or even powerful binoculars, shows ν Sco as a wide double, with blue-white components of mags. 4.0 and 6.3. Telescopes of 75 mm and above reveal at high magnification that the fainter ▶

73
SCORPIUS
Sco · Scorpii

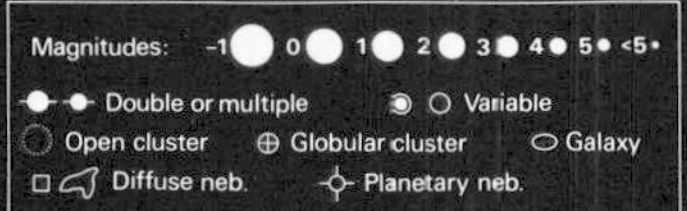

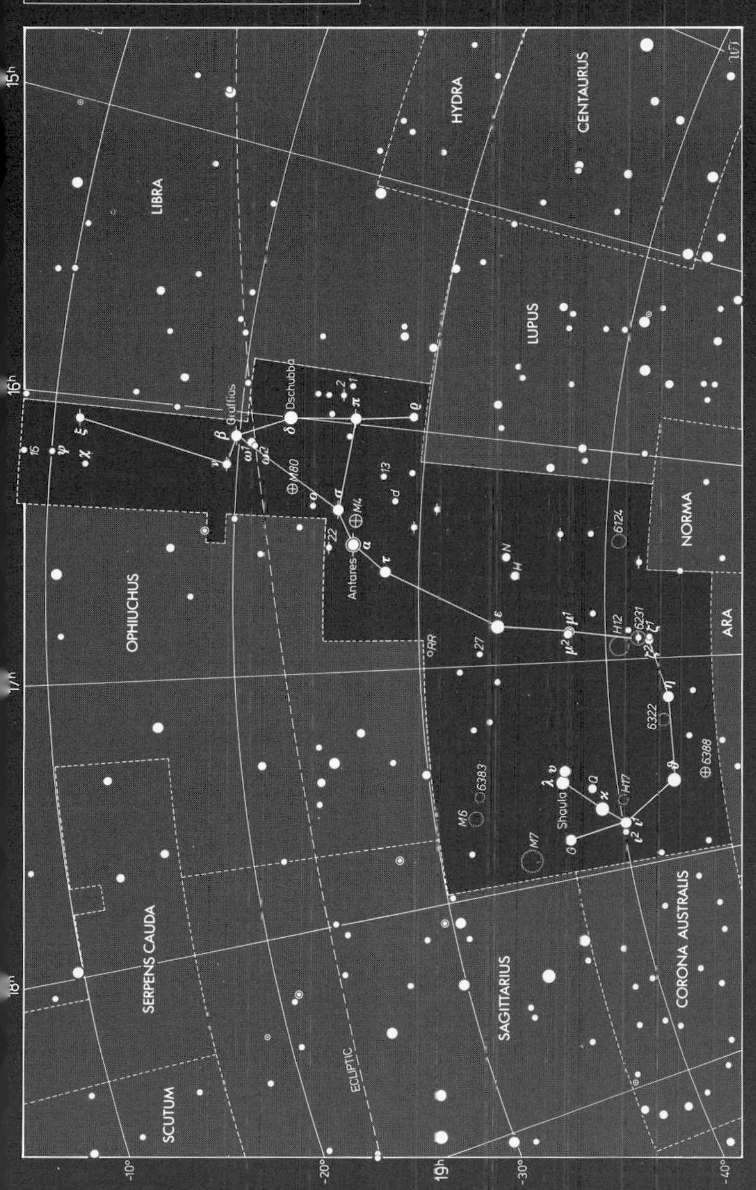

The glittering star clusters M6 (top) and M7 in Scorpius are ideal objects for observation with binoculars. M7 lies in front of dense Milky Way starfields. Chris Floyd.

◀ star is itself a close double of mags. 6.8 and 7.8. The brighter star is an even closer double of mags. 4.3 and 6.8, requiring an aperture of 150 mm to split.

ξ (xi) Sco, 16h 04m –11°, 90 l.y. away, is a celebrated multiple star. A small telescope shows it as a white star of mag. 4.2 with a mag. 7.3 orange companion; also visible in the same field is a fainter and wider pair, called Struve 1999, composed of mag. 7.4 and 8.1 stars that are gravitationally connected to ξ Sco. Therefore, at first sight, ξ Sco looks like another Double Double. But the brighter star is itself a close double, consisting of yellow-white stars of mags. 4.8 and 5.1 orbiting each other every 46 years. At their closest in 1996 they are impossible to split in amateur telescopes, but after the year 2005 they should be divisible in 250 mm aperture, and 100 mm should separate them by 2015.

ω (omega) Sco, 16h 07m –21°, is a pair of unrelated stars distinguishable with the naked eye, consisting of a blue-white star of mag. 4.0, 720 l.y. away, and an orange star of mag. 4.3, 330 l.y. away.

M4 (NGC 6121), 16h 24m –27°, is a 6th-mag. globular cluster appearing almost as large as the full Moon. It appears like a woolly ball in binoculars but is not as easy to spot as its magnitude would suggest because its light is spread over a large area. In 100-mm telescopes individual stars are resolved and there is a noticeable bar of stars across its centre. M4 is more loosely scattered than many globulars, and does not have a strong central condensation. It is one of the closest globulars to us, 6800 l.y. away.

M6 (NGC 6405), 17h 40m –32°, is an impressive 4th-mag. open cluster of about 80 stars arranged in radiating chains, sometimes called the Butterfly Cluster. Binoculars resolve some of its stars, the brightest of which is the orange giant semi-regular variable BM Sco which ranges from 5th to 7th mag. every 27 months or so. M6 lies 2000 l.y. away.

M7 (NGC 6475), 17h 54m –35°, is a huge, scattered 3rd-mag. cluster of 80 or so stars, visible to the naked eye as a brighter knot in the Milky Way with an apparent diameter over twice that of the full Moon. It is easily resolved in binoculars, and M6 at the edge of the same binocular field makes this an exceptionally rich sight. The brightest stars of M7 are of 6th mag. The central group is arranged in an X-shape with scattered outliers that form triangular surroundings like a Christmas tree, all against the backdrop of a dense star cloud. An outstanding cluster and a classic for small apertures. M7 lies 800 l.y. away.

M80 (NGC 6093), 16h 17m –23°, is a small, 7th-mag. globular cluster visible in binoculars or a small telescope, appearing like the fuzzy head of a comet. It lies 27,000 l.y. away.

NGC 6231, 16h 54m –42°, is a bright naked-eye cluster of 120 or so stars in a rich area of the Milky Way that is well worth sweeping with binoculars. The brightest stars of the group are of 6th mag. and give the impression of a mini-Pleiades in binoculars and small telescopes. The mag. 4.7 blue supergiant ζ^1 (zeta^1) Sco is probably an outlying member of this cluster, which lies 6000 l.y. away. NGC 6231 is connected to a larger, scattered cluster of fainter stars visible in binoculars, called H 12, which lies 1° to the north. The chain of stars linking NGC 6231 and H 12 outlines one of the spiral arms of our Galaxy.

SCULPTOR The Sculptor

One of the faint and half-forgotten constellations introduced in the 1750s by the French astronomer Nicolas Louis de Lacaille to fill in the southern skies. It represents a sculptor's studio. Although Sculptor's stars are of little interest, it does contain the South Pole of our Galaxy, which lies 90° from the Milky Way. Here, then, we can look out into deep space, unobscured by stars or dust, and see many faint galaxies. Among these is a faint dwarf elliptical member of our own Local Group, detectable only on long-exposure photographs taken through large telescopes.

α (alpha) Sculptoris, 0h 59m –29°, mag. 4.3, is a blue-white giant 590 l.y. away.

β (beta) Scl, 23h 33m –38°, mag. 4.4, is a blue-white star 260 l.y. away.

γ (gamma) Scl, 23h 19m –33°, mag. 4.4, is a yellow giant 225 l.y. away.

δ (delta) Scl, 23h 49m –28°, mag. 4.6, is a blue-white star 200 l.y. away.

ε (epsilon) Scl, 1h 46m –25°, 95 l.y. away, is a binary of mags. 5.4 and 8.6, visible in small telescopes. The estimated orbital period is over 1000 years.

κ^1 (kappa^1) Scl, 0h 09m –28°, 88 l.y. away, is a tight pair of white stars, of mags. 6.1 and 6.2, at the limit of resolution of a 75-mm telescope.

R Scl, 1h 27m –33°, is a deep-red semi-regular variable star that fluctuates between mags. 5.8 and 7.7 with a period of about a year.

NGC 55, 0h 15m –39°, is an 8th-mag. spiral galaxy seen nearly edge-on, so it appears elongated. It is similar in size and shape to NGC 253 (see below), although not quite as bright. Its distance is estimated at 6 million l.y.

NGC 253, 0h 48m –25°, is a 7th-mag. spiral galaxy seen nearly edge-on, hence appearing cigar-shaped. Nearly ½° long, it can be picked up in binoculars but requires at least 100 mm aperture to distinguish the mottling caused by dust clouds in its spiral arms. Its estimated distance is 9 million l.y. (see page 206).

In addition to NGC 55 and NGC 253, Sculptor contains the 9th-mag. galaxy NGC 300, which photographs show to be a loose spiral. Anglo-Australian Telescope Board.

SCULPTOR
Scl · Sculptoris

Magnitudes: -1 0 1 2 3 4 5 <5

Double or multiple · Variable

Open cluster · Globular cluster · Galaxy

Diffuse neb. · Planetary neb.

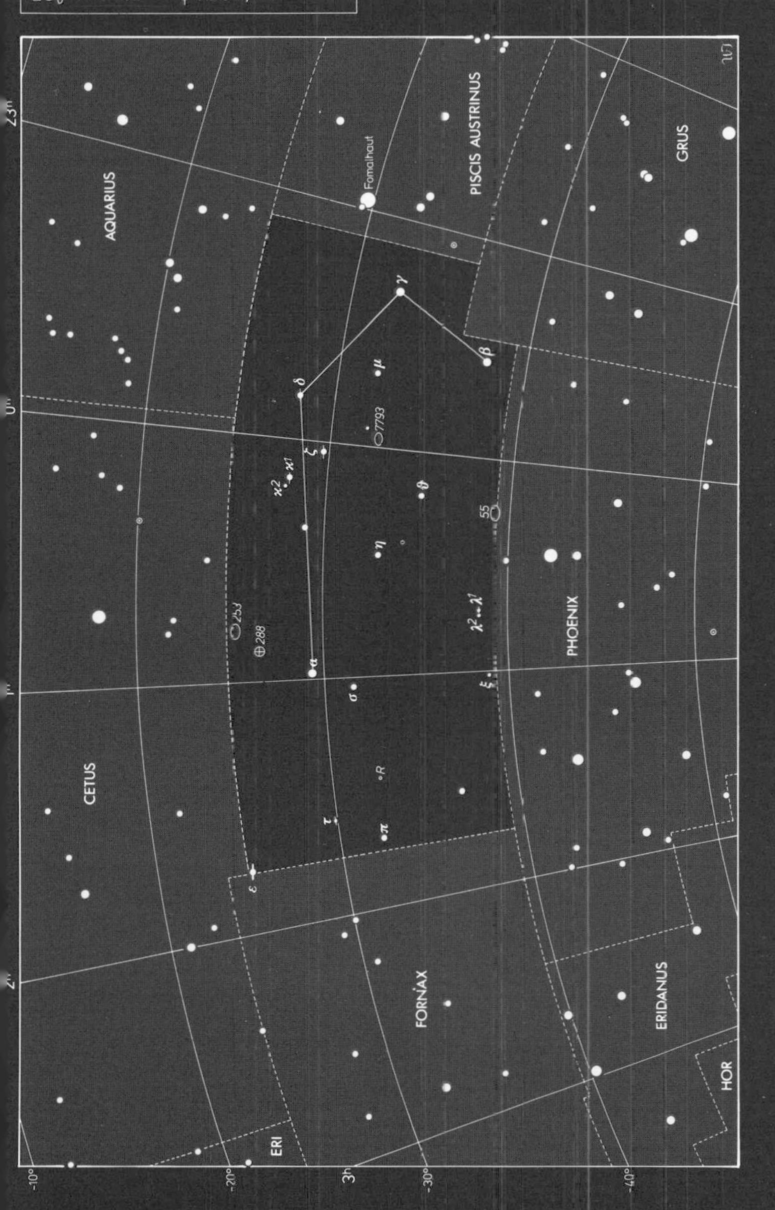

SCUTUM The Shield

A faint constellation between Aquila and Serpens, introduced in 1684 by the Polish astronomer Johannes Hevelius under the title Scutum Sobiescianum, Sobieski's Shield, in honour of his patron, King John III Sobieski. Rich star clouds of the Milky Way are the main attraction in Scutum.

α (alpha) Scuti, 18h 35m –8°, mag. 3.9, is an orange giant 140 l.y. away.

δ (delta) Sct, 18h 42m –9°, 255 l.y. distant, is the prototype of a rare class of variable stars that pulsate in size every few hours, producing small-amplitude brightness changes. δ Scuti itself varies from mag. 4.7 to 4.8 with a period of 4 hours 40 minutes.

M11 (NGC 6705), 18h 51m –6°, the Wild Duck Cluster, is a showpiece open cluster of about 200 stars, half the apparent width of the full Moon. The cluster gets its name from its noticeable fan shape, resembling a flight of ducks. At 6th mag. it is at the limit of naked-eye visibility, but binoculars show it as a misty patch. In a telescope with a magnification of around ×100 it breaks up into a sparkling field of faint stardust. An 8th-mag. star, slightly brighter than the rest, lies at the fan's apex, with a double star nearby. M11 is 5600 l.y. away.

Among the rich Milky Way starfields of Scutum lies M11, the Wild Duck Cluster. In small instruments it is noticeably fan-shaped. Royal Observatory Edinburgh.

SCUTUM
Sct · Scuti

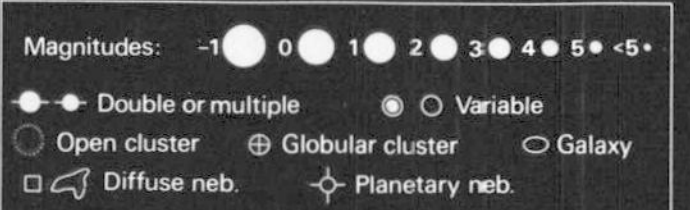

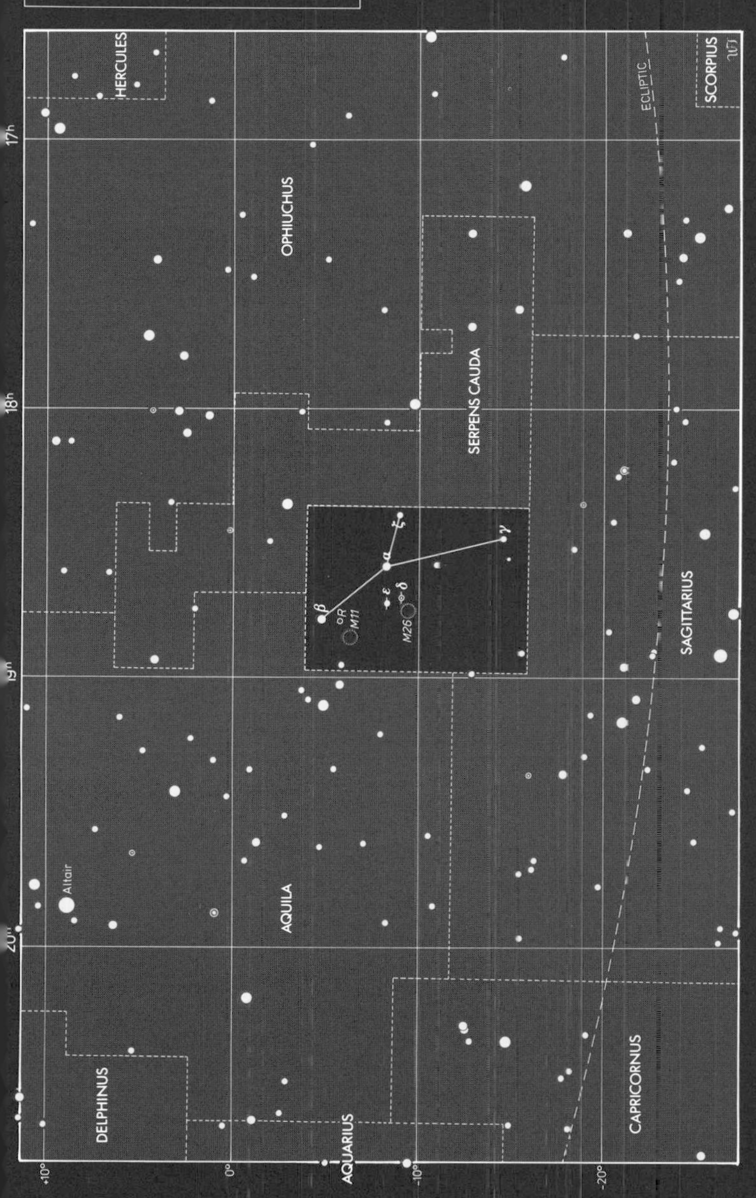

SERPENS The Serpent

An ancient constellation, representing a snake wound around the body of Ophiuchus. Serpens is actually split into two halves, one either side of Ophiuchus: Serpens Caput, the head, which is the larger and more prominent half; and Serpens Cauda, the tail. It is the only constellation to be split in half, but both halves count as one constellation.

α (alpha) Serpentis, 15h 44m +6°, (Unukalhai, 'the serpent's neck'), mag. 2.7, is an orange giant star 72 l.y. away.

β (beta) Ser, 15h 46m +15°, 135 l.y. distant, is a blue-white star in the serpent's head of mag. 3.7, with a 10th-mag. companion visible in small telescopes. An unrelated 7th-mag. star is visible nearby in binoculars.

γ (gamma) Ser, 15h 56m +16°, mag. 3.9, is a white star 42 l.y. away.

δ (delta) Ser, 15h 35m +11°, 85 l.y. away, is white star of mag. 4.2 with a close mag. 5.2 companion visible in a small telescope at high magnification.

ϑ (theta) Ser, 18h 56m +4°, (Alya), 105 l.y. distant, is an elegant pair of white stars of mags. 4.6 and 5.0, easily split in the smallest telescopes.

ν (nu) Ser, 17h 21m –13°, 130 l.y. away, is a blue-white star of mag. 4.3 with a wide mag. 8.3 companion visible in binoculars or small telescopes.

τ^1 (tau^1) Ser, 15h 26m +15°, a red giant of mag. 5.2, is the brightest member of a loose scattering of eight stars of 6th mag. near β Ser, all visible in binoculars.

M5 (NGC 5904), 15h 19m +2°, is a 6th-mag. globular cluster 25,000 l.y. away, visible in binoculars or small telescopes. It is regarded as one of the finest globulars in the northern sky, second only to the famous M13 in Hercules. A telescope of 100 mm aperture reveals its brilliant, condensed centre and mottled outer regions, with chains of stars radiating outwards. Close to M5 lies 5 Ser, a mag. 5.1 yellow star with a mag. 10 companion.

M16 (NGC 6611), 18h 19m –14°, is a hazy-looking star cluster over 8000 l.y. distant, the apparent size of the full Moon, embedded in the Eagle Nebula. The cluster is a grouping of about 60 stars of 8th mag. and fainter at the limit of resolution in binoculars, more condensed at the northern edge. The surrounding Eagle Nebula adds a touch of haziness to the cluster when seen in binoculars. The nebula is too faint to be seen well in amateur telescopes, but shows up beautifully on long-exposure photographs (see pages 230 and 263).

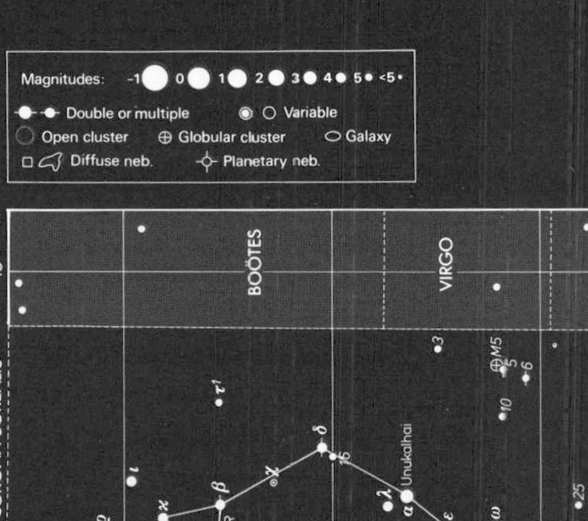

76 SERPENS

Ser · Serpentis

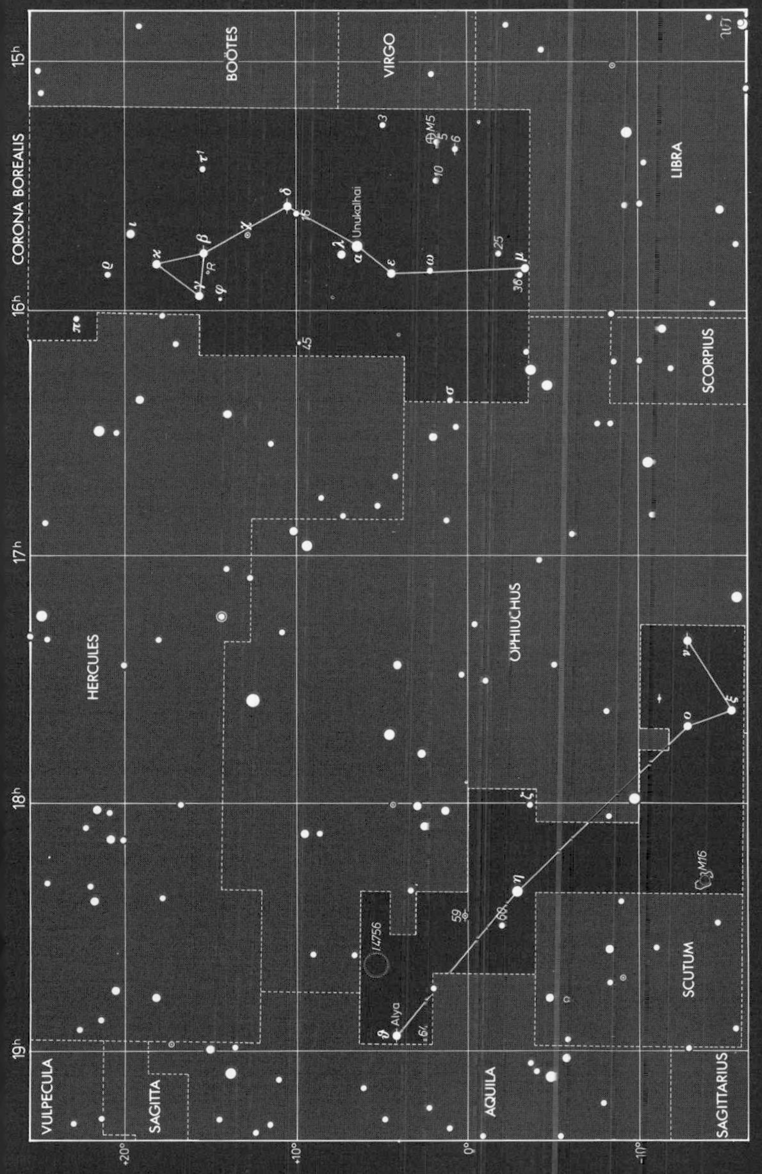

SEXTANS The Sextant

A barren constellation south of Leo, introduced in 1687 by the Polish astronomer Johannes Hevelius. It commemorates the instrument he used for measuring star positions. Hevelius continued to use his sextant for making naked-eye sightings of star positions long after telescopes were available.

α (alpha) Sextantis, 10h 08m 0°, mag. 4.5, is a blue-white giant star 330 l.y. away.

β (beta) Sex, 10h 30m –1°, mag. 5.1, is a blue-white star 520 l.y. away.

γ (gamma) Sex, 9h 53m –8°, mag. 5.1, is a white star 185 l.y. away.

δ (delta) Sex, 10h 30m –3°, mag. 5.2, is a blue-white star 300 l.y. away.

NGC 3115, 10h 05m –8°, is a 9th-mag. elliptical galaxy known as the Spindle Galaxy, 14 million l.y away. Moderate-sized amateur telescopes show its elongated outline and brighter centre.

M16 and the surrounding Eagle Nebula in Serpens, a visually spectacular combination of star cluster and gas cloud, with intruding fingers of cold, dark dust and gas. *(For a description see page 228.)* Hale Observatories photograph.

SEXTANS

Sex · Sextantis

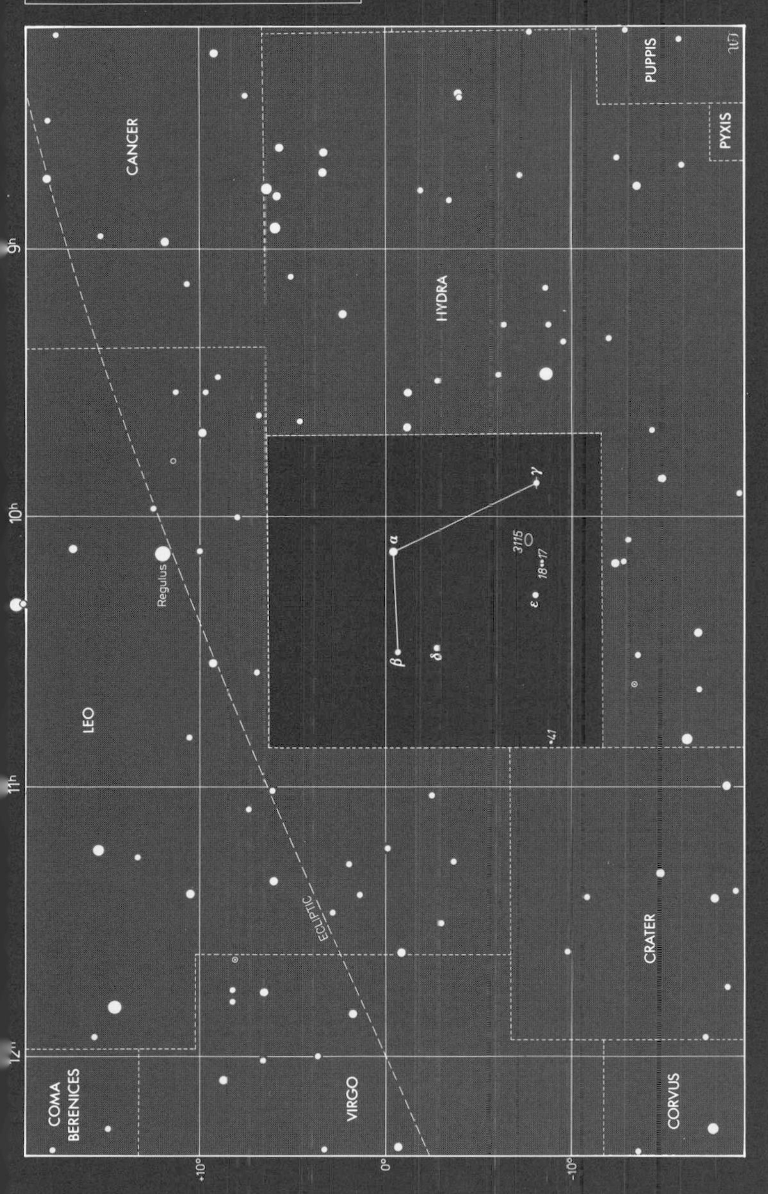

TAURUS The Bull

One of the most ancient constellations, recognized since the dawn of civilization. In Greek mythology, Taurus represents the animal disguise adopted by Zeus to carry off Princess Europa to Crete. Only the front half of the bull is depicted in the sky, its face being formed by the V-shaped cluster of stars known as the Hyades. Its glinting red eye is marked by the star Aldebaran, and its long horns are tipped by the stars β (beta) and ζ (zeta) Tauri. In addition to the Hyades, Taurus contains the celebrated star cluster of the Pleiades, or Seven Sisters. In Taurus occurred the famous supernova that was seen from Earth in 1054 and gave rise to the Crab Nebula, M1. At 4h 22.0m, +19° 32′, lies the faint Hind's Variable Nebula, NGC 1555, discovered in the 19th century by the English astronomer John Russell Hind; within this nebula lies the star T Tauri, prototype of a class of irregular variables believed to be stars in the process of formation. Each year, the Taurid meteors radiate from near ε (epsilon) Tauri, reaching a maximum of about 10 per hour around November 4. The Sun passes through the constellation from mid-May to late June.

α (alpha) Tauri, 4h 36m +17°, (Aldebaran, 'the follower', i.e. of the Pleiades) is a red giant irregular variable that fluctuates between about mags. 0.75 and 0.95. Although it appears to be part of the Hyades cluster, it is in fact an unrelated foreground star, 68 l.y. away.

β (beta) Tau, 5h 26m +29°, (Elnath, 'the butting one'), mag. 1.7, is a blue-white giant 145 l.y. away.

ζ (zeta) Tau, 5h 38m +21°, mag. 3.0, is a blue giant 520 l.y. away.

ϑ^1 ϑ^2 (theta^1 theta^2) Tau, 4h 29m +16°, 150 l.y. away, is a naked-eye or binocular double in the Hyades, consisting of white and yellow giants of mags. 3.8 and 3.4 respectively. ϑ^2 is the brightest member of the Hyades.

$\varkappa$ (kappa) Tau, 4h 25m +22°, 150 l.y. away, is a white star of mag. 4.2 that forms a naked-eye or binocular duo with 67 Tau, mag. 5.3. Both are outlying members of the Hyades.

λ (lambda) Tau, 4h 01m +12°, 330 l.y. away, is an eclipsing binary of the Algol type, varying between mags. 3.4 and 3.9 every 4 days.

σ^1 σ^2 (sigma^1 sigma^2) Tau, 4h 39m +16°, is a wide binocular double of white stars, mags. 5.1 and 4.7 respectively. σ^2, the brighter of the pair, is thought to be a member of the Hyades, 150 l.y. away, but the cluster membership of σ^1 is uncertain.

φ (phi) Tau, 4h 20m +27°, 280 l.y. away, is an optical double divisible in small telescopes, consisting of a mag. 5.0 orange giant and a white star of mag. 8.4.

χ (chi) Tau, 4h 23m +26°, 360 l.y. away, is a double star for small telescopes, with blue and gold components of mags. 5.4 and 7.6. ▶

78
TAURUS
Tau · Tauri

Magnitudes: -1 0 1 2 3 4 5 <5
Double or multiple
Variable
Open cluster
Globular cluster
Galaxy
Diffuse neb.
Planetary neb.

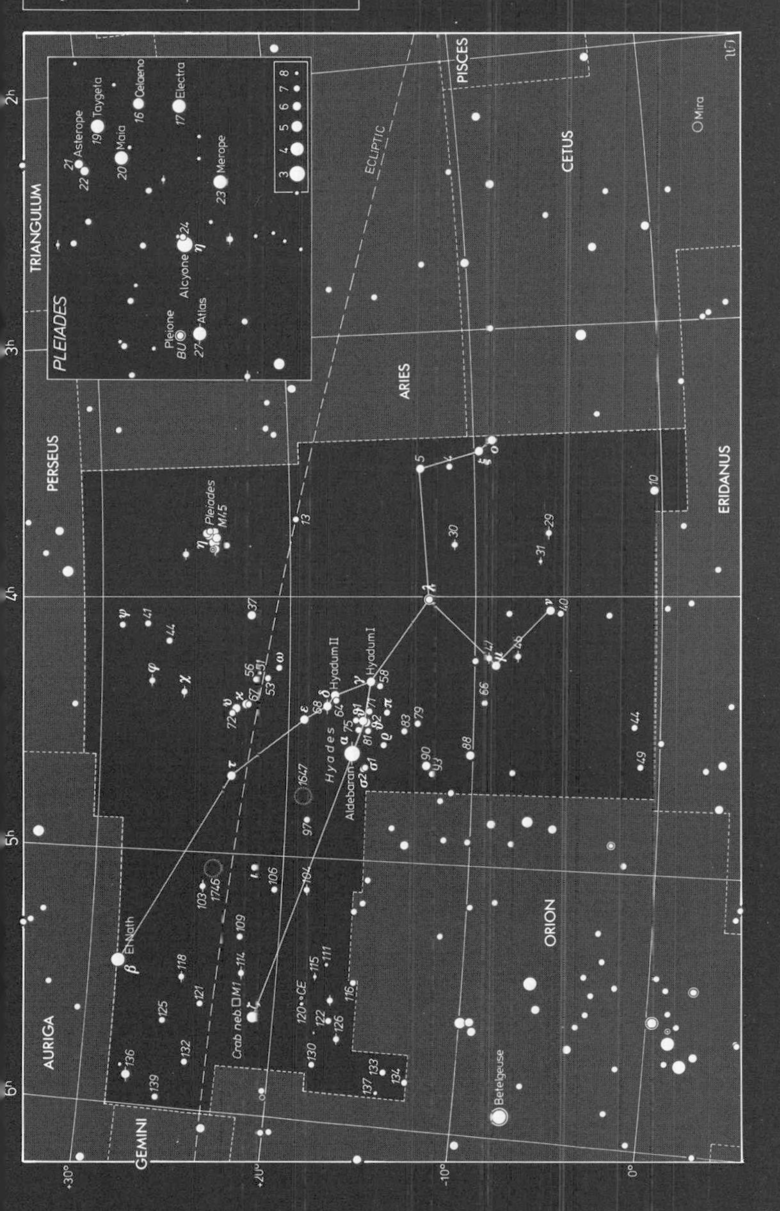

TELESCOPIUM The Telescope

A constellation invented in the 1750s by the Frenchman Nicolas Louis de Lacaille to honour the most important of astronomical instruments. As with so many of Lacaille's constellations, it is faint and contrived, containing little to interest owners of small telescopes.

α (alpha) Telescopii, 18h 27m –46°, mag. 3.5, is a blue-white star 460 l.y. away.

δ^1 δ^2 (delta^1 delta^2) Tel, 18h 32m –46°, is a pair of blue-white stars of mags. 4.9 and 5.1, visible separately in binoculars. They are unrelated to each other, being at 650 and 1300 l.y. from us respectively.

ε (epsilon) Tel, 18h 11m –46°, mag. 4.5, is a yellow giant 175 l.y. away.

ζ (zeta) Tel, 18h 29m –49°, mag. 4.1, is an orange giant 180 l.y. away.

◀ The Hyades is a large and bright cluster of about 200 stars covering over 5° of sky. The brightest members form a distinctive V-shape, easily visible to the naked eye. In mythology, the Hyades were the daughters of Atlas and Aethra, and half-sisters of the Pleiades. Because of its size, the cluster is best studied with binoculars rather than a telescope. The bright star Aldebaran is not a member of the Hyades, but is superimposed on it by chance; the brightest true member is actually ϑ^2 (theta^2) Tauri. The cluster lies 150 l.y. away; its distance is important, for it marks the first step in our distance scale of the Galaxy.

M1 (NGC 1952), 5h 35m +22°, is the celebrated Crab Nebula, the remains of a star that exploded as a supernova. It can be glimpsed through binoculars on clear, dark nights. Despite its fame the Crab Nebula is a disappointing object for small telescopes, appearing as an elliptical 8th-mag. wisp of nebulosity. At the centre of the nebula, beyond the reach of amateur telescopes, is a 16th-mag. star, the remains of the star that exploded. This faint star is now known to be a pulsar. The Crab Nebula and pulsar lie about 6500 l.y. away.

M45, 3h 47m +24°, the Pleiades, is the brightest and most famous star cluster in the sky; it is popularly termed the Seven Sisters, after a group of mythological nymphs, the daughters of Atlas and Pleione. Approximately seven stars are visible to the naked eye, covering three full Moon widths of sky; binoculars bring dozens more into view. About 100 stars belong to the cluster, which lies 410 l.y. away. Unlike the stars of the Hyades, which are older and more evolved, the Pleiades formed within the last 50 million years and include many young blue giants. The brightest member of the Pleiades is η (eta) Tauri (Alcyone), mag. 2.9. Other prominent members are 16 Tau (Celaeno), mag. 5.5; 17 Tau (Electra), mag. 3.7; 19 Tau (Taygeta), mag. 4.3; 20 Tau (Maia), mag. 3.9; 21 Tau (Asterope), mag. 5.8; 23 Tau (Merope), mag. 4.2; 27 Tau (Atlas), mag. 3.6; and BU Tau (Pleione), a shell star that throws off rings of gas at irregular intervals, causing it to fluctuate unpredictably between mags. 4.8 and 5.5. The whole of the Pleiades is embedded in a faint nebulosity, the remains of the cloud from which the stars formed. This nebula is noticeable on long-exposure photographs, and under very clear conditions the brightest part of the nebula, around Merope, may be glimpsed in binoculars or small telescopes.

TELESCOPIUM
Tel · Telescopii

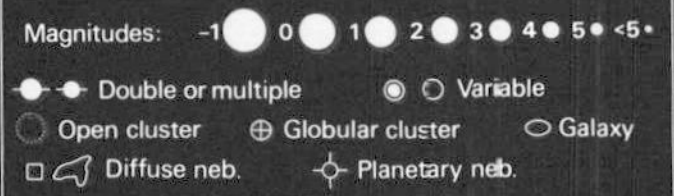

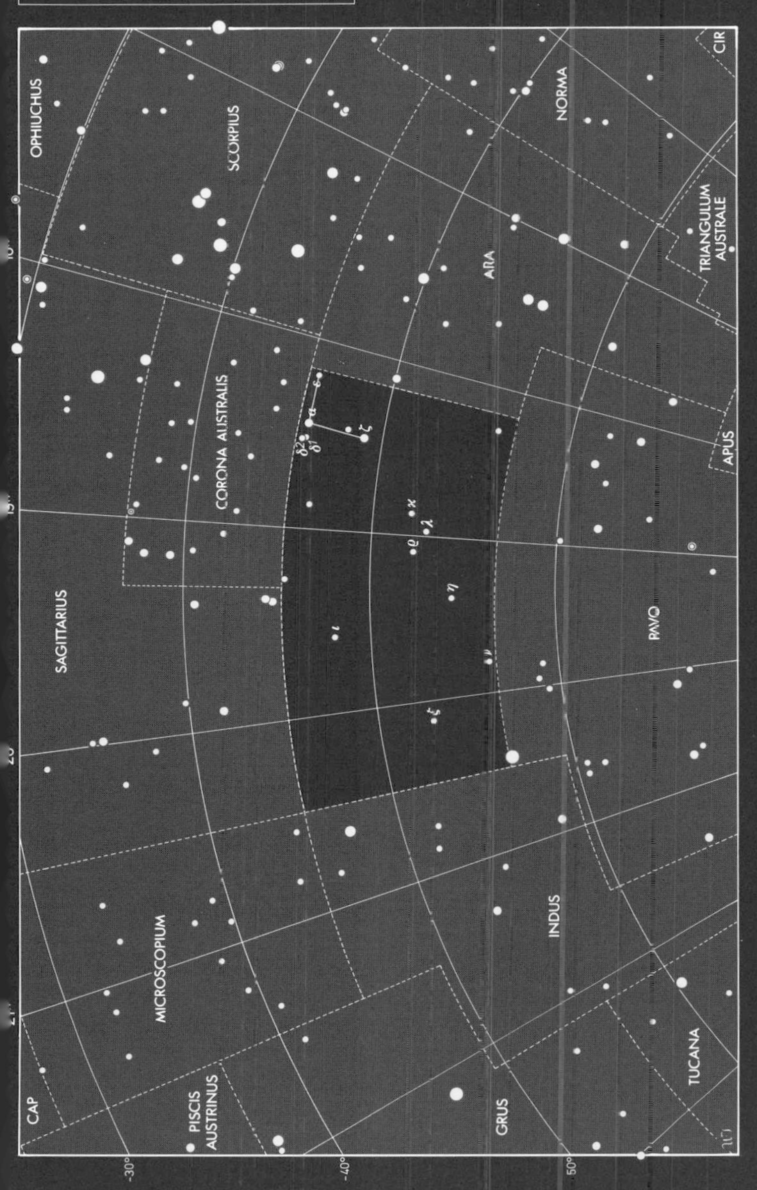

TRIANGULUM The Triangle

A small but distinctive constellation lying between Andromeda and Aries, consisting of three main stars that form a thin delta shape; the Greeks referred to it as Deltoton. Its most important feature is the spiral galaxy M33, the third-largest member of our Local Group of galaxies, after the Andromeda Galaxy and our own Milky Way.

α (alpha) Trianguli, 1h 53m +30°, mag. 3.4, is a white star 65 l.y. away.

β (beta) Tri, 2h 10m +35°, mag. 3.0, is a white giant 135 l.y. away.

γ (gamma) Tri, 2h 17m +34°, mag. 4.0, is a blue-white star 120 l.y. away.

6 Tri, 2h 12m +30°, 280 l.y. away, is a mag. 5.0 golden-yellow giant with a close mag. 6.9 bluish companion visible in a small telescope.

M33 (NGC 598), 1h 34m +31°, is a spiral galaxy 2.7 million l.y. away in our Local Group. Presented almost face-on, it covers a larger area of sky than the full Moon. Despite its size and proximity it is not prominent visually because its light is spread over such a large area. M33 is best picked up on a dark night in binoculars or a small telescope with low power to enhance the contrast. Unlike most galaxies, it does not have a noticeably stellar nucleus. Quite large amateur telescopes are needed to trace the spiral arms.

The spiral galaxy M33 in Triangulum, a member of our Local Group of galaxies. Hale Observatories photograph.

TRIANGULUM
Tri · Trianguli

Magnitudes: -1 0 1 2 3 4 5 <5
Double or multiple · Variable
Open cluster · Globular cluster · Galaxy
Diffuse neb. · Planetary neb.

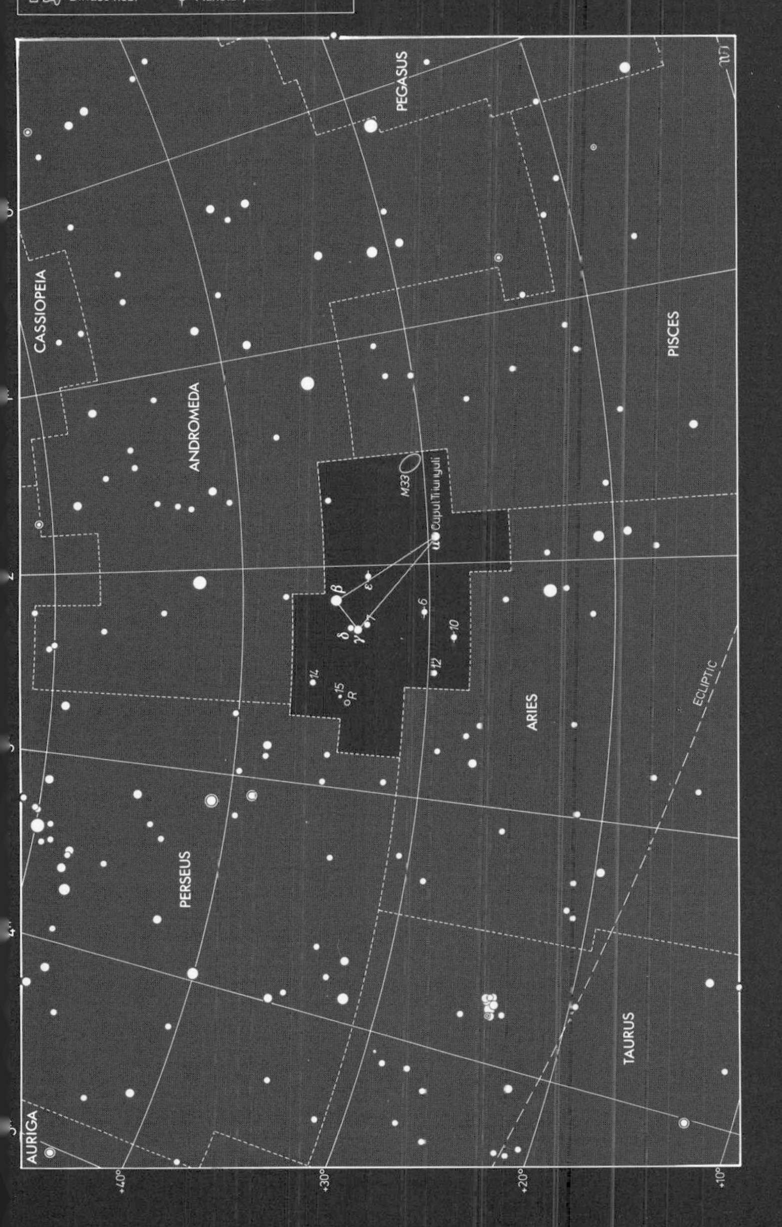

TRIANGULUM AUSTRALE The Southern Triangle

A small but readily distinguishable constellation near α (alpha) Centauri, introduced at the end of the 16th century by the Dutch navigators Pieter Dirkszoon Keyser and Frederick de Houtman. Its three main stars are brighter than those of its northern equivalent, Triangulum.

α (alpha) Trianguli Australis, 16h 49m –69°, mag. 1.9, is an orange giant star 105 l.y. away.

β (beta) TrA, 15h 55m –63°, mag. 2.9, is a giant white star 42 l.y. away.

γ (gamma) TrA, 15h 19m –69°, mag. 2.9, is a blue-white star 72 l.y. away.

NGC 6025, 16h 04m –61°, is a 5th-mag. binocular star cluster of about 60 members of mag. 7 and fainter, 2700 l.y. away.

Johann Bayer (1572–1625)

Johann Bayer was a German lawyer in Augsburg and an amateur astronomer who, in 1603, published the first star atlas that covered the entire sky, the *Uranometria*. He based his coverage of the northern heavens mainly on the observations made by the great Danish astronomer Tycho Brahe, while his information on the southern skies came from the work of the Dutch navigator Pieter Dirkszoon Keyser. In addition to the 48 constellations known since ancient times, Bayer showed the 12 new constellations around the southern celestial pole invented by Keyser and his countryman Frederick de Houtman: Apus, Chamaeleon, Dorado, Grus, Hydrus, Indus, Musca, Pavo, Phoenix, Triangulum Australe, Tucana and Volans. Bayer's most important legacy to astronomers was the system of identifying stars by Greek letters. In each constellation the brightest stars were assigned letters of the Greek alphabet, usually, but not always, in approximate order of brightness (Gemini, Orion and Sagittarius are prominent examples of constellations in which the star labelled α (alpha) is not the brightest). Before Bayer's time, stars without proper names were identified by the cumbersome descriptions of the Greek astronomer Ptolemy such as 'in the left forearm of the advance twin', a reference to the 4th-magnitude star that we now refer to as ϑ (theta) Geminorum. Clearly, to make sense of such descriptions astronomers had to know their constellation figures well, and even then there was considerable room for confusion. The system of so-called Bayer letters was a vast improvement, and remains in use to this day.

TRIANGULUM AUSTRALE
TrA · Trianguli Australis

Magnitudes: -1 0 1 2 3 4 5 <5
Double or multiple
Variable
Open cluster
Globular cluster
Galaxy
Diffuse neb.
Planetary neb.

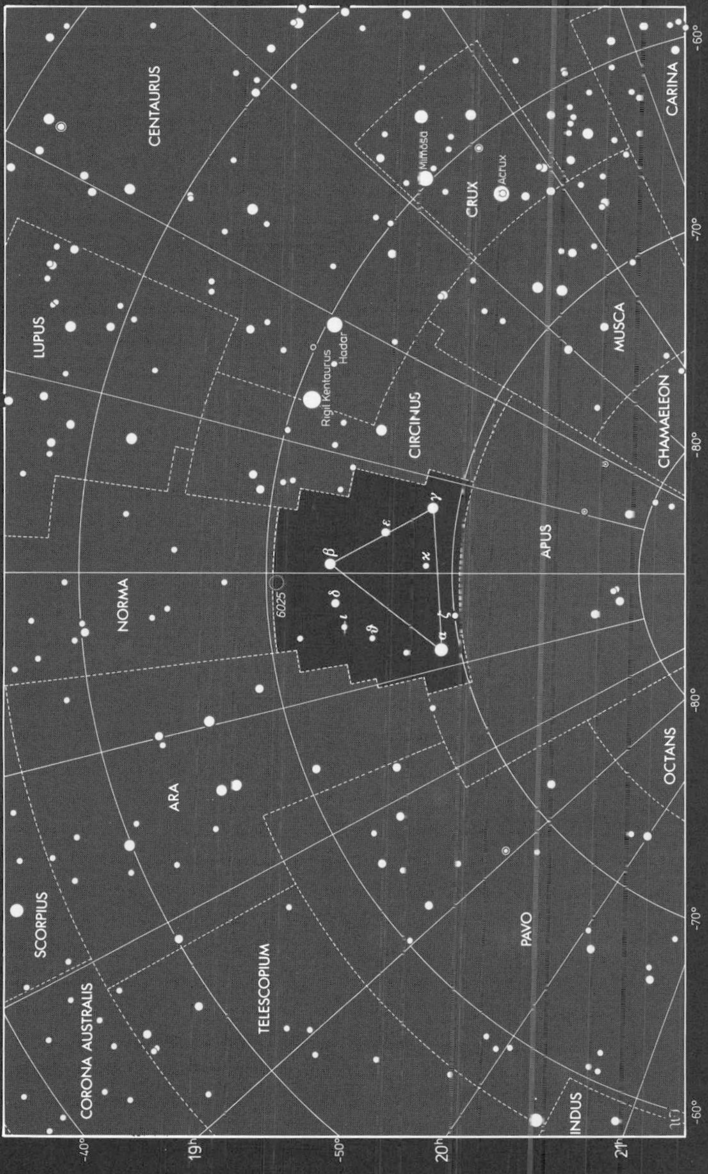

CENTAURUS
LUPUS
CRUX
Mimosa
Acrux
Hadar
Rigil Kentaurus
CIRCINUS
MUSCA
CARINA
CHAMAELEON
APUS
NORMA
6025
ARA
SCORPIUS
CORONA AUSTRALIS
TELESCOPIUM
PAVO
OCTANS
INDUS
-60°
-70°
-80°
-40°
-50°
19h
20h
21h

TUCANA The Toucan

A constellation near the south pole of the sky, introduced in the late 16th century by the Dutch navigators Pieter Dirkszoon Keyser and Frederick de Houtman. It represents a toucan, the South American bird with the large beak. Its most notable features are the Small Magellanic Cloud and the globular cluster known as 47 Tucanae.

α (alpha) Tucanae, 22h 19m –60°, mag. 2.9, is an orange giant star 120 l.y. away.

β (beta) Tuc, 0h 32m –63°, is a complex multiple star. Binoculars or small telescopes show that it consists of two almost identical blue-white stars, β^1 and β^2, mags. 4.4 and 4.5; β^2 is itself a close binary with a period of 44 years, requiring a telescope above 200 mm to divide. Nearby lies a mag. 5.1 white star, β^3. All three stars share the same proper motion through space, but their estimated distances are somewhat discordant, being 275, 150 and 240 l.y. respectively.

γ (gamma) Tuc, 23h 17m –58°, mag. 4.0, is a giant white star 140 l.y. away.

δ (delta) Tuc, 22h 27m –65°, mag. 4.5, is a blue-white star 270 l.y. away with a mag. 9 companion visible in small telescopes.

κ (kappa) Tuc, 1h 16m –69°, 78 l.y. away, is a double star consisting of components of mags. 4.9 and 7.5 visible in small telescopes. This pair moves through space with a wide mag. 7.2 star, itself a close binary with an orbital period of 86 years, divisible in 150 mm aperture.

47 Tuc (NGC 104), 0h 24m –72°, is a large and brilliant globular cluster the size of the full Moon, visible to the naked eye as a fuzzy 4th-mag. star – hence, on early charts, it was actually catalogued as a star and given a star's designation. Among globular clusters it is second only to ω (omega) Centauri in size and brightness. Telescopes of 100 mm aperture begin to resolve 47 Tuc, and even binoculars show its brilliant central blaze. It is among the closest globulars to us, 15,000 l.y. away. (See the photograph on page 158.)

Small Magellanic Cloud (SMC) is a satellite galaxy of the Milky Way, as is its larger brother in Dorado. The Small Magellanic Cloud appears to the naked eye as a nebulous, tadpole-shaped patch 3½° across. Binoculars and small telescopes resolve it into clusters and glowing gas clouds. It lies 190,000 l.y. away. (See the photograph on page 158.)

NGC 362, 1h 03m –71°, is a 7th-mag. globular cluster visible in binoculars near the edge of the Small Magellanic Cloud, but not associated with it. NGC 362 actually lies 29,000 l.y. away, in our own Galaxy.

TUCANA
Tuc · Tucanae

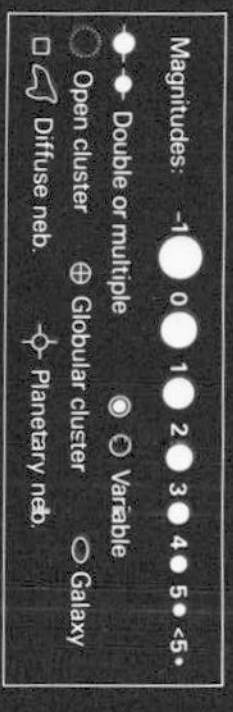

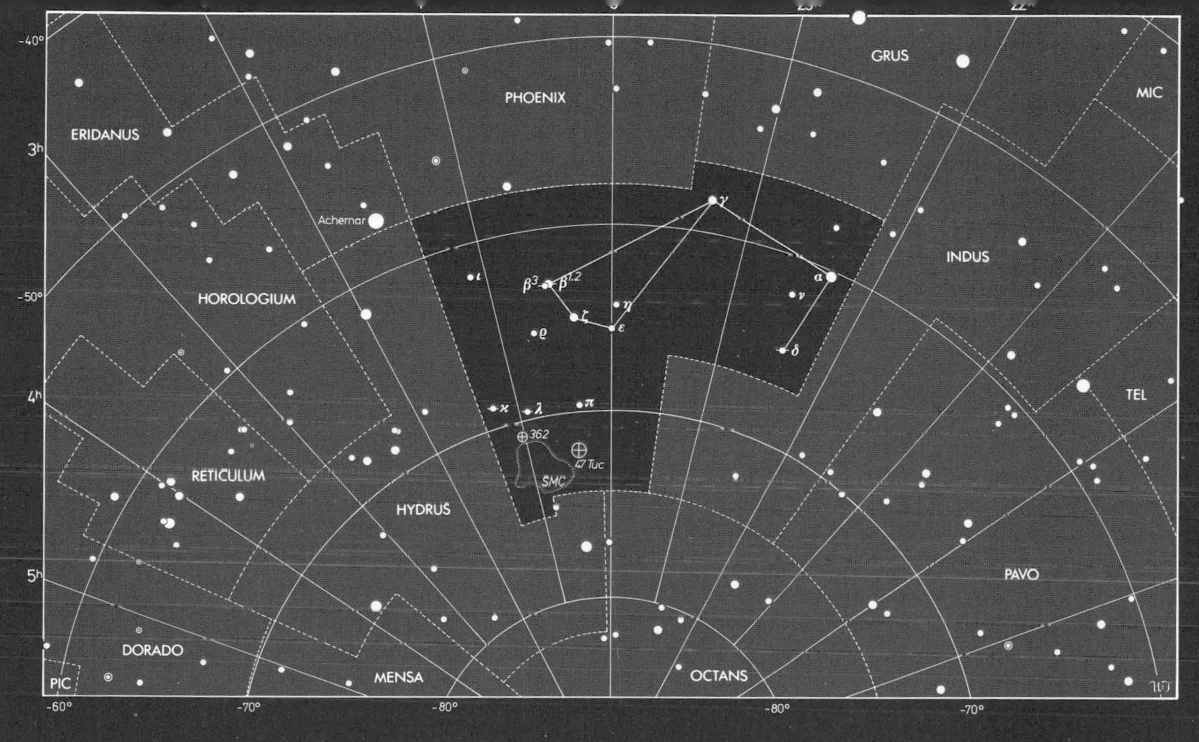

URSA MAJOR The Great Bear

The third-largest constellation in the sky. Its central feature is the seven stars that make up the familiar shape variously called the Plough or the Big Dipper, the best known of all star patterns, although why so many people, including the North American Indians, visualized this group as a bear remains a mystery. In Europe the pattern was seen as a wagon or chariot. Others, notably the Arabs, viewed the dipper shape not as a bear, but as a bier or coffin. In Greek mythology the bear represented Callisto, who was turned into a bear in punishment for her illicit love affair with Zeus. The stars of the Big Dipper, except Alkaid and Dubhe, are moving together through space. Two stars in the bowl of the dipper, Merak and Dubhe, act as a pointer to Polaris, the North Pole Star in neighbouring Ursa Minor. The handle of the Big Dipper points to the bright star Arcturus in Boötes. At 11h 03.3m, +35° 58′, lies the mag. 7.5 red dwarf Lalande 21185, which is the Sun's fourth-closest stellar neighbour, 8.2 l.y. away. Its name comes from its number in a catalogue drawn up by the 18th-century French astronomer Joseph Lalande. Ursa Major contains numerous galaxies, but only a few of them are easily visible in amateur telescopes.

α (alpha) Ursa Majoris, 11h 04m +62°, (Dubhe, 'the bear'), mag. 1.8, is a yellow giant 108 l.y. away. It has a close mag. 4.8 companion that orbits it every 45 years. The two can be split in 220 mm aperture, apart from the few years either side of 2011, when they are at their closest.

β (beta) UMa, 11h 02m +56°, (Merak, 'flank'), mag. 2.4, is a white star 78 l.y. away.

γ (gamma) UMa, 11h 54m +54°, (Phecda, 'thigh'), mag. 2.4, is a white star 88 l.y. away.

δ (delta) UMa, 12h 15m +57°, (Megrez, 'root of the tail'), mag. 3.3, is a white star 62 l.y. away.

ε (epsilon) UMa, 12h 54m +56°, (Alioth), mag. 1.8, is a white star with a peculiar spectrum and of uncertain distance.

ζ (zeta) UMa, 13h 24m +55°, (Mizar), mag. 2.3, is a celebrated multiple star. Keen eyesight, or binoculars, reveal its mag. 4.0 companion Alcor. Mizar is 60 l.y. from Earth and Alcor 90 l.y. away, too far apart to make this a genuine binary. However, a small telescope reveals that Mizar has another mag. 4.0 companion closer to it, which definitely is related. This star was first seen by the Italian astronomer Giovanni Riccioli in 1650, making Mizar the first double star to be discovered telescopically. Mizar was also the first star discovered to be a spectroscopic binary, by the American astronomer E. C. Pickering in 1889. The companion of Mizar is another spectroscopic binary, as is Alcor, making this a highly complex group. (See the diagram on page 245.)

η (eta) UMa, 13h 48m +49°, (Alkaid or Benetnasch, both from the Arabic for 'leader of the mourners'), mag. 1.9, is a blue-white star about 100 l.y. away. ►

83
URSA MAJOR
UMa · Ursae Majoris

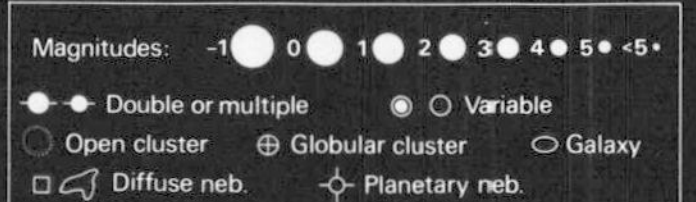

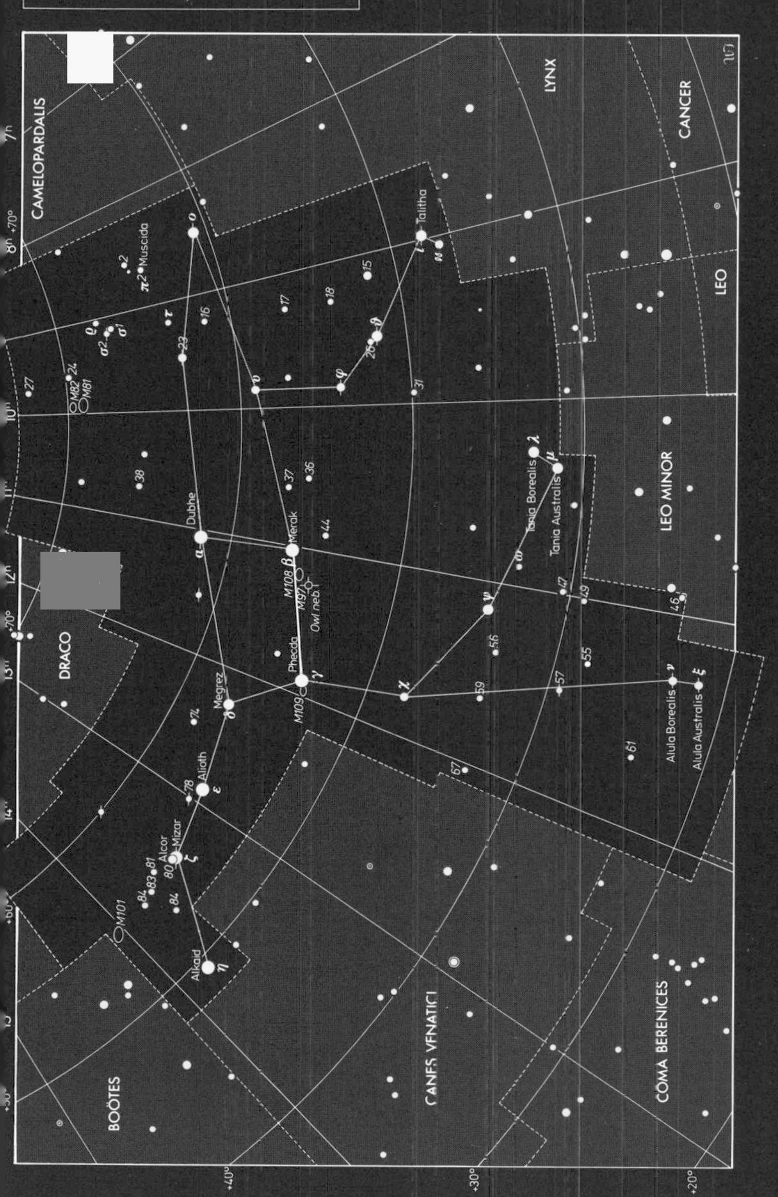

M81 is a near-perfect spiral galaxy in Ursa Major, tilted at a slight angle to our line of sight. Hale Observatories photograph.

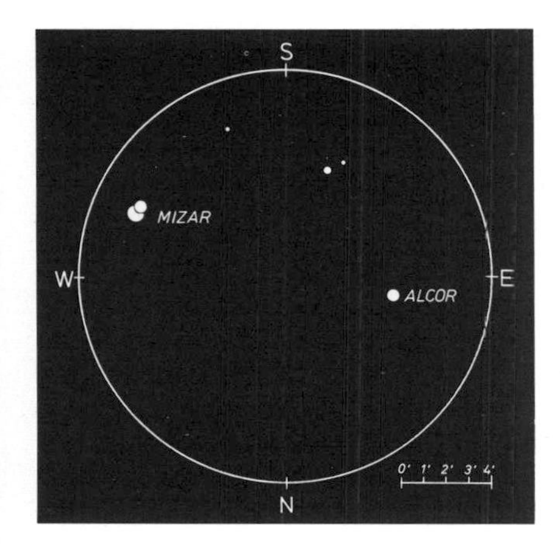

Mizar and Alcor as seen through a telescope.
Wil Tirion.

◀ ξ (xi) UMa, 11h 18m +32°, 26 l.y. away, was the first double star to have its orbit computed. Its two yellow components (both spectroscopic binaries), of mags. 4.3 and 4.8, orbit each other every 60 years. They were closest in 1992, when 150 mm aperture was needed to separate them, but are now moving apart and should be divisible in 75 mm by the year 2000.

M81 (NGC 3031), 9h 56m +69°, is a beautiful 7th-mag. spiral galaxy, one of the brightest in the sky, visible in binoculars. A small telescope shows it as a roundish, softly glowing patch noticeably brighter towards the centre. Being tilted at an angle to us it appears somewhat elliptical in outline, covering over half a Moon's width at its longest. In the same telescopic field of view ½° to the north is M82; the two galaxies are about 10 million l.y. from us.

M82 (NGC 3034), 9h 56m +70°, is a neighbour galaxy of M81, a quarter as bright and less than half the size but still visible in binoculars. In a small telescope it appears as an elongated blur and can actually appear more prominent than M81 because of its higher surface brightness. Long-exposure photographs show that M82 is actually an edge-on spiral galaxy interacting with a huge cloud of dust; it is officially described as 'peculiar'.

M97 (NGC 3587), 11h 15m +55°, is an elusive 11th-mag. planetary nebula known as the Owl Nebula because of two dark patches like eyes that give it the appearance of an owl's face when seen through a large telescope. In moderate apertures it appears as only a pale disk over three times the size of Jupiter and will probably need an aperture of at least 75 mm to be seen at all. The Owl lies 1300 l.y. away.

M101 (NGC 5457), 14h 03m +54°, is a spiral galaxy visible in binoculars as a pale, rounded smudge almost as large as the full Moon; because of its large size it is less prominent than its quoted 8th mag. would suggest. Long-exposure photographs show it as a face-on galaxy with tightly wound spiral arms, but these are not apparent in small telescopes, nor is there any strong central brightening. M101 lies 23 million l.y. away.

URSA MINOR The Little Bear

A constellation said to have been introduced in about 600 BC by the Greek astronomer Thales. Ursa Minor contains the present North Celestial Pole, within 1° of which lies the conveniently placed 2nd-magnitude star we call Polaris, α (alpha) Ursae Minoris. Precession will bring Polaris closest to the pole around AD 2100, after which it will start to move away again. Ursa Minor is also termed the Little Dipper because its seven brightest stars outline a shape like a smaller version of the Big Dipper in Ursa Major. The stars β (beta) and γ (gamma) Ursae Minoris in the Little Dipper's bowl are called the Guardians of the Pole. The stars in Ursa Minor are useful for checking atmospheric transparency, because their brightnesses run in one-magnitude steps from 2nd to 5th magnitude; the faintest member of Ursa Minor visible on a given night is a guide to the limiting magnitude.

α (alpha) Ursae Minoris, 2h 32m +89°, (Polaris), mag. 2.0, is a yellow supergiant star about 650 l.y. away. It is listed in catalogues as a Cepheid variable of small amplitude, but during the 20th century its pulsations became smaller until by 1990 it had almost ceased to vary at all. Polaris is also a double star, with a mag. 8.2 companion visible in a small telescope.

β (beta) UMi, 14h 51m +74°, (Kochab), mag. 2.1, is an orange giant 95 l.y. away.

γ (gamma) UMi, 15h 21m +72°, (Pherkad), mag. 3.0, is a giant white star 225 l.y. away. The mag. 5.0 orange giant 11 UMi that appears near it, as seen by the naked eye or in binoculars, is unrelated, lying 360 l.y. away.

δ (delta) UMi, 17h 32m +87°, mag. 4.4, is a white star 140 l.y. away.

ε (epsilon) UMi, 16h 46m +82°, mag. 4.2, 200 l.y. away, is a yellow giant eclipsing binary that varies every 39.5 days by under 0.1 mag., not discernible to the naked eye.

ζ (zeta) UMi, 15h 44m +78°, mag. 4.3, is a white star 110 l.y. away.

η (eta) UMi, 16h 18m +76°, mag. 5.0, is a white star 65 l.y. away. A wide mag. 5.5 companion, 19 UMi, is an unrelated background object.

URSA MINOR

UMi · Ursae Minoris

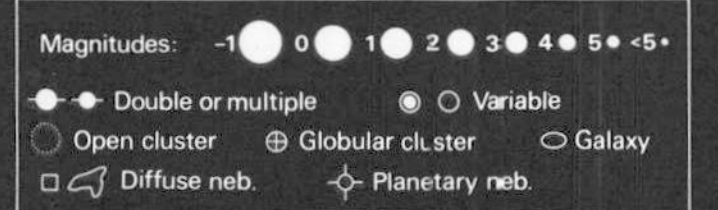

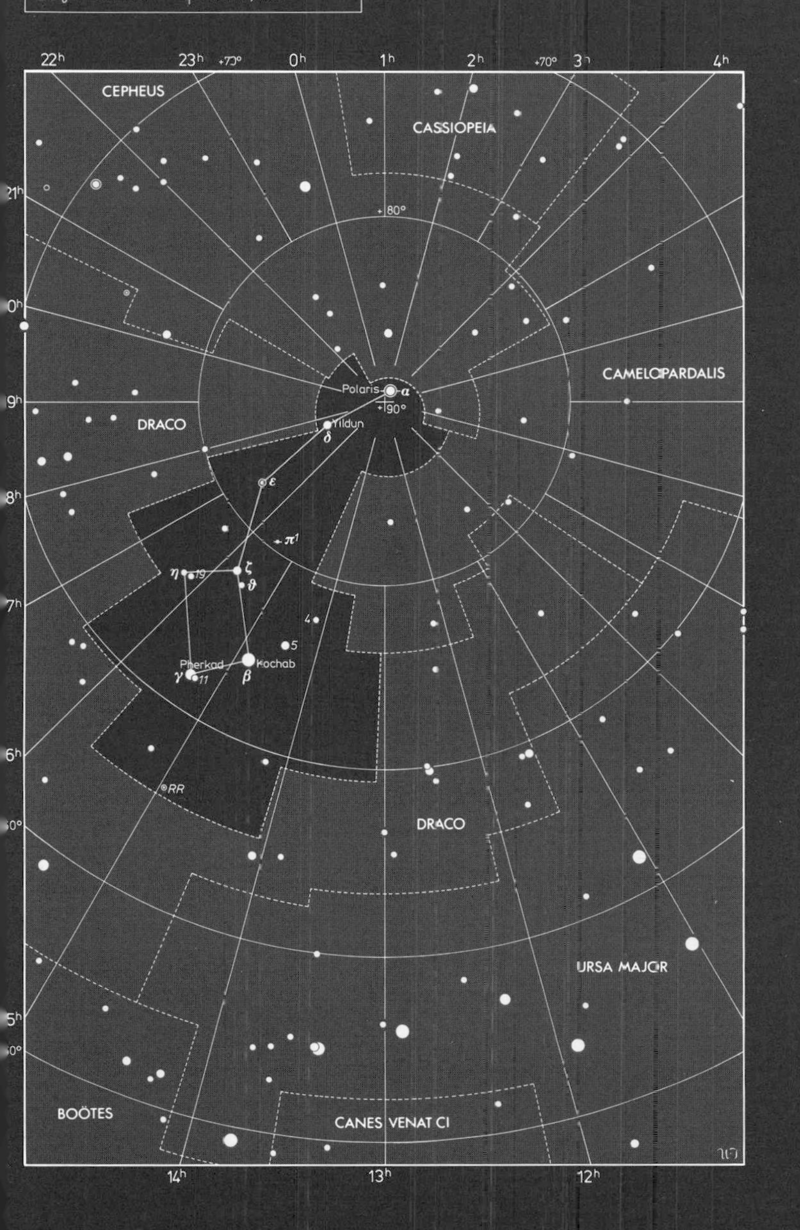

VELA The Sails

Formerly part of the ancient constellation Argo Navis, representing the ship of Jason and the Argonauts, until made separate by Nicolas Louis de Lacaille in 1763. Since Vela is only a subdivision of a once-larger constellation, it has no stars labelled α (alpha) or β (beta). The stars $\varkappa$ (kappa) and δ (delta) Velorum, in conjunction with ι (iota) and ε (epsilon) Carinae, form a shape known as the False Cross, sometimes mistaken for the real Southern Cross. Vela lies in a part of the Milky Way rich with faint nebulosity, visible on long-exposure photographs. This nebulosity is known as the Gum Nebula, after the Australian astronomer Colin S. Gum who drew attention to it in 1952. The Gum Nebula is believed to be the remains of one or more supernovae that occurred in Vela. Another remnant of a supernova in the constellation is the Vela Pulsar, which flashes 11 times per second, one of the few pulsars that can be seen flashing visually as well as at radio wavelengths.

γ (gamma) Velorum, 8h 09m –47°, is an interesting multiple star. Binoculars or small telescopes show that it consists of two apparently unrelated blue-white components of mags. 1.8 and 4.3. The brighter of these is the brightest known Wolf–Rayet star, a rare class of stars with very hot surfaces which seem to be ejecting gas. Its distance is uncertain, but the fainter star is estimated to lie 1600 l.y. away. There are also two wider companions, of 8th and 9th mags.

δ (delta) Vel, 8h 45m –55°, is a white star 75 l.y. away of mag. 2.0 with a mag. 5.1 companion requiring 100 mm aperture to see.

$\varkappa$ (kappa) Vel, 9h 22m –55°, mag. 2.5, is a blue-white star 460 l.y. away.

λ (lambda) Vel, 9h 08m –43°, mag. 2.2, is an orange supergiant 300 l.y. away.

H Vel, 8h 56m –53°, 520 l.y. away, is a neat double of mags. 4.8 and 7.4, difficult in the smallest telescopes because of the magnitude contrast.

NGC 2547, 8h 11m –49°, 1300 l.y. away, is a cluster of about 80 stars of mag. 6.5 and fainter, just visible to the naked eye and best seen in binoculars.

NGC 3132, 10h 08m –41°, is a relatively large and bright planetary nebula of 8th mag., in a small telescope appearing larger than Jupiter, with a central star of 10th mag. It lies 2600 l.y. away.

NGC 3228, 10h 22m –52°, is a cluster of about 15 faint stars for binoculars and small telescopes, 1600 l.y. away.

IC 2391, 8h 40m –53°, is a large naked-eye cluster of 50 stars, 590 l.y. away, scattered around the mag. 3.6 blue-white star o Vel, a small-amplitude variable of the β Cephei type. About 1° from it is the binocular cluster NGC 2669.

IC 2395, 8h 41m –48°, is a binocular cluster of 40 stars, 2800 l.y. away. A mag. 5.5 star seems to be the brightest member, although it may actually be a foreground star. Also visible ½° to the south is the 8th-mag. cluster NGC 2670.

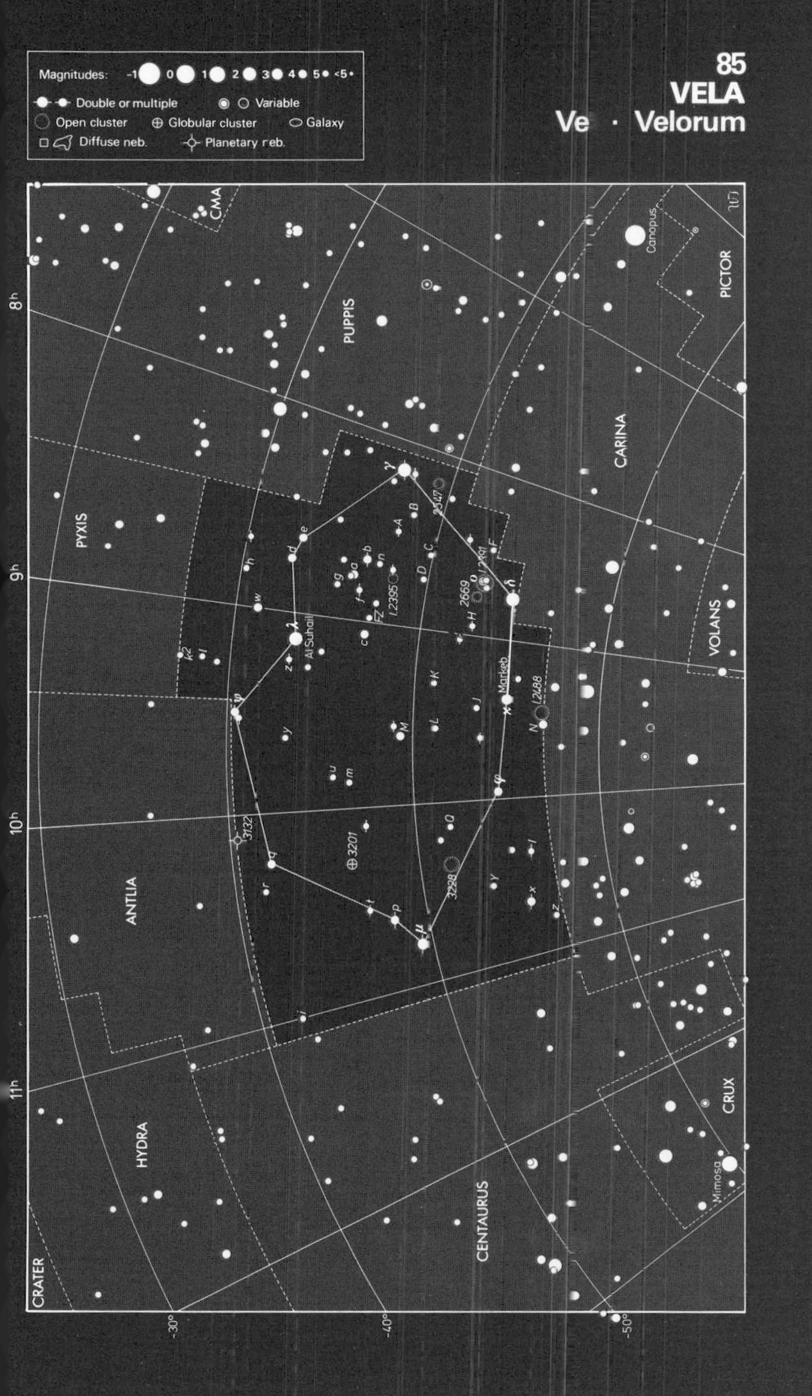
85
VELA
Ve · Velorum
Magnitudes: -1 0 1 2 3 4 5 <5
Double or multiple
Variable
Open cluster
Globular cluster
Galaxy
Diffuse neb.
Planetary neb.
CMA
PUPPIS
Canopus
PICTOR
CARINA
PYXIS
VOLANS
Al Suhail
Markeb
ANTLIA
HYDRA
CENTAURUS
CRUX
Mimosa
CRATER
8h
9h
10h
11h
-30°
-40°
-50°
2547
2669
3132
3201
3228
I.2391
I.2395
I.2488

VIRGO The Virgin

The second-largest constellation. Virgo is usually identified as the goddess of justice, the scales of justice being represented by neighbouring Libra. But another legend sees her as Demeter, the corn goddess, and in the sky she is pictured holding an ear of wheat (the star Spica). The Sun passes through the constellation from mid-September to early November, and thus is within Virgo's boundaries at the time of the autumnal equinox around September 23 each year, when the Sun moves south of the celestial equator. Virgo contains the nearest major cluster of galaxies to us, which spills over into neighbouring Coma Berenices; the area is sometimes known as 'the realm of the galaxies'. The Virgo Cluster lies about 45 million l.y. away and contains about 3000 members, several dozen of which are visible in apertures of 150 mm or so, although they appear as little more than hazy patches of light. Some of the brightest members of the cluster are mentioned below. Virgo contains the brightest quasar, 3C 273, located at 12h 29.1m, +2° 03′. It is unrelated to the Virgo galaxy cluster. Optically, 3C 273 appears as a 13th-mag. blue star. It is estimated to lie about 3000 million l.y. away.

α (alpha) Virginis, 13h 25m –11°, (Spica, 'ear of wheat'), mag. 1.0, is a blue-white star 280 l.y. away. It is a spectroscopic binary, tidally distorted by its companion so that it varies slightly as it rotates, although by less than 0.1 mag.

β (beta) Vir, 11h 51m +2°, mag. 3.6, is a yellow star 33 l.y. away.

γ (gamma) Vir, 12h 42m –1°, (Porrima), 36 l.y. away, is a celebrated double star. Together, the stars shine as mag. 2.8. But a small telescope reveals γ Vir to consist of a matching pair of yellow-white stars, each of mag. 3.5. They orbit each other in 169 years and are now closing together. By the year 2000 they will need a 100-mm telescope to split them, and at their closest around the year 2005 they will require 250 mm. They then rapidly move apart again, becoming visible in 100 mm by 2010, and in 60 mm after 2012, remaining divisible in small telescopes for the rest of the 21st century.

δ (delta) Vir, 12h 56m +3°, mag. 3.4, is a red giant 260 l.y. away.

ε (epsilon) Vir, 13h 02m +11°, (Vindemiatrix, 'grape gatherer'), mag. 2.8, is a yellow giant about 100 l.y. away.

ϑ (theta) Vir, 13h 10m –6°, 170 l.y. away, is a double star visible in a small telescope, consisting of blue-white components of mags. 4.4 and 8.0.

τ (tau) Vir, 14h 02m +2°, mag. 4.3, a white star 130 l.y. away, forms a wide optical double for small telescopes with a mag. 9.6 companion.

φ (phi) Vir, 14h 28m –2°, 75 l.y. away, is a yellow giant of mag. 4.8 with a mag. 9.3 companion, difficult to see in the smallest telescopes because of the magnitude contrast. ▶

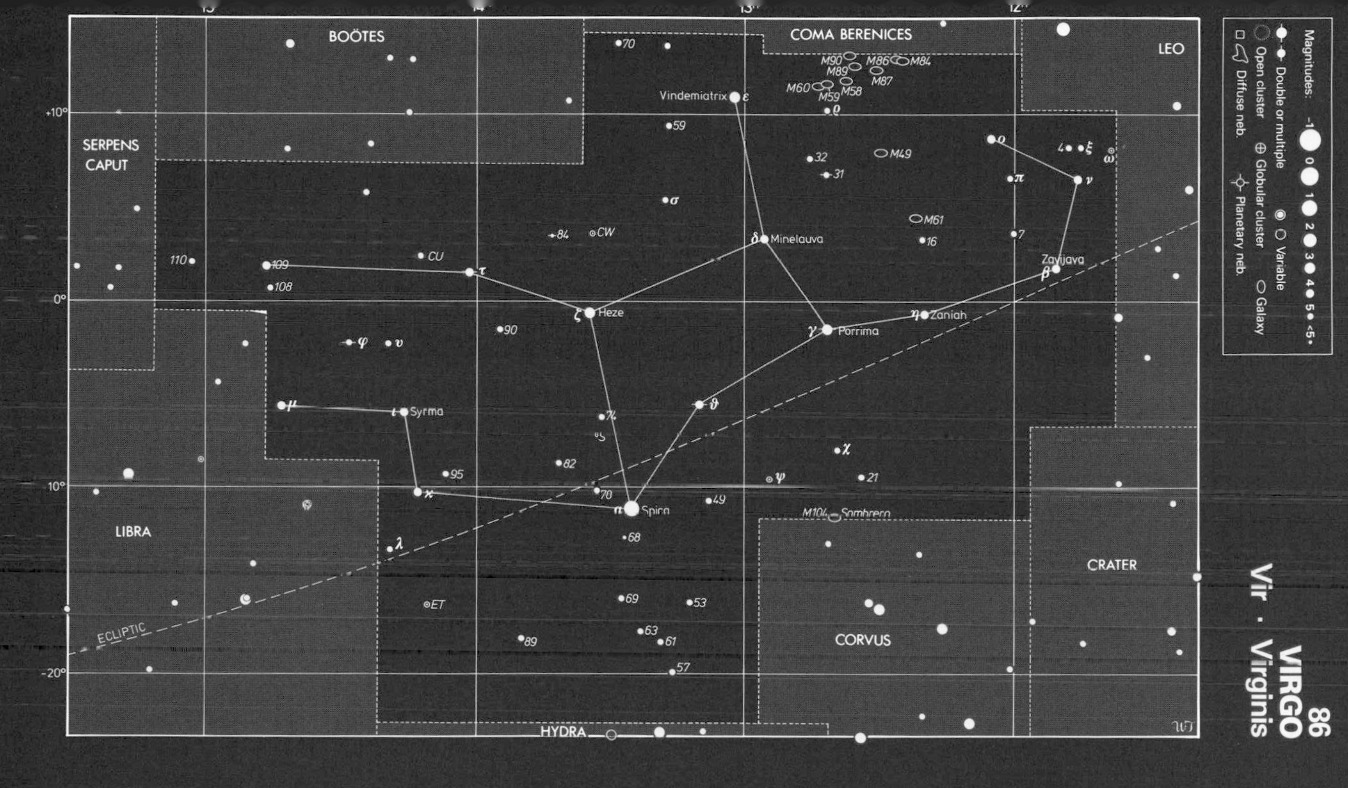
86
VIRGO
Vir · Virginis
Magnitudes: -1 0 1 2 3 4 5 <5
Double or multiple
Variable
Open cluster
Globular cluster
Galaxy
Diffuse neb.
Planetary neb.
LEO
COMA BERENICES
BOÖTES
SERPENS CAPUT
CRATER
CORVUS
HYDRA
LIBRA
ECLIPTIC
Zavijava
Zaniah
Porrima
Minelauva
Vindemiatrix
Heze
Spica
Syrma
Sombrero
M104
M61
M49
M84
M86
M87
M89
M90
M58
M59
M60
CW
CU
ET
+10°
0°
-10°
-20°

VOLANS The Flying Fish

This figure was invented at the end of the 16th century by the Dutch navigators Pieter Dirkszoon Keyser and Frederick de Houtman under the name Piscis Volans. None of its stars is particularly bright, but there are two fine double stars for small telescopes.

α (alpha) Volantis, 9h 02m –66°, mag. 4.0, is a white star 75 l.y. away.

β (beta) Vol, 8h 26m –66°, mag. 3.8, is an orange giant about 100 l.y. away.

γ (gamma) Vol, 7h 09m –70°, 200 l.y. away, is a pair of gold and cream stars of mags. 3.8 and 5.7, beautiful in a small telescope.

δ (delta) Vol, 7h 17m –68°, mag. 4.0, is a yellow giant 330 l.y. away.

ε (epsilon) Vol, 8h 08m –69°, 550 l.y. away, is a mag. 4.4 blue-white star with a mag. 8.0 companion visible in a small telescope.

◀ M49 (NGC 4472), 12h 30m +8°, is an 8th-mag. elliptical galaxy visible as a rounded glow in a 75 mm telescope under low power. It is one of the largest and brightest members of the Virgo cluster of galaxies.

M58 (NGC 4579), 12h 38m +12°, is a 10th-mag. spiral galaxy with a noticeably brighter core.

M59 (NGC 4621), 12h 42m +12°, is a 10th-mag. elliptical galaxy lying about a quarter of the way from M60 to M58.

M60 (NGC 4649), 12h 44m +12°, is a 9th-mag. elliptical galaxy, one of the most prominent members of the Virgo Cluster, detectable in 75 mm aperture.

M84 (NGC 4374), 12h 25m +13°, and M86 (NGC 4406), 12h 26m +13°, are a pair of 9th-mag. elliptical galaxies appearing in the same telescopic field as fuzzy patches with noticeably brighter cores. M86 is slightly the larger of the two and noticeably elongated, whereas M84 appears round.

M87 (NGC 4486), 12h 31m +12°, is a celebrated giant elliptical galaxy. It is both a radio source, known as Virgo A, and an X-ray source. Photographs taken through large telescopes show a jet of matter emerging from M87 as though ejected in an explosion (see page 120). In amateur telescopes, M87 appears as a rounded, 9th-mag. glow with a noticeable nucleus.

M90 (NGC 4569), 12h 37m +13°, is a large 9th-mag. spiral galaxy, tilted at an angle to us so that it appears elongated.

M104 (NGC 4594), 12h 40m –12°, is an 8th-mag. spiral galaxy seen edge-on so that it appears elongated. It is popularly known as the Sombrero Hat because of its distinctive appearance on long-exposure photographs (see page 124). With its bulging nucleus ringed by tightly coiled spiral arms, its appearance is reminiscent of Saturn. Apertures above 150 mm reveal a dark lane of dust along its rim, silhouetted against the brighter arms and nucleus. The Sombrero is not in the Virgo Cluster but lies somewhat closer, about 35 million l.y. away.

87 VOLANS
Vol · Volantis

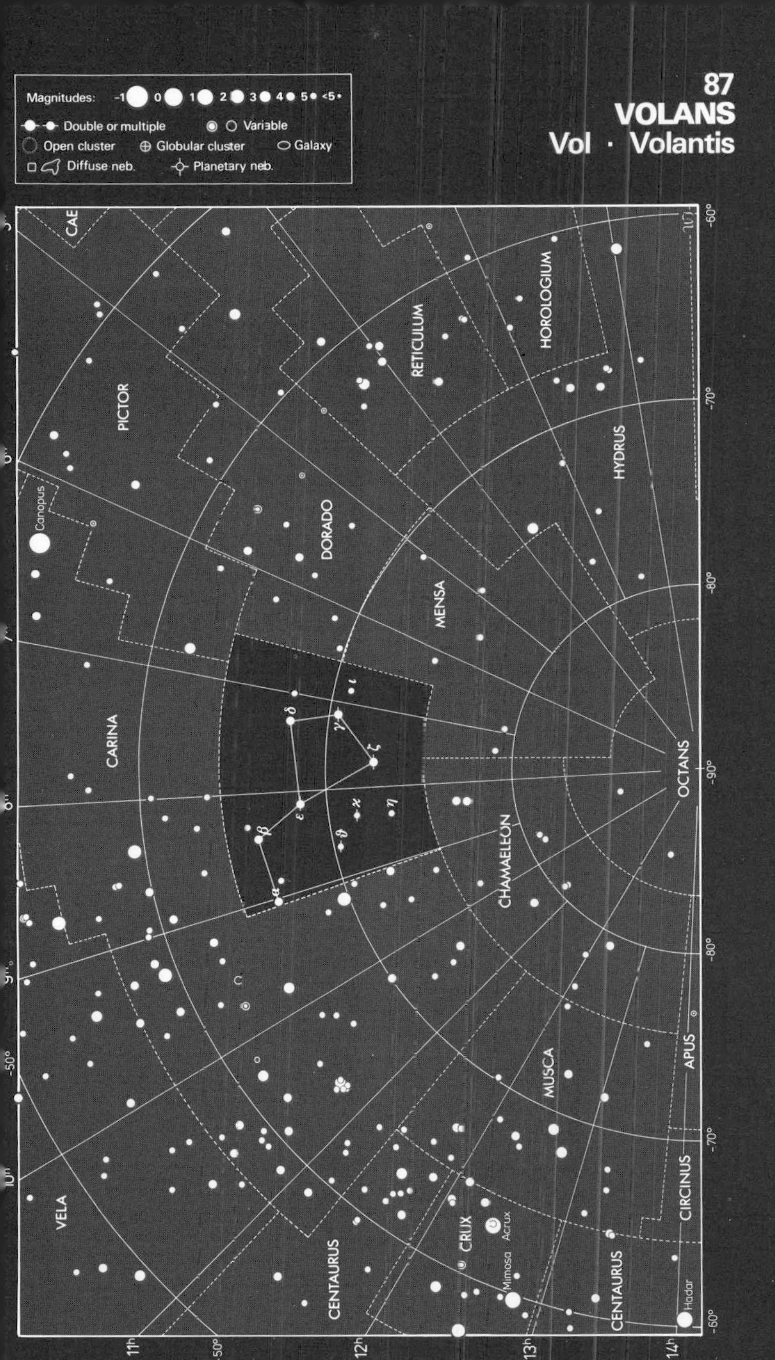

VULPECULA The Fox

A faint constellation at the head of Cygnus. It originated in 1687 with the Polish astronomer Johannes Hevelius, who called it Vulpecula cum Anser, the Fox and Goose. Since then the goose has fled, leaving the fox. In 1967 this diminutive constellation was the site of an astounding discovery – the first pulsar, or flashing radio source, which was detected by radio astronomers in Cambridge, England. On Vulpecula's border with Sagitta is a notable little group of 6th- and 7th-magnitude stars, the brightest of which is 4 Vulpeculae, known popularly as the Coathanger but more formally as Collinder 399 or Brocchi's Cluster, striking in binoculars. The Coathanger's most remarkable feature is an almost straight line of six stars; a curve of stars, forming the Coathanger's hook, extends from the centre of this line.

α (alpha) Vulpeculae, 19h 29m +25°, mag. 4.4, is a red giant about 250 l.y. away. An unrelated mag. 5.8 orange giant, 8 Vul, is seen nearby in binoculars.

M27 (NGC 6853), 20h 00m +23°, the Dumbbell Nebula, is a large and bright planetary nebula, reputedly the most conspicuous of its kind, visible in binoculars but better seen through a telescope. M27 is of 8th mag. and covers an area one-quarter the diameter of the full Moon. Visually it appears as a dumbbell-shaped misty green glow. M27 is 1000 l.y. away. (For photographs see pages 214 and 266.)

NGC 2442, an 11th-mag. barred spiral galaxy in Volans. Anglo-Australian Telescope Board.

VULPECULA
Vul · Vulpeculae

Magnitudes: −1 0 1 2 3 4 5 <5

Double or multiple — Variable

Open cluster — Globular cluster — Galaxy

Diffuse neb. — Planetary neb.

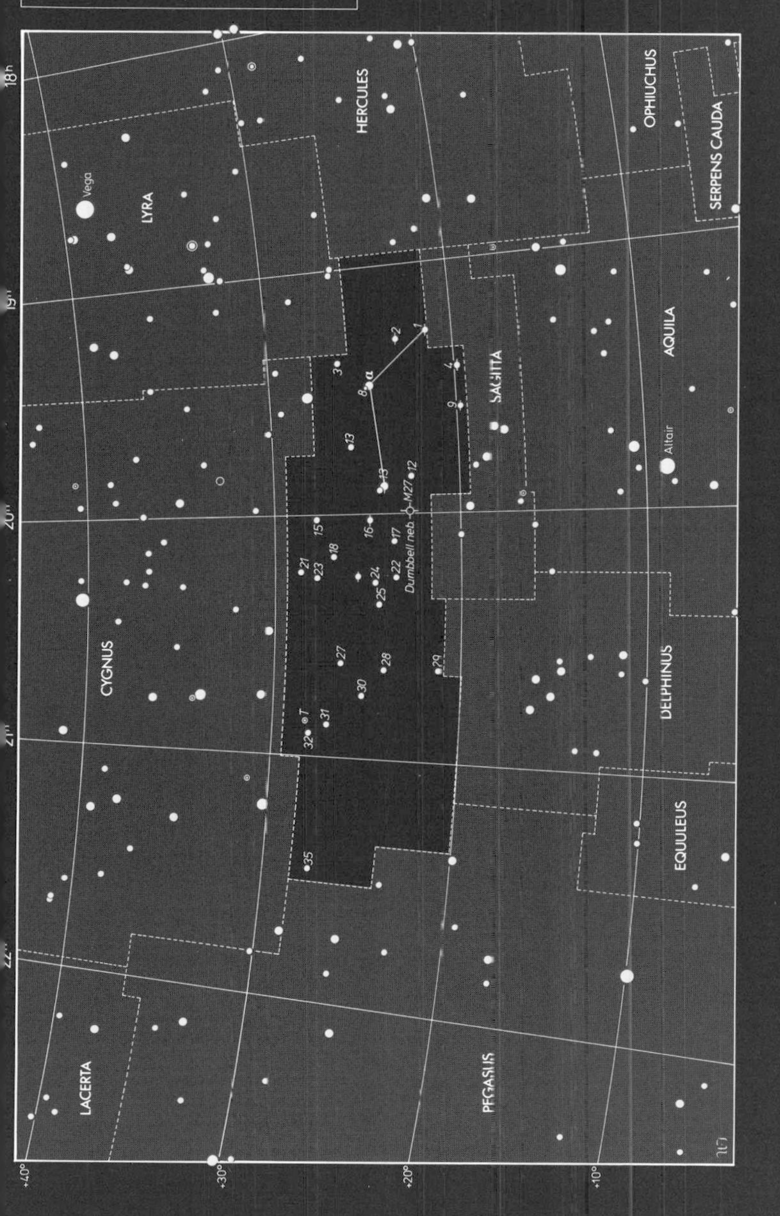

The Pleiades (popularly known as the Seven Sisters, representing the daughters of Atlas in Greek mythology) is a famous open star cluster in the constellation Taurus, embedded in faint nebulosity. *(See page 234.)* Colin Hunt.

SECTION II

Stars

Stars are balls of gas made incandescent by energy from nuclear reactions deep in their interiors. They come in a wide range of sizes and brightnesses, from faint dwarfs a hundredth of the Sun's diameter to dazzling supergiants hundreds of times the size of the Sun. They range in temperature from intensely hot blue-white stars (more than 20,000°C) to cool red stars (3000°C). The Sun, which is a medium-temperature yellow star, turns out to be pretty average in all respects.

Stars are born from massive clouds of gas and dust within our Galaxy. An interstellar gas cloud is termed a *nebula* (plural: nebulae), from the Latin meaning 'cloud'. A nebula is not uniformly distributed in space, but contains denser knots – the seeds of future stars. If the knot is dense enough it begins to contract under the inward pull of its own gravity. As it gets smaller and denser it heats up, until conditions of temperature and pressure at the centre of the shrinking blob become so extreme that nuclear reactions begin. The gas blob has switched on to become a true star, generating its own heat and light for millions of years.

Several star-spawning clouds are well within reach of observation by amateurs. Most famous is the Orion Nebula, marking the sword in the constellation of Orion the Hunter. This nebula is visible as a hazy green glow to the naked eye; binoculars show it more clearly. At the centre of the Orion Nebula is a star called ϑ^1 (theta^1) Orionis, which small telescopes show consists of four component stars. Energy emitted by the brightest of these four stars makes the nebula shine. But behind the bright, visible part of the cloud is an even larger, still-dark area where stars are being born at this moment. The Orion Nebula is estimated to contain enough matter to produce 10,000 stars: it is a star cluster in the making. Another famous stellar birthplace is the Tarantula Nebula in the southern constellation Dorado, which dwarfs the Orion Nebula and is in fact the largest nebula known.

One celebrated young cluster of stars is the Pleiades in the constellation Taurus, the Bull. Long-exposure photographs show that the stars of the Pleiades are still surrounded by wisps of the nebula from which they formed. At least five members of the Pleiades can be distinguished by normal eyesight; binoculars and small telescopes bring dozens more members into view. The whole cluster is estimated to contain about a hundred stars. The brightest and youngest of these formed no more than 2 million years ago, making them extremely youthful by astronomical standards.

The Pleiades is an example of a type of cluster referred to as an *open cluster* or *galactic cluster*. About a thousand are known to astronomers,

Left: *The Jewel Box, NGC 4755, also known as the* κ *(kappa) Crucis cluster, a glittering star cluster in the southern constellation Crux, the Southern Cross. Most of its stars are blue-white, but at its centre is a red giant.* *(See page 128.)* Anglo-Australian Telescope Board.

Below: *The spidery shape of the Tarantula Nebula, NGC 2070, glows brightly in the constellation Dorado. The nebula is part of the Large Magellanic Cloud, as are the other nebulae near it in this photograph.* *(See page 136.)* Royal Observatory Edinburgh.

A true-colour photograph of the Orion Nebula, M42, whose extensive swirls of ghostly glowing gas make it one of the most celebrated objects in the entire heavens. *(See pages 192–3.)* Anglo-Australian Telescope Board.

and many of them are listed in this book. Near the Pleiades in Taurus is a larger and older open cluster, the Hyades, which is estimated to be about 500 million years old. Being older than the Pleiades, the stars have had more time to drift apart. Eventually, most open clusters disperse completely. The Sun was probably a member of such a cluster when it was born 4600 million years ago. A different type of cluster is a globular cluster, described on page 278.

Nebulae are made of a 10:1 mixture of hydrogen and helium, the primary constituents of the Universe, so, naturally enough, stars have the same composition. Stars get their energy from nuclear reactions which transform hydrogen into helium. In the reaction, four hydrogen atoms are crushed together to make one atom of helium; an uncontrolled version of the same reaction occurs in a hydrogen bomb.

There are certain limits on the size of a star. A gas blob with less than 6 per cent of the Sun's mass cannot become a star, because conditions in its interior will not become sufficiently extreme for nuclear reactions to begin. This 6 per cent limit may be considered the dividing line between a planet and a star. If the gaseous planet Jupiter in our Solar System had been about 60 times more massive than it actually is, it would have become a small star. At the other end of the scale, the largest stars have masses of about a hundred times that of the Sun. It was once thought that stars more massive than this would produce so much energy that they would literally disintegrate, but this may not be true in all cases. A few stars are known that seem to have masses greater than a hundred Suns, one example being η (eta) Carinae.

A star's most vital statistic is its mass, for this factor affects everything else about it: its temperature, its brightness and its lifetime. The stars with the least mass are, not surprisingly, the coolest; they are known as *red dwarfs*. A typical red dwarf such as Barnard's Star, the second-closest star to the Sun, has a mass about a tenth that of the Sun and glows a dull red with a surface temperature of about 3000°C. Even though Barnard's Star is only six light years away, it is too faint to be seen with the naked eye. Surprisingly enough, stars with the lowest mass live the longest. Their nuclear fires burn so slowly that they can survive for as much as a million million years, a hundred times as long as the Sun. The Sun itself, which by definition is of one solar mass, has a surface temperature of 5500°C, and is expected to live for about 10,000 million years. It is currently in the prime of its life.

Moving up the scale, a star such as Sirius, which is twice the Sun's mass, can live for only about 1000 million years, a tenth of the Sun's age. The surface temperature of Sirius is a blue-white 11,000°C. Larger and hotter still, the star Spica in the constellation Virgo has a mass of about 11 Suns and a surface temperature of around 24,000°C. The lifetime of this intensely hot, highly luminous star is less than 1 per cent of the lifetime of the Sun.

A star's colour is a direct indicator of its temperature. The most precise way to measure a star's temperature is to study the spectrum of its light, which is done by splitting the light up in a device called a spectroscope. Stars are classified into a sequence of so-called *spectral types* according to their temperature. The bluest and hottest stars are classified as spectral types O and B (Spica is a B-type star). Then come the cooler blue-white A-type stars, which include Sirius, and then F-type stars, which appear white; Procyon is an F-type star. G-type stars appear yellow-white; they include the Sun, α (alpha) Centauri and τ (tau) Ceti. Cooler still are K-type stars, such as ε (epsilon) Eridani, which appear orange. Coolest of all are the red stars with an M-type spectrum, of which Barnard's Star is an example. Each spectral type is subdivided into ten steps from 0 to 9; on this more precise scale, the Sun ranks as a G2 star.

The seemingly haphazard lettering sequence for the spectral types is the result of a previous classification scheme which was rearranged and

Stellar Spectral Types

Type	*Colour*	*Temperature range (°C)*	*Examples*
O	Blue	40,000–25,000	ζ Puppis (supergiant)
B	Blue	25,000–11,000	Spica (main sequence) Regulus (main sequence) Rigel (supergiant)
A	Blue-white	11,000–7500	Vega (main sequence) Sirius (main sequence) Deneb (supergiant)
F	White	7500–6000	Canopus (supergiant) Procyon (subgiant) Polaris (supergiant)
G	Yellow-white	6000–5000	Sun (main sequence) α Centauri (main sequence) τ Ceti (main sequence) Capella (giant)
K	Orange	5000–3500	ε Eridani (main sequence) Arcturus (giant) Aldebaran (giant)
M	Red	3500–3000	Barnard's Star (main sequence) Antares (supergiant) Betelgeuse (supergiant)

The largest and most beautiful planetary nebula, the Helix Nebula, NGC 7293, in Aquarius, is a shell of gas thrown off by a dying star. The red colour of its outer part is caused by nitrogen and hydrogen, the blue-green of the central part by oxygen. *(See page 76.)* Anglo-Australian Telescope Board.

shortened to produce the present system. The sequence of stellar spectral types is remembered by the mnemonic: 'Oh Be A Fine Girl, Kiss Me'.

When the spectral type of stars is plotted against their actual luminosity (absolute magnitude), all stars that are in stable, hydrogen-burning middle age lie in a well-defined band across the graph known as the *main sequence*. A star's position along the main sequence is fixed by its mass, with the least massive stars at the bottom and the most

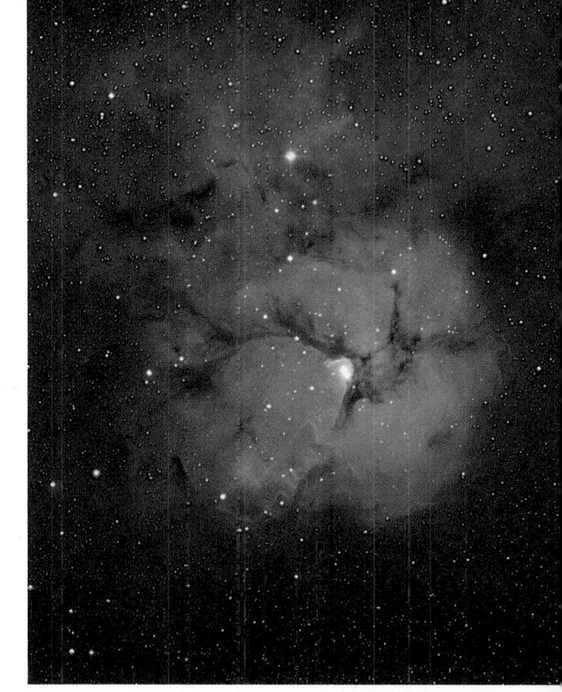

The Trifid Nebula, M20, in Sagittarius, is a cloud of gas and dust in which stars are forming. In one part of the nebula the light from newborn stars makes hydrogen gas glow pinkish red. Another part of the nebula appears blue, caused by starlight reflected from dust grains. Dark lanes of dust trisect the nebula, giving rise to its popular name. *(See page 219.)*
Anglo-Australian Telescope Board.

M16 in Serpens is a cluster of stars embedded in a faint nebula called the Eagle. The cluster is easily seen in amateur instruments, but the nebula shows up well only on long-exposure photographs such as this. *(See page 228.)*
Anglo-Australian Telescope Board.

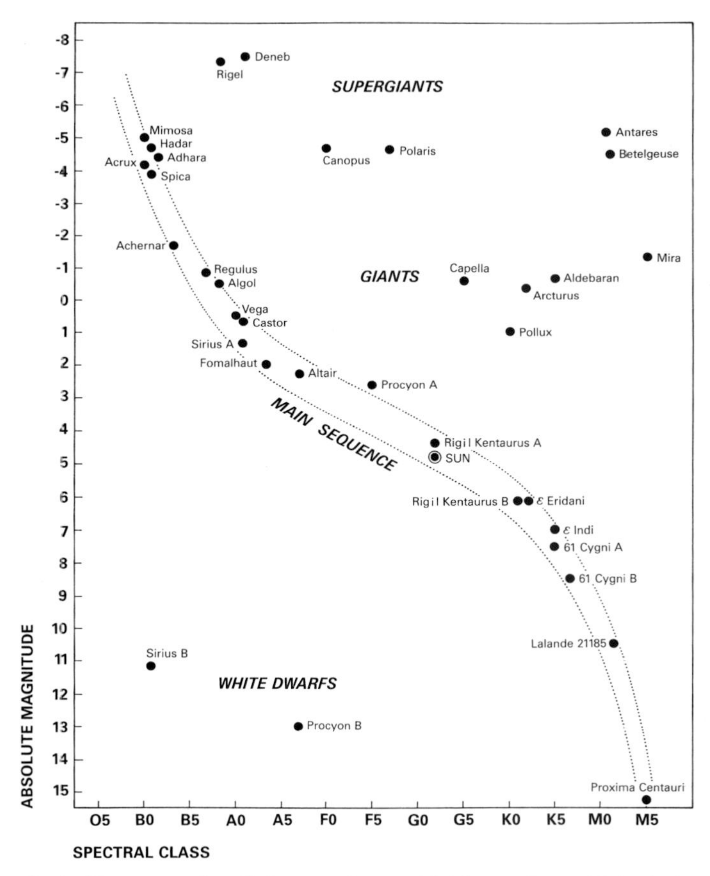

In the Hertzsprung–Russell diagram, the actual brightness of stars (their absolute magnitude) is plotted against their temperature (spectral class). Stars of different types fall in different areas of the Hertzsprung–Russell diagram. The band running from top left to bottom right is the main sequence, on which the Sun falls. Giants and supergiants lie above the main sequence; white dwarfs lie below and to the left of it. Wil Tirion.

massive stars at the top. The Sun, as befits its middle-of-the-road nature, lies about halfway along the main sequence. Such a plot of star brightness against spectral type is known as a Hertzsprung–Russell diagram, after the Danish astronomer Ejnar Hertzsprung and the American Henry Norris Russell who devised it in 1911–13.

Although most stars lie on the main sequence, a number of particularly bright stars lie above and to the right of it, while a few faint stars lie below and to the left of it. These stars are all in late stages of evolution. We can best understand what is happening to them by following the future evolution predicted for the Sun.

As we have seen, the Sun formed about 4600 million years ago and is about halfway through its expected lifespan. In a few thousand million years, though, it will start to run out of hydrogen at its core. In search of more hydrogen to use as fuel, the nuclear reactions inside the Sun will start to move outwards, releasing more energy. Eventually, when surrounded by a shell of burning hydrogen, even the helium atoms in the Sun's core will enter into nuclear reactions of their own, fusing together to form carbon atoms. With all this extra energy being given off, the Sun will become much brighter than it is today and will start to swell alarmingly in size. But as the Sun's outer layers expand they will also cool, becoming redder in colour, and the Sun will turn into a *red giant*, similar to the bright stars Aldebaran and Arcturus. At its largest, the red giant Sun will grow to at least a hundred times its present diameter, engulfing Mercury, Venus and perhaps even the Earth within its distended outer layers. Needless to say, all life on our planet will long since have become extinct.

On the Hertzsprung–Russell diagram, the Sun's increase in brightness will move it upwards off the main sequence, and its change in spectral type will in addition move it to the right. Stars at the top end of the main sequence become so big and bright at this stage of their evolution that they are referred to not as mere giants but as *supergiants*. Prominent examples of red supergiants are Betelgeuse and Antares, both of which are hundreds of times larger than the Sun. Other stars which have not yet evolved enough to become red in colour but are nevertheless firmly in the supergiant bracket are Rigel, Deneb and Polaris.

Stellar Luminosity Classes

Ia	Bright supergiant
Iab	Less bright supergiant
Ib	Supergiant
II	Bright giant
III	Giant
IV	Subgiant
V	Main sequence
VI	Subdwarf
VII	White dwarf

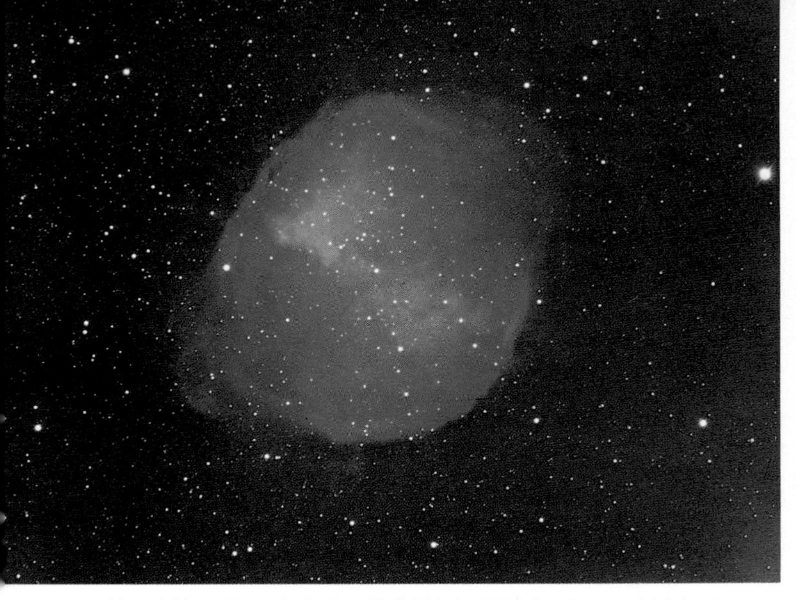

Shaped like a figure 8, the Dumbbell Nebula, M27, is a large and bright planetary nebula in Vulpecula. Photograph by Laird Thompson with the 3.6-metre Canada–France–Hawaii Telescope. (See page 254.) University of Hawaii.

M57, the Ring Nebula in Lyra, photographed by Laird Thompson with the Canada–France–Hawaii Telescope. The red colour is produced by hydrogen; the yellow and blue come from oxygen. (See page 176.) University of Hawaii.

The Horsehead Nebula, a cloud of dark dust looking in silhouette like an interstellar chess piece against the redness of glowing hydrogen gas. At left is the overexposed image of ζ (zeta) Orionis. Royal Observatory Edinburgh.

The Crab Nebula, M1, the remains of a star that exploded as a supernova. The blue colour comes from electrons spiralling in the Crab's intense magnetic field. Canada–France–Hawaii Telescope photograph by Laird Thompson. University of Hawaii.

To distinguish whether a star is, say, a giant or supergiant, or lies on the main sequence, astronomers assign stars a luminosity class (see the table on page 265) in addition to the spectral type. (Astronomers tend to think of stars as either dwarfs or giants, depending on whether they are on the main sequence or have evolved off it. Main sequence stars are frequently referred to as dwarfs, even though the most massive of them may be several times larger than the Sun.) Taken together, the spectral type and luminosity class define the main properties of a star as it exists at present. But those properties change as the star ages.

Stars spend only a few per cent of their total lifetime in the red giant phase, which in the case of stars like the Sun amounts to no more than a few hundred million years. A red giant is a star that has grown old and is about to die. Once a red giant has swollen to its maximum size, its distended outer layers drift off into space, forming a stellar smoke ring known rather confusingly as a *planetary nebula*, even though it has nothing to do with planets. The name was first used in 1785 by William Herschel who said that they looked like the small, rounded disks of planets as seen through his telescope. Probably the best-known of all planetary nebulae is the Ring Nebula in Lyra, though it is not the easiest to see. Much larger is the Dumbbell Nebula in Vulpecula, which can be picked up in binoculars on a clear, dark night. Two small but bright planetary nebulae for amateur telescopes are NGC 6826 in Cygnus and NGC 7662 in Andromeda.

At the centre of a planetary nebula, the core of the former red giant is exposed as a small, intensely hot star. Once the surrounding gases of the planetary nebula have dispersed, usually after thousands of years, the central star remains as a so-called *white dwarf*. A white dwarf is only about the diameter of the Earth, but contains most of the matter of the original star; only about 10 per cent of the star's mass is lost in the planetary nebula stage. White dwarfs are therefore exceptionally dense bodies. A teaspoonful of white dwarf material would have a mass of thousands of kilograms. Over thousands of millions of years, white dwarfs slowly cool off and fade into oblivion.

Being so small, white dwarfs are very faint. Not one is visible to the naked eye. Both the nearby bright stars Sirius and Procyon have white dwarf companions, but Procyon's companion is too close to its parent to be distinguishable in amateur telescopes, and the companion of Sirius can be glimpsed only under the most favourable conditions. The easiest white dwarf to see is a companion of the star o^2 (omicron^2) Eridani (also known as 40 Eridani); a small telescope will show it. Of added interest is a fainter third member of this system, a red dwarf, also visible in amateur telescopes.

Our Sun, it seems, is destined to go through the stage of being a planetary nebula before fading away as a white dwarf. But stars with several times the Sun's mass, towards the top end of the main sequence, suffer a far more spectacular end. As we have seen, they first become dazzling supergiants rather than mere giants. They do not get a chance to reach the planetary nebula stage. So massive are they that the nuclear reactions at their centres continue in runaway fashion until the star becomes unstable and explodes. Such an explosion is known as a *supernova*.

In a supernova eruption a star's brightness increases millions of times, so that for a few days the star can rival the brilliance of an entire galaxy. The shattered outer layers of the star are thrown off into space at speeds of around 5000 km per second. In 1054 astronomers on Earth saw a star erupt as a supernova in the constellation Taurus. The star became brighter than Venus and was visible in daylight for three weeks. It finally faded below naked-eye visibility more than a year after it had first appeared.

At the site of that explosion lies one of the most famous objects in the heavens: the Crab Nebula, the shattered remains of the star that erupted as a supernova. The Crab Nebula is visible as a smudgy patch in amateur telescopes, but is best seen on long-exposure photographs taken with large instruments. Over the next 50,000 years or so the gases of the Crab Nebula will disperse into space, forming delicate traceries like those of the Veil Nebula in Cygnus, itself the remains of a former supernova.

The last supernova observed in our Galaxy was in 1604. This star, in the constellation Ophiuchus, reached a maximum magnitude of just over –2, as bright as Jupiter. It was studied by the German astronomer Johannes Kepler, and is often known as Kepler's Star in his honour. Hundreds of supernovae in other galaxies have been seen telescopically since then, but only one has become bright enough to be visible to the naked eye. That was Supernova 1987A, which erupted in the Large Magellanic Cloud, a near neighbour of the Milky Way. It was first spotted on February 24, 1987, by astronomers in the southern hemisphere and rose to a maximum magnitude of 2.9 in late May, eventually fading from naked-eye view by the end of that year. Another supernova in our Galaxy is long overdue. When it comes, it should be a spectacular sight. Many astronomers dream of seeing it outshining the other stars in the night sky, dazzling the eye and casting shadows.

A star might not blow itself entirely to bits in a supernova explosion. Sometimes the central core of the exploded star is left as an object even smaller and denser than a white dwarf, known as a *neutron star*. In a neutron star, the protons and electrons of the star's atoms have been crushed by the tremendous forces of the supernova so that they combine to form the particles known as neutrons. A typical neutron star is a mere 20 km in diameter, but contains as much mass as one or two Suns. Being so minute, neutron stars can spin very rapidly without flying apart. Each time they spin we see a flash of radiation like a

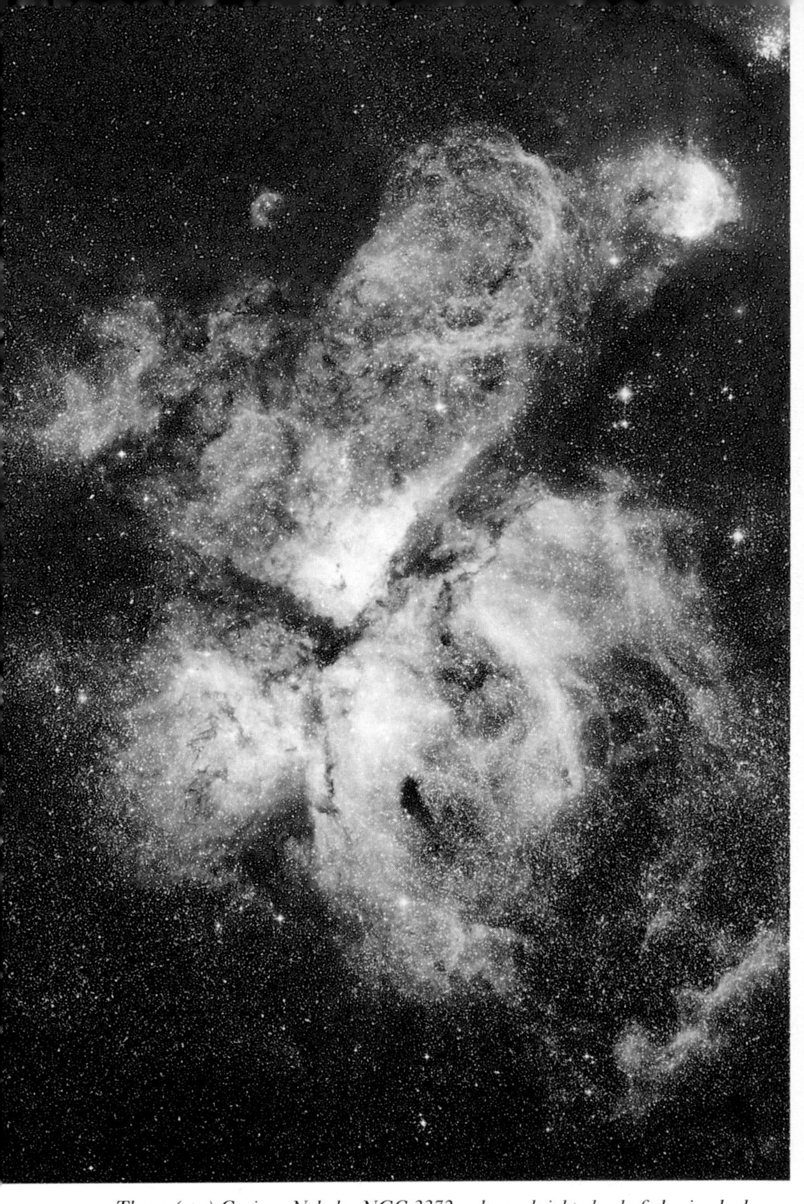

The η (eta) Carinae Nebula, NGC 3372, a large, bright cloud of glowing hydrogen gas containing clusters of hot young stars. The nebula is named after η Carinae itself, a massive star that lies in the brightest central part of the nebula, near a dark notch called the Keyhole. η Carinae could become a supernova within the next 10,000 years. *(See page 102.)* Royal Observatory Edinburgh.

Twisted filaments of gas veining the constellation Vela are the dismembered remains of a star that erupted in a supernova explosion about 10,000 years ago. The core of the star was left behind as a faint pulsar; this flashes 11 times per second. The Vela supernova remnant is part of the much larger Gum Nebula, named after the astronomer Colin S. Gum. Royal Observatory Edinburgh.

lighthouse beam. Astronomers have detected radio pulses from several hundred such sources, which they term *pulsars*; one lies at the centre of the Crab Nebula. The Crab Pulsar flashes 30 times per second; others pulse more slowly, down to once every four seconds. Most neutron stars are too faint to be seen optically, but the pulsar in the Crab Nebula has been seen flashing in step with the radio pulses.

If the core of the exploded star has a mass of more than three Suns, then even a neutron star is not the end for it. Instead, it becomes something still more bizarre: a *black hole*. No force can shore up a dead star weighing more than three solar masses against the inward pull of its own gravity. It continues to shrink, becoming ever smaller and denser until its gravity becomes so great that nothing can escape from it, not even its own light. It has dug its own grave, a black hole. Since a black hole is, by definition, invisible, it is of only academic interest to amateur observers. However, professional astronomers have detected X-ray emissions from space which they believe are being given off by hot gas plunging into the bottomless pits of black holes. The best-known candidate for a black hole is Cygnus X-1; it lies near a visible 9th-magnitude star in the constellation Cygnus.

Double and Multiple Stars

To the naked eye, stars appear as solitary, isolated objects. But the great majority of stars – over 75 per cent – actually have one or more companion stars, too faint or too close to be seen separately with the naked eye but which can be distinguished telescopically. Many attractive double and multiple stars are within the range of binoculars and small telescopes.

There are two sorts of double star. In one type, the two stars are not actually related but happen to lie in the same line of sight by chance; in this arrangement, termed an *optical double*, one star may be hundreds of light years farther from us than the other. Optical doubles are comparatively rare. Most double stars are physically linked to each other by gravity, forming a genuine *binary* system. The two stars in an optical binary orbit their mutual centre of gravity, which can take centuries. In systems of more than two stars, the orbits can become very complicated.

Twins or triplets are most common among stars, but some stellar families can be much larger. One celebrated multiple star is the so-called Double Double, ε (epsilon) Lyrae. Binoculars show it to be a wide double, but modest-sized amateur telescopes reveal that each of these stars is itself double, making a quadruple system. Even more remarkable is Castor, a system of six stars all linked by gravity. Amateur telescopes show that Castor consists of two bright blue-white stars close together, with a fainter third star some way off. Professional observations have revealed that each of these three stars is a *spectroscopic binary*. In a spectroscopic binary the two stars are too close together to be seen individually in any telescope, but analysis of the light from the star reveals the companion's presence. The two brightest stars of Castor take about four centuries to orbit each other, whereas their spectroscopic companions whirl around them in only a few days. Sometimes the members of a binary system eclipse each other as seen from Earth. These eclipsing binaries are dealt with in the next section, on variable stars.

For amateurs, the attraction of observing double stars lies in separating (or 'splitting') close doubles and in comparing their brightnesses and colours. Some pairs have beautiful colour contrasts, such as the yellow and blue-green components of Albireo, β (beta) Cygni. The closer together two stars are, the larger the aperture of telescope needed to split them. If one star is much fainter than the other it will be swamped by the primary star's light, and so will be more difficult to see than a companion that is of similar brightness to the main star. The constellation notes in this book include guides to the likely aperture needed to split various double stars. But there can be no hard-and-fast rules. So much depends on the observer's eyesight, the quality of telescope and the observing conditions. The only way to find out for sure is to look yourself.

Variable Stars

Certain stars vary in brightness, and are known as *variable stars*. Amateur astronomers can make valuable observations of them. An observer estimates the brightness of the variable star by comparing it with nearby stars of known magnitude. The observations are plotted on a graph to form what is known as a *light curve*, which can reveal much about the nature of the star under study.

The usual cause of a star's variation is actual changes in its light output, but in some cases the star is a member of a binary system in which one star periodically eclipses the other. One famous *eclipsing binary* star, the first of its type to be noticed, is Algol in the constellation Perseus. Algol consists of a blue dwarf, from which most of the light comes, orbited by a fainter yellow subgiant. Every 2.87 days the magnitude of Algol drops from 2.1 to 3.4 as the fainter star eclipses the brighter one; the eclipse lasts about 10 hours. Compare it at maximum with α (alpha) Persei, magnitude 1.8, and at minimum with δ (delta) Persei, magnitude 3.0. There is also a secondary minimum when the brighter star eclipses the fainter one, but the drop in light is too small for the eye to notice. Most variable star observers begin by observing

Algol; regular predictions of its eclipses are issued by astronomical societies. Another eclipsing binary of note is β (beta) Lyrae, which varies between magnitudes 3.3 and 4.3 every 12.9 days. Compare it with nearby γ (gamma) Lyrae, constant at magnitude 3.2, and κ (kappa) Lyrae, magnitude 4.3.

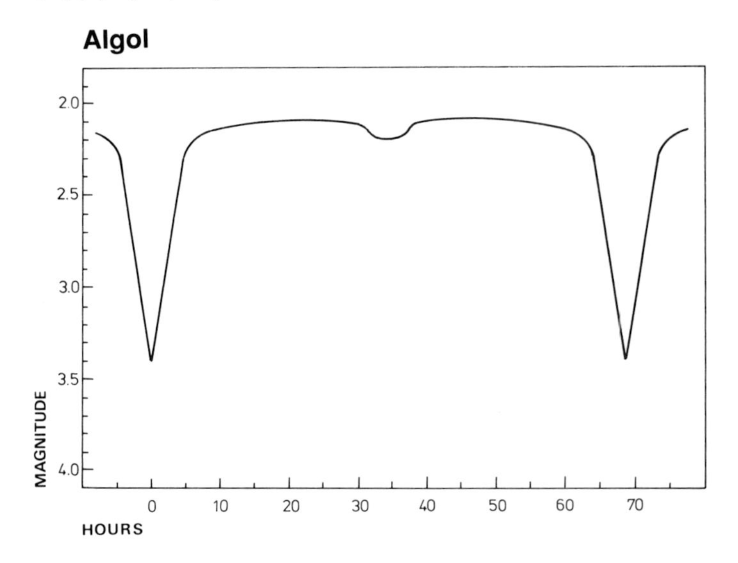

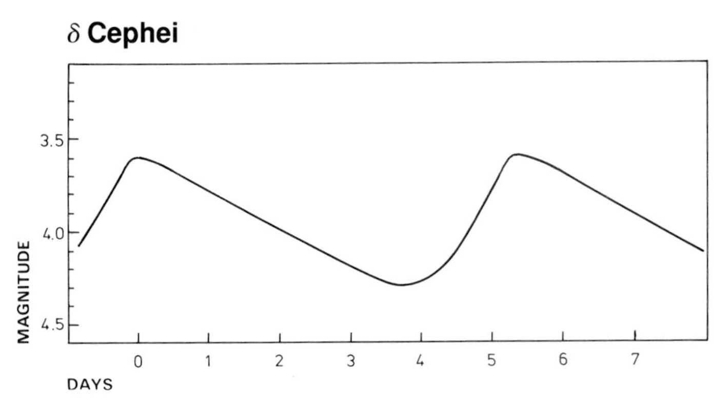

The brightness of Algol in Perseus and δ (delta) Cephei rises and falls in a predictable fashion every few days, as shown by these light curves. Note that the rise of δ Cephei to maximum is quicker than its subsequent decline. The small secondary minimum of Algol is undetectable to the naked eye. Wil Tirion.

Of the stars that vary intrinsically in brightness, most do so because of changes in their size. These are known as *pulsating variables* (not to be confused with pulsars). Of particular importance to astronomers are the so-called *Cepheid variables*, named after their prototype δ (delta) Cephei. Cepheid variables are yellow supergiant stars that go through one cycle of pulsations in periods ranging from about 2 to 40 days, varying in brightness by up to one magnitude as they do so. δ Cephei itself varies between magnitudes 3.5 and 4.4 every 5.4 days, and is thus an easy object for amateur observation. Good comparison stars are ε (epsilon) Cephei, magnitude 4.2, and ζ (zeta) Cephei, magnitude 3.4.

The importance of Cepheid variables is that the length of their light cycle is directly related to their absolute magnitude: the brighter the Cepheid, the longer it takes to complete one cycle of variation. Astronomers therefore have an easy way of finding these stars' absolute magnitude, simply by observing their period of fluctuation; when the absolute magnitude is compared with the apparent magnitude, the star's distance is easily computed. Cepheid variables thus make important distance indicators in astronomy.

Related pulsating stars are the *RR Lyrae variables*. These are old blue stars frequently found in globular clusters, and they vary by about 0.5 to 1.5 magnitudes in less than a day. The prototype, RR Lyrae itself, varies from magnitude 7.1 to 8.1 every 0.57 days. Minor additional types of pulsating variables include β (beta) Cephei stars and δ (delta) Scuti stars, both of which have short periods of only a few hours and ranges of variation too small to be noticeable to the naked eye.

Each type of variable star falls in a specific area on the Hertzsprung–Russell diagram (page 264), and represents stars of different mass at various stages of evolution. Variability of one kind or another seems to be an inevitable consequence of the ageing process of a star.

Red giants and red supergiants are old stars that frequently prove to be variable. They pulsate, but not with anything like the same regularity as the types of stars mentioned above. The most abundant variables known are the *long-period variables*, which have periods ranging from about three months to two years, and amplitudes of several magnitudes. Their prototype is Mira, *o* (omicron) Ceti, in the constellation Cetus, the Whale, a red giant whose period averages 332 days during which it ranges between about magnitudes 3 and 9; the exact period and amplitude differ slightly from cycle to cycle. Another variable of the same type is χ (chi) Cygni.

More erratic are the *semi-regular variables*, typically with periods of about 100 days and amplitudes of 1 to 2 magnitudes. The *irregular variables* have little or no discernible pattern at all to their fluctuations. All these stars are red giants and red supergiants that have reached a stage of instability, oscillating in size and brightness. Sometimes it is difficult to decide which class a star should be placed in. Examples of semi-regular and irregular variables are Antares, Betelgeuse, α (alpha) Herculis and μ (mu) Cephei.

Most spectacular of all variable stars are the *novae*, which suddenly and unexpectedly erupt by perhaps 10 magnitudes or more (10,000 times or more in brightness), sometimes becoming visible to the naked eye where no star appeared before. Their name comes from the Latin meaning 'new', for they were once thought to be genuinely new stars. Now we know that they are merely old, faint stars undergoing a temporary outburst. Despite their name, they are not related to supernovae; those stars explode for different reasons.

According to current theory, novae are close double stars, one member of which is a white dwarf. Gas spilling from the companion star onto the white dwarf is thrown off in an eruption. The star does not disrupt itself in a nova outburst. In fact, some novae have undergone more than one recorded outburst, and perhaps all novae recur, given time.

A nova rises to maximum brightness in a few days. Amateur astronomers are often the first to spot such eruptions and to notify professional observatories. After a few days or weeks at maximum comes a slow decline over several months, as the nova slowly sinks back to its previous obscurity, sometimes punctuated by occasional minor additional outbursts. Following the progress of a nova is one of the most fascinating aspects of variable star observation. Unfortunately, prominent naked-eye novae occur only about once a decade, but many more are visible in binoculars.

Dense Milky Way starfields, blotted by dark nebulosity, towards the centre of our Galaxy in Sagittarius. *(See page 278.)* Hale Observatories photograph.

Milky Way, Galaxies and the Universe

Our Sun and all the stars visible in the night sky are members of a vast aggregation of stars known as the Galaxy (given a capital G to distinguish it from any other galaxy). Our Galaxy is spiral in shape, with arms composed of stars and nebulae that wind outwards from a central bulge of stars. It is about 100,000 light years in diameter; the Sun lies in a spiral arm 30,000 light years from the Galaxy's centre. Astronomers estimate that the Galaxy contains at least 100,000 million stars.

Most of the stars in the Galaxy lie in a disk about 2000 light years thick. Seen from our position within the Galaxy, this disk of stars appears as a faint, hazy band crossing the sky on clear, dark nights. We call the band of stars the Milky Way, and the term Milky Way is often used as a name for our entire Galaxy. The starfields of the Milky Way are particularly dense in the region of Sagittarius, which is the direction of the Galaxy's centre. Note that the plane of the Milky Way is tilted at 63° with respect to the celestial equator. This results from a combination of the tilt of the Earth's axis, and the fact that the plane of the Earth's orbit as a whole is tilted with respect to the plane of the Galaxy.

Around our Galaxy are dotted more than a hundred ball-shaped clusters of stars known as *globular clusters*, several of which are visible with the naked eye or binoculars. Globular clusters contain from a hundred thousand to several million stars, all bound together by gravity.

The brightest globular clusters are ω (omega) Centauri and 47 Tucanae, both in the southern hemisphere; in the northern hemisphere the best example is M13 in Hercules. To the naked eye and in binoculars, these objects appear as softly glowing patches of light. Moderate-sized telescopes start to resolve some of the individual red giant stars, giving the cluster a speckled appearance. Globular clusters are believed to have formed early in the history of the Galaxy. They contain some of the most ancient stars known, 12,000 million or so years old, over twice the age of the Sun.

Our Galaxy has two small companion galaxies called the Magellanic Clouds. To the naked eye they appear like detached portions of the Milky Way, in the southern constellations Dorado and Tucana. The Large Magellanic Cloud contains about 10,000 million stars (a tenth of the number in our Galaxy) and lies 169,000 light years from us. The Small Magellanic Cloud has about a fifth of the number of stars in the Large Magellanic Cloud, and lies somewhat farther away at about 190,000 light years. Both Clouds contain numerous star clusters and bright nebulae, and are rich territories for sweeping with instruments of all sizes. One can but imagine the magnificent view of our Galaxy that any astronomers living in the Magellanic Clouds would have.

Milky Way: star clouds and dust lanes extend northwards from Sagittarius to Scutum and Serpens. The central pink patch is M8 (the Lagoon Nebula). Two smaller patches above it are M17 (the Omega Nebula) and M16 (the Eagle Nebula). David Malin.

Countless other galaxies are dotted like islands in the Universe as far as the largest telescopes can see. Most galaxies are members of clusters containing up to thousands of galaxies. Our own Milky Way is the second-largest member of a small cluster of about 30 galaxies known as the Local Group. The largest galaxy in the Local Group is visible to the naked eye as a fuzzy, elongated patch in the constellation Andromeda. The Andromeda Galaxy is estimated to contain about twice as many stars as are in our own Galaxy, and to be about 25 per cent greater in diameter. It lies 2.4 million light years away. Long-exposure photographs reveal that the Andromeda Galaxy is a spiral, tilted so that we see it almost edge-on. If our own Galaxy were viewed from outside, it would look like this. The Andromeda Galaxy has two small companion galaxies, its equivalent of the Magellanic Clouds, which are visible in amateur telescopes.

The only other member of the Local Group within easy reach of amateur instruments is M33 in Triangulum, another spiral galaxy, farther from us than the Andromeda Galaxy and containing considerably fewer stars. It can be picked up in binoculars under clear, dark skies. The nearest rich cluster of galaxies to the Local Group lies in the constellation of Virgo, part of it spilling over into neighbouring Coma Berenices. Of its 3000 or so known members, dozens are within the reach of amateur telescopes.

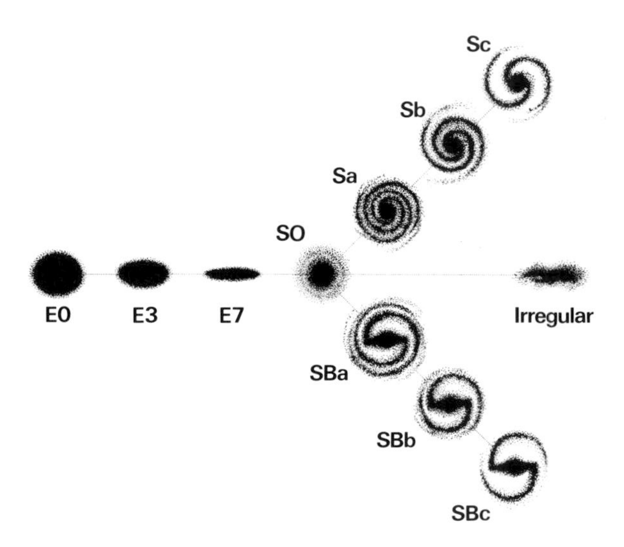

Galaxies are classified according to shape: elliptical, type E; spiral, type S; barred spiral, type SB; and irregular. Wil Tirion.

Astronomers classify galaxies into three main types: elliptical, spiral and barred spiral. *Elliptical galaxies* range in shape from virtually spherical, known as E0, to flattened lens shapes, known as E7. They include both the largest and the smallest galaxies in the Universe. Supergiant ellipticals composed of up to 10 million million stars are the most luminous galaxies known. An example is M87 in the Virgo Cluster. At the other end of the scale, dwarf ellipticals resemble large globular clusters. Dwarf ellipticals may be the most abundant type of galaxy in the Universe, but their faintness makes them difficult to see.

Spiral galaxies (type S), such as our own Milky Way and the Andromeda Galaxy, have arms winding out from a central bulge. There are usually two arms, but sometimes more. In *barred spirals* (type SB), the arms emerge from the ends of a bar of stars that runs across the galaxy's centre. Spiral and barred spiral galaxies are subdivided according to how tightly their arms are wound: types Sa and SBa have the tightest-wound arms, while types Sc and SBc have the loosest-wound arms. Until recently, our own Galaxy was thought to fall midway between Sb and Sc, but there is growing evidence that it might instead be a barred spiral; M31 in Andromeda is type Sb.

Most of the galaxies observed in the Universe are large, bright spirals but, as mentioned above, they may actually be outnumbered by dwarf galaxies that are too faint to see over great distances. In addition to these three main types, there are certain galaxies classified as irregular; the Magellanic Clouds fall into this class, although there is some semblance of spiral structure in the Large Magellanic Cloud.

Being faint, fuzzy objects, galaxies are best seen where skies are clear and dark, away from city haze and stray light. Low magnification is best, to increase the contrast of the galaxy against the sky background. Through a telescope you should see the nucleus of a galaxy as a star-like point surrounded by the misty halo of the rest of the galaxy. Do not expect to see spiral arms, as on a long-exposure photograph. Galaxies present themselves to us at all angles, so even spirals can seem to be elliptical in shape when viewed edge-on.

Peculiar things are going on within some galaxies. For example, certain galaxies give off vast amounts of energy as radio waves; these are called *radio galaxies*, and include the supergiant ellipticals M87 in Virgo and NGC 5128 in Centaurus. Some spiral galaxies have unusually bright nuclei. These are termed *Seyfert galaxies*, after the astronomer Carl Seyfert, who was the first to draw attention to them, in 1943. M77 in Cetus is the brightest Seyfert galaxy.

Most peculiar of all are the objects known as *quasars*, which emit as much energy as hundreds of normal galaxies from an area less than a light year in diameter. Despite their exceptional nature, quasars are not at all exciting visually, which is why they were overlooked until 1963. The brightest quasar, 3C 273, appears as an unimposing

The beautiful Whirlpool Galaxy, M51, in Canes Venatici. At the end of an arm lies its irregular companion galaxy, NGC 5195. Photographed with the 3.6-metre Canada–France–Hawaii Telescope by Laird Thompson. University of Hawaii.

A dark lane of dust rings the star-packed core of the giant elliptical galaxy NGC 5128 in the constellation Centaurus. This odd galaxy is also known as the radio source Centaurus A. *(See page 107.)* Anglo-Australian Telescope Board.

13th-magnitude star in Virgo. Most astronomers now believe that quasars are actually the active centres of galaxies so far off in the Universe that their outer regions are too faint to be seen. Radio galaxies, Seyferts and quasars are probably all related in some way; perhaps they are all young galaxies seen at different stages of their early evolution. One favourite theory is that their central powerhouse is a massive black hole that gobbles up stars and gas from the surrounding galaxy.

In 1929 the American astronomer Edwin Hubble discovered that the galaxies are moving apart from one another as though the Universe were expanding, like a balloon being inflated. (But clusters of galaxies such as the Local Group are not expanding – they are held together by their mutual gravitational attraction.) Hubble's discovery that the Universe is expanding came from a study of the spectrum of each galaxy's light. This revealed that the light from the galaxies was being lengthened in wavelength as a result of high-speed recession (which is called the Doppler effect). Such a lengthening of wavelength is called a *red shift*, because the light from the galaxy is moved towards the red (longer-wavelength) end of the spectrum. Incidentally, the red shift does not make galaxies actually look redder, because the blue end of the spectrum is filled in by light that was formerly at ultraviolet wavelengths.

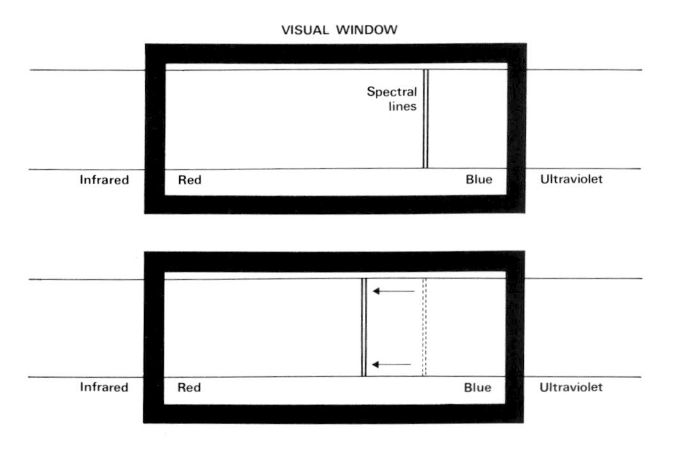

The red shift in the light from a distant galaxy is measured by the change in position of lines in its spectrum. The change in position of the lines is here shown with reference to the visual window, the range of wavelengths to which the human eye is sensitive. All the light from the galaxy is shifted by the same amount, so while some of the galaxy's light moves out of sight beyond the red end of the visual window, other light moves into the visible region from the ultraviolet. Wil Tirion.

Hubble found that the amount of red shift in a galaxy's light is directly related to its distance, the most distant galaxies having the greatest red shifts. Therefore, by measuring the red shift of a galaxy astronomers can tell how far away it is. Quasars, for instance, exhibit such enormous red shifts that they must be the most distant objects visible in the Universe, up to 15,000 million light years away.

Since the Universe is expanding, it is logical to conclude that it was once smaller and more densely packed than it is now. According to the most widely accepted theory, the entire Universe was originally a compressed, superdense blob which, for some unknown reason, exploded in a cataclysm known as the Big Bang. The galaxies are the fragments from that explosion, still flying outwards. As far as anyone can tell, the Universe will continue to expand for ever. According to the best estimates the Big Bang took place between 10,000 million and 20,000 million years ago; that is the age of the Universe as we know it. It is impossible to tell what, if anything, happened before the Big Bang.

NGC 4151 in Canes Venatici is an example of a Seyfert galaxy, a type of spiral galaxy with a bright but variable centre. Seyfert galaxies are thought to be closely related to quasars. Hale Observatories photograph.

The Sun

Our Sun is a glowing sphere of hydrogen and helium gas 1.4 million km in diameter, 109 times the diameter of the Earth and 745 times as massive as all the planets put together. The Sun is vital to all life on Earth, because it provides the heat and light without which our planet would be uninhabitable. It is important to astronomers because it is the only star that we can observe in close-up. Studying the Sun tells us a lot that we could never otherwise know about stars in general.

Whereas most celestial objects present problems for observers because they are so faint, the Sun provides us with entirely the opposite problem: it is so bright that it is dangerous to look at. Anyone glancing for an instant at the Sun through any form of optical instrument, be it binoculars or telescope, risks instant blindness. Even staring with the naked eye at the Sun for long periods can permanently damage your eyesight. There is only one safe way to observe the Sun, and that is by

The Sun on July 4, 1974, at a time of relatively low solar activity. The large sunspot group could engulf the Earth several times over. Note the effect of limb darkening, against which brighter faculae may be seen. Kitt Peak National Observatory.

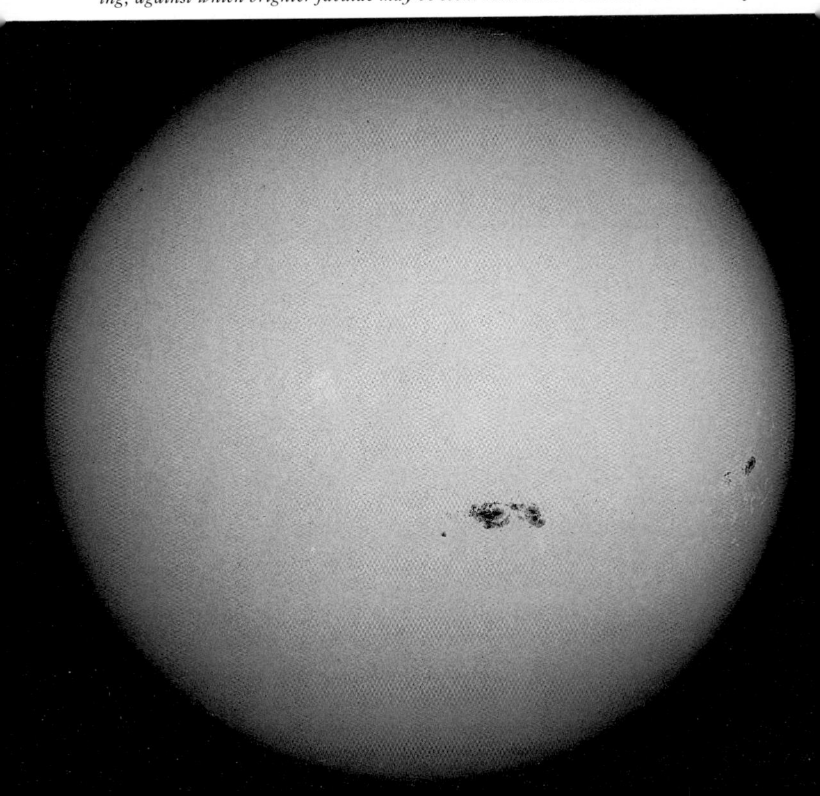

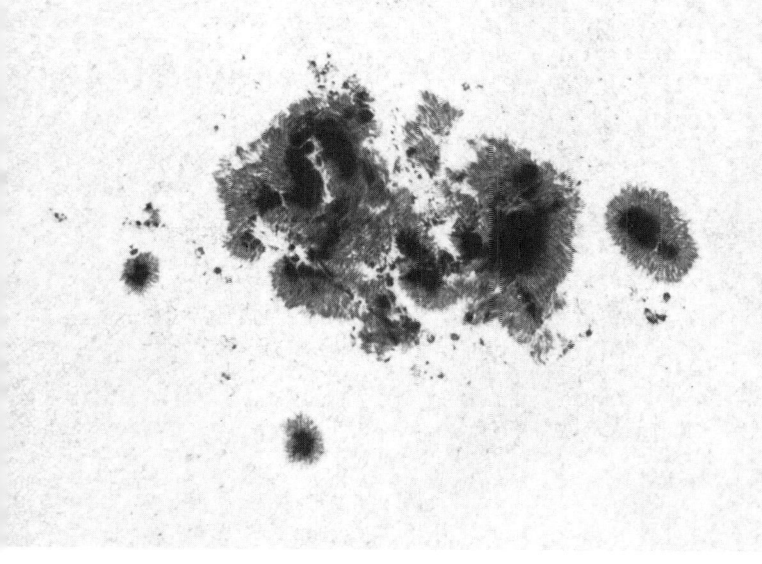

A large group of sunspots 200,000 km long as it appeared on May 17, 1951. Filamentary structure in the outer penumbra of the spots gives them a sunflower-like appearance. Note also the granulated appearance of the surrounding solar photosphere. Mount Wilson Observatory photograph.

projecting its image onto a piece of white card. Some telescopes come equipped with dark Sun filters to place over the eyepiece. These should never be used, for they can crack during observation, with inevitably disastrous results.

A view of the projected image of the Sun displays the brilliant solar surface, or *photosphere* ('sphere of light'), consisting of seething gases at a temperature of 5500°C. Although this is intensely hot by terrestrial standards, it is cool by comparison with the Sun's core, where the energy-generating nuclear reactions take place; there, the temperature is calculated to be about 15 million °C.

The photosphere exhibits a mottled, rice-grain effect termed *granulation*, caused by cells of hot gas bubbling up in the photosphere like water boiling in a pan. Granules range from about 300 km to 1500 km in diameter. Look carefully at a projected image of the Sun, and you will notice that the edges of the Sun appear fainter than the centre of the disk, an effect known as *limb darkening*. This is caused by the fact that the gases of the photosphere are somewhat transparent, so that at the centre of the disk we are looking more deeply into the Sun's interior than at the limbs. Brighter patches known as *faculae* may be visible against the limb darkening. Faculae are areas of higher temperature on the photosphere. A number of prominent dark markings known as *sunspots* will probably also be visible. They are areas of cooler gas that appear dark by contrast against the brighter photosphere.

Sunspots are temporary features that occur where magnetic fields on the Sun burst through the photosphere. Apparently, the presence of a strong magnetic field blocks the outward flow of heat from inside the Sun, producing the cooler spot. Sunspots have a dark centre known as the *umbra*, of temperature about 4000°C, surrounded by a brighter *penumbra* at about 5000°C. (These temperatures, incidentally, are about the same as on the surface of a red giant such as Aldebaran.)

Sunspots range in size from small pores no bigger than a large granule to enormous, complex patches 100,000 km or so across. The largest sunspots tend to form in groups that could span the distance from the Earth to the Moon. Spots that big are visible to the naked eye when the Sun is dimmed by the atmosphere, shortly after sunrise or before sunset. A major sunspot takes about a week to develop to full size, then slowly dies away again over the next fortnight or so.

As the Sun rotates, new spots are brought into view at one limb while old groups disappear around the other limb. Spot groups seldom last more than one or two solar rotations (a month or two). Usually, a spot group contains two main components, oriented east–west. The spot that leads across the disk as the Sun rotates is termed the p (preceding) spot, and is usually larger than its follower, the f spot. The p and f spots have opposite magnetic polarities, like the ends of a horseshoe magnet. The horseshoe is completed by invisible lines of force looping between the spots.

Sometimes the intense magnetic fields in a complex sunspot group become entangled, releasing a sudden flash of energy known as a *flare* that may last from a few minutes to an hour. Atomic particles are spewed out into space by the eruption of a flare. These particles reach Earth about a day later, causing effects in the upper atmosphere such as radio interference and the glowing displays known as *aurorae*. Unfortunately, aurorae occur only near the Earth's magnetic poles, so you will not be able to see them often unless you live in far northerly or southerly latitudes. An auroral display is an ethereal sight: the sky seems to glow with coloured light that can take the form of folded drapery or arches, shimmering and changing shape for hours on end.

By observing the progress of spots as they are carried across the Sun's face, we can measure the rotation period of the Sun. Being gaseous and not solid, the Sun does not rotate at the same rate at all latitudes. It spins most quickly at its equator, in 25 days; this falls to about 28 days at latitude 45°; and near the poles it is slowest of all, 34 days. The average figure usually quoted, 25.38 days, refers to the rotation rate at latitude 17°. As a result of the Earth's orbital motion around the Sun, a spot takes about two days extra to return to the same place on the Sun as seen from Earth.

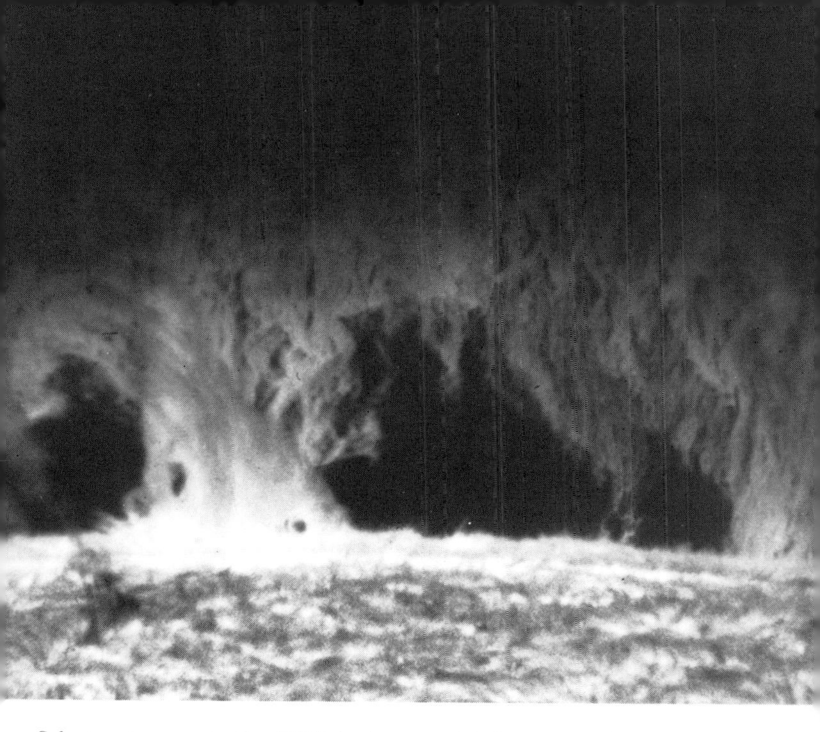

Solar prominences extending 65,000 km into space look like enormous burning trees at the edge of the Sun. Photograph taken in the red light of hydrogen. Big Bear Solar Observatory.

The number of sunspots on view waxes and wanes in a cycle lasting 11 years on average, although the length of individual cycles has been as short as 8 years and as long as 16 years. At times of minimum activity the Sun may be spotless for days on end, whereas at solar maximum over a hundred spots may be visible at any one time. However, the level of activity varies considerably from cycle to cycle; the average number of sunspots visible at maximum has ranged from 40 to 180, and even when the Sun is supposedly quiet, some large spots and flares can break out. Solar activity is notoriously unpredictable, which adds to its fascination for observers.

A few general rules can be deduced, however. The first spots of each new cycle appear at latitudes of about 30–35° north and south of the Sun's equator. As the cycle progresses, spots tend to form closer to the equator. Sunspot numbers build up to a peak, then start to decay again. As minimum approaches, the last spots of the cycle are found at latitudes between 5° and 10° north and south of the equator. At the same time – that is, around solar minimum – the first spots of the next cycle are forming at higher latitudes. Sunspots are seldom found on the Sun's equator, and spots at latitudes greater than 40° are exceedingly rare.

Above the photosphere is a tenuous layer of gas known as the *chromosphere*, about 10,000 km deep. So faint is the chromosphere that it is normally invisible except with special instruments. It can, however, be seen for a few seconds at a total eclipse, when it appears as a pinkish crescent just before and after the Moon completely covers the face of the Sun. Its pinkish colour is caused by light from hydrogen gas, and gives rise to its name, which means 'colour sphere'.

Also visible around the edge of the Sun at total eclipses are huge clouds of gas extending from the chromosphere into space, known as *prominences*. They have the same characteristic rosy-pink colour as the chromosphere, caused by emission from hydrogen. Like so many features of the Sun, prominences are controlled by magnetic fields. So-called quiescent prominences extend for 100,000 km or more across the Sun, sometimes forming graceful arches tens of thousands of kilometres high. When seen silhouetted against the brighter background of the photosphere, they are termed filaments. Quiescent prominences can last for months. At the other end of the scale are eruptive prominences, with lifetimes of only a few hours. They are flares seen at the edge of the Sun, ejecting material into space at speeds of up to 1000 km per second. All forms of solar activity – sunspots, flares and prominences – follow the 11-year solar cycle.

The Sun's crowning glory is its *corona*, a faint halo of gas that comes into view only when the brilliant photosphere is blotted out at a total solar eclipse. The corona is composed of exceptionally rarefied gas at a temperature of 1 to 2 million °C. Petal-like streamers of coronal gas extend from the Sun's equatorial zone, while shorter, more delicate plumes fan out from the polar regions. The shape of the corona changes during the solar cycle; at solar maximum, when there are more active areas on the Sun, the corona appears more regular in shape than at solar minimum.

Gas from the corona is continually flowing away from the Sun out into the Solar System, forming what is known as the *solar wind*. Atomic particles of the solar wind are detected streaming past the Earth at a speed of about 400 km per second. The most obvious effect of the solar wind is to make comet tails point away from the Sun. The solar wind extends outwards beyond the orbit of the most distant planet, finally merging with the thin gas between the stars. In a sense, therefore, all the planets of the Solar System can be said to lie within the outer reaches of the Sun's corona.

Right: *The spectacular corona of the Sun exhibited extensive streamers at the total solar eclipse of July 11, 1991, photographed from Mexico.* Armagh Planetarium.

The Solar System

In one sense our Sun is unusual, for it is not accompanied by another star but has a family of nine planets, assorted moons and countless smaller lumps of debris. (Any planets around other stars are too faint to be seen, at least with optical telescopes on Earth.) This retinue of bodies, all held captive by the Sun's gravity, is called the Solar System.

All the objects in the Solar System shine by reflecting the light coming from the Sun. Several of the planets when at their brightest can equal or outshine the brightest stars. All the planets orbit the Sun in approximately the same flat plane, so they are always to be found near the ecliptic. A bright 'star' that disturbs a familiar constellation pattern along the ecliptic is therefore almost certainly a planet (although it could, just possibly, be a nova).

As seen from above the Sun's north pole, the planets orbit the Sun in an anticlockwise direction. Their orbits are slightly elliptical in shape, so that each planet's distance from the Sun varies somewhat during one orbit. In order of average distance from the Sun, the nine planets are: Mercury, Venus, Earth, Mars, Jupiter, Saturn, Uranus, Neptune and Pluto. The four inner planets are rocky, relatively small bodies. Then come four giant planets made mostly of gas. Pluto is a small, frozen oddity on the outer limits.

Venus is the planet with the most circular orbit; its distance from the Sun varies by little more than 1.5 million km. The planet with the most elliptical orbit is distant Pluto, which at times actually strays across the orbit of Neptune. Some of the minor bodies, or asteroids, in the Solar System have orbits that are more elliptical than this, and the comets have highly distended orbits that can take them from the innermost part of the Solar System out to way beyond Pluto. The closest point of a celestial body's orbit to the Sun is known as the *perihelion*; the farthest point is the *aphelion*.

The basic unit of distance in the Solar System is the *astronomical unit*, the average distance of the Earth from the Sun; it is equivalent to 149,597,870 km. Light takes 499 seconds (8.3 minutes) to cross this distance, so we on Earth see the Sun as it actually appeared over 8 minutes ago. Clearly, although the astronomical unit is large compared with everyday distances on Earth, it is insignificant in comparison with a light year. There are 63,240 astronomical units in a light year.

How long a planet takes to orbit the Sun depends on its distance: the closest planets orbit the quickest, and the farthest orbit the slowest. Each planet's orbital period is technically its 'year', although the period is usually expressed in terms of Earth days and years. The

Top right: *The sizes of the planets to scale, compared with a section of the Sun.*

Bottom right: *The orbits of the planets to scale.* Wil Tirion.

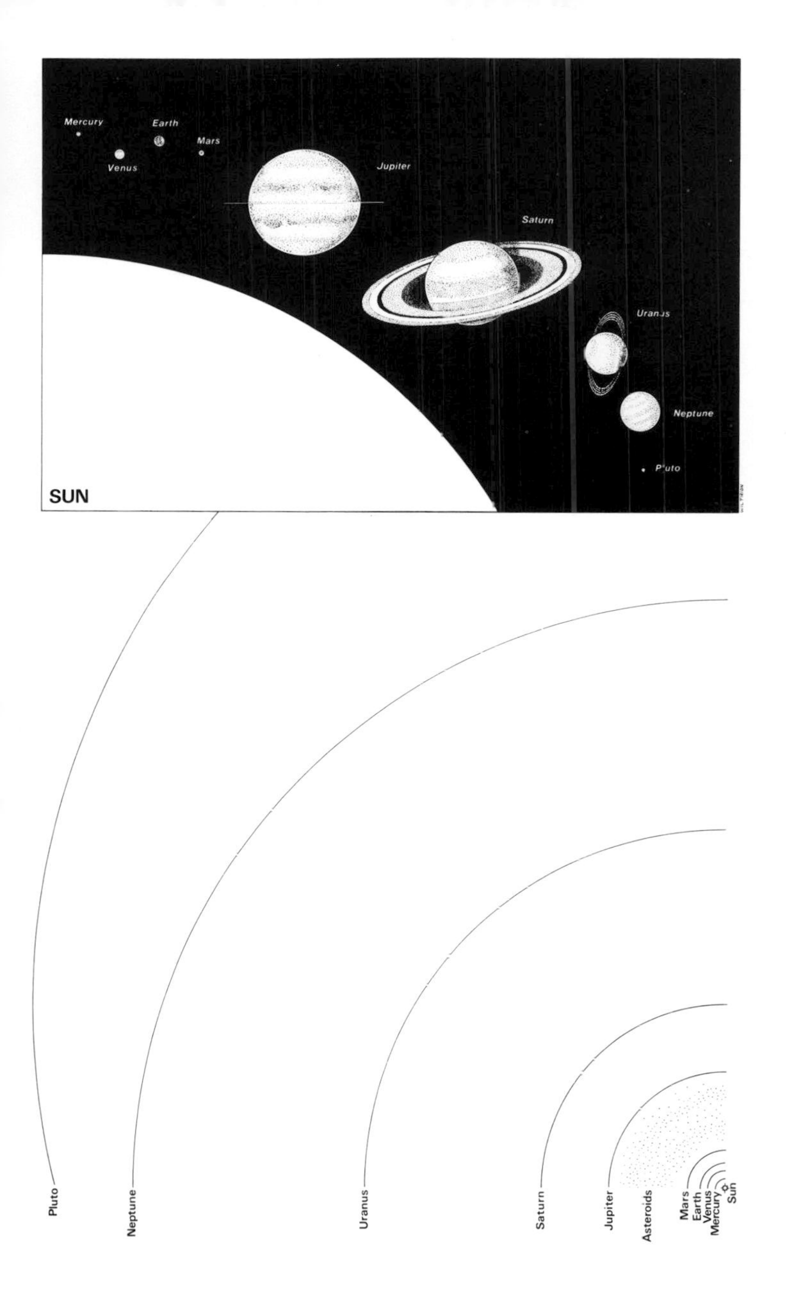
Mercury
Earth
Mars
Venus
Jupiter
Saturn
Uranus
Neptune
Pluto
SUN
Pluto
Neptune
Uranus
Saturn
Jupiter
Asteroids
Mars
Earth
Venus
Mercury
Sun

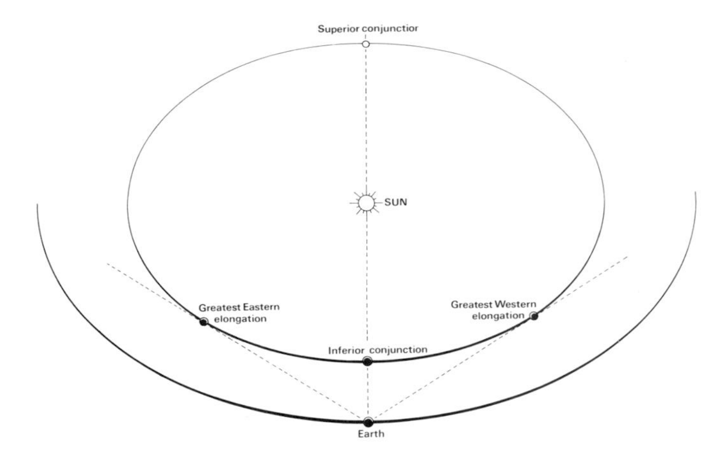

When Venus or Mercury lies directly between the Earth and the Sun it is said to be at inferior conjunction; on the far side of the Sun it is at superior conjunction. Maximum angular separation from the Sun is called greatest elongation. Wil Tirion.

orbital periods of the planets range from 88 days for Mercury to 250 years for Pluto. These orbital periods are known technically as *sidereal periods*, for they are measured with respect to the distant stars.

From time to time, as Mercury and Venus move around their orbits, they come between us and the Sun; when that happens, they are said to be at *inferior conjunction*. That is also the time when Mercury and Venus are closest to the Earth, although this fact is of little use to observers since the planets are lost in the Sun's glare and their illuminated hemispheres are in any case turned away from the Earth.

The orbits of Mercury and Venus are inclined by 7° and 3.4° respectively to the orbit of the Earth, but that is sufficient to ensure that they usually pass above or below the disk of the Sun at inferior conjunction. But occasionally Mercury or Venus does actually cross the face of the Sun as seen from Earth. At such an event, called a *transit*, the planet appears as a tiny dark dot like a small sunspot against the glaring solar surface. Transits of Mercury are more common than those of Venus: the next are due in 1999, 2003, 2006 and 2016, whereas the transits of Venus in 2004 and 2012 will be the only ones for over a century.

When Mercury or Venus lies on the far side of the Sun from us it is said to be at *superior conjunction*. When Mars or any of the planets beyond it lies on the far side of the Sun it is simply said to be at conjunction – being farther from the Sun than we are, they cannot possibly come between Earth and the Sun, so there is no ambiguity about which sort of conjunction is meant. Planets at conjunction are invisible in the Sun's glare.

The best time to view the two inner planets is when they are at their maximum possible angular separation from the Sun, known as *greatest elongation*. At greatest elongation, Mercury and Venus are at exactly half-phase. An eastern elongation means that the planets are setting after the Sun in the evening sky; at western elongation, the planets rise before the Sun in the morning sky. The time between one greatest eastern elongation of Mercury and the next, or one greatest western elongation and the next, is 116 days. With Venus, greatest western elongations or greatest eastern elongations recur every 584 days, although Venus is so prominent that it can be adequately observed well away from greatest elongation. Incidentally, the time between one given appearance of a planet and the next – be it conjunction, elongation or whatever – is known as its *synodic period*. This differs from the sidereal period because our observation platform, the Earth, is moving in orbit around the Sun.

The best opportunities to see Mars and the outer planets is when they are directly opposite the Sun in the sky; this is termed *opposition*. A planet at opposition appears due south for northern hemisphere observers (due north for observers in the southern hemisphere) at midnight local time, or 1 a.m. if daylight saving time is in operation. At opposition, Mars, Jupiter, Saturn, Uranus, Neptune and Pluto lie closest to Earth (Mercury and Venus, being between the Earth and the Sun, cannot come to opposition). These outer planets therefore appear biggest and brightest around the time of opposition.

Mars, Venus and Jupiter as seen from Earth. On Mars the dark wedge of Syrtis Major is visible, as is the southern polar cap. The photograph of Venus was taken in ultraviolet light to bring out the cloud features. Lowell Observatory.

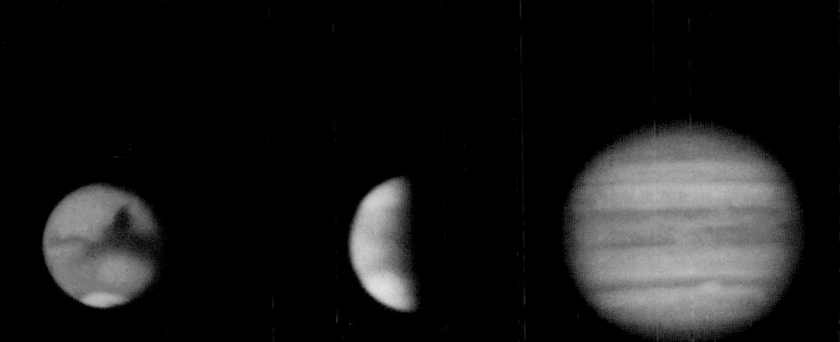

The Moon

The Moon, the Earth's natural satellite and nearest celestial neighbour, is an object of perennial fascination for observation with instruments of all sizes. Despite its small size – 3476 km in diameter, roughly a quarter that of the Earth – it is so close, on average 384,400 km, that even ordinary binoculars reveal a wealth of detail on its cratered surface. Some of the most interesting objects to look out for are shown in the maps on pages 316–327.

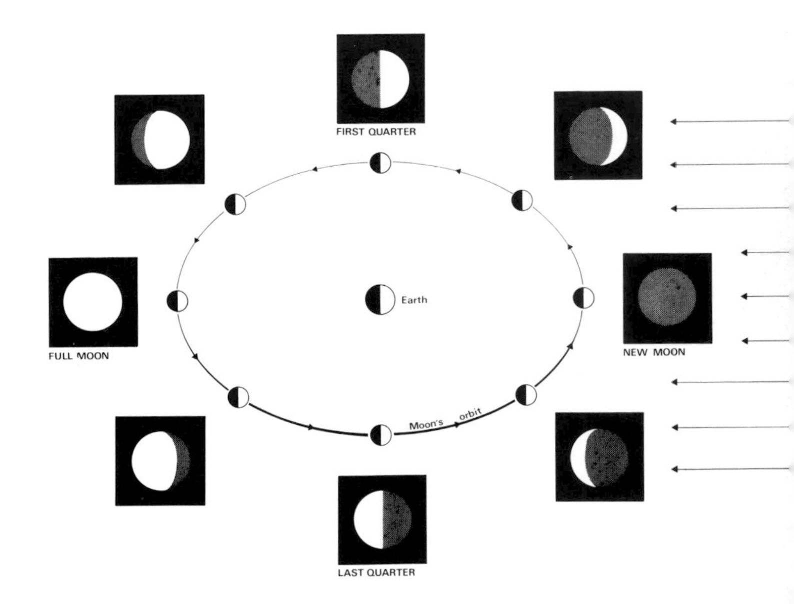

The Moon passes through its cycle of phases as it orbits the Earth and we thus see differing proportions of its illuminated side. Wil Tirion.

In the course of a month the Moon goes through a cycle of phases from new (unilluminated) through waxing half (first quarter), full, waning half (last quarter), back to new again. Strictly speaking, there are two types of month. The first lasts 27.3 days, the time the Moon takes to complete one orbit of the Earth relative to a fixed point such as the distant stars; this is known as a *sidereal month*. But the Earth also moves relative to the Sun in that time, so the Moon must complete rather more than one orbit to return to the same phase as seen from Earth. The time the Moon takes to complete one cycle of phases,

The Moon at gibbous phase, 10 days after new. Note Sinus Iridum on the terminator at the top. Most of the features on this photograph are visible in small telescopes or even binoculars, if mounted firmly. Lick Observatory.

29.5 days, is called a *synodic month*. Each spot on the Moon is subjected to two weeks of daylight, during which surface temperatures reach the boiling point of water (100°C), followed by a two-week night when temperatures plummet to –170°C.

The Moon spins on its axis in 27.3 days, the same time it takes to complete one orbit of the Earth, so that the same face of the Moon is always turned towards us; this is known as a *captured rotation*. In practice, though, we can see slightly more than half the Moon's surface. The Moon's equator is tilted at about 6½° to the plane of its orbit, so that at times we can see as much as 6½° over the Moon's north or south pole; this is known as *libration in latitude*. In addition, the Moon's speed of motion along its elliptical orbit changes rhythmically as it approaches and recedes from the Earth, while its axial rotation remains uniform. The Moon therefore seems to rock slightly to west and east as it orbits the Earth, so that we can see up to 7½° or so around each edge; this is known as *libration in longitude*. The net effect of these librations is that we can see 59 per cent of the Moon's surface.

The line dividing the lit and unlit portions of the Moon is known as the *terminator*. Objects near the terminator are thrown into sharp relief by the low angle of illumination, so that craters and mountains appear particularly rugged. As the Sun rises higher over the moonscape, details become more washed out. Near full Moon many individual formations are difficult to pick out. An exception is those craters that have bright ray systems, apparently made of pulverized rock thrown out from the crater during its formation; the rays become more prominent under high illumination. It is easiest to pick out the difference in contrast between the bright highlands and the dark lowland plains known as *maria* (singular: mare) at full Moon.

After looking at the brilliant full Moon it comes as a surprise to realize that the lunar surface rocks are actually dark grey in colour; on average, the Moon's surface reflects only 7 per cent of the light that hits it. If the Moon were, for instance, covered in clouds like those of Venus, it would be over ten times brighter.

Formations on the Moon bear a variety of curious names. The dark lowland plains are known as maria, Latin for 'seas', because the first observers imagined them to be stretches of water; the name persists, even though it has been clear for centuries that the Moon is both airless and waterless. Thus we have, for instance, the Oceanus Procellarum (Ocean of Storms), Mare Imbrium (Sea of Rains) and Mare Tranquillitatis (Sea of Tranquillity). Less prominent lowlands are characterized as bays (Sinus), marshes (Palus) or lakes (Lacus). Mountains on the Moon are named after terrestrial ranges; hence we have the lunar Alps and the lunar Apennines. The craters have been named after philosophers and scientists of the past, though it is fair to say that the selection has been somewhat arbitrary.

Partners in space: the Moon and Earth, photographed in 1992 by the Galileo space probe, on its way to Jupiter. The Moon is in the foreground, moving from left to right. On the Earth, Antarctica is visible through clouds at the bottom.
NASA

We owe the modern system of lunar nomenclature to an Italian astronomer, Giovanni Riccioli, who in 1651 published a map bearing many of the names now universally accepted. Riccioli allocated himself a prominent crater near one edge of the Moon, with a large neighbouring formation named after his pupil, Francesco Grimaldi. Less favoured colleagues did not fare so well. Galileo, one of the greatest scientists of all time, but who fell out with the Church, is given an insignificant 15 km diameter crater in Oceanus Procellarum.

Riccioli's system has been followed to the present day, but now that the near and far sides of the Moon have been mapped in detail by spacecraft, the ground rules have been extended to include names from areas of human endeavour other than astronomy. Thus we now find Freud, H. G. Wells and Montgolfier commemorated on the Moon. Finding suitable names for features on celestial bodies has become a pressing problem now that many planets and their moons have been surveyed by space probe.

Lunar observation began in 1609, when Galileo turned his first telescope towards the Moon. His drawings, published in 1610, are crude

Mare Imbrium, a large, rounded lowland plain on the Moon. The dark-floored crater Plato pierces the mountains on its northern shore. The mare's southern reaches are splashed with Copernicus's bright rays. Mount Wilson Observatory.

by modern standards but served to show that the Moon's surface is rugged and mountainous. Debate raged for the next three and a half centuries over the origin of the Moon's features. One school of thought held that they were caused by volcanic action; the opposition believed the craters and maria to have been blasted out by massive impacts of meteorites and asteroids. There is no point in recapitulating the arguments here. Suffice it to say that the impact theory has emerged as the undisputed winner, although it is accepted that there has also been a certain amount of volcanic activity on the Moon.

Not until the late 1960s, when space probes and astronauts first reached the Moon, was the controversy resolved. On July 31, 1964, an American probe called Ranger 7 headed towards a corner of the Mare Nubium. As it zoomed close in it sent back a stream of photographs showing features as small as a metre across, a hundred times better than what is visible through Earth-based telescopes. Rangers 8 and 9 followed the next year. The lesson from these probes was that even the apparently flattest parts of the Moon's surface were pitted with small craters caused by aeons of meteorite bombardment. Landing sites for future manned spacecraft would therefore have to be chosen with particular care, to prevent the landing craft from toppling into a crater or hitting a boulder (this nearly happened on the Apollo 11 descent).

Ranger's cursory examination was followed by a two-pronged attack: the Surveyors, a series of craft which soft-landed automatically to give an astronaut's-eye view of the lunar surface (the Rangers had simply crashed), and the Lunar Orbiters which, as the name implies, photographed the Moon from close orbit. Together, these two series of probes revolutionized our knowledge of the Moon between 1966 and 1968, and paved the way for the manned Apollo landings. The Surveyors showed that the Moon's topsoil is made of compacted dust,

Oblique view of the twin craters Aristarchus (bright, with rays) and Herodotus, taken with a mapping camera aboard an Apollo spacecraft. From a small crater near Herodotus emerges the snake-like Schröter's Valley. Fairchild.

Apollo 17 astronaut Harrison Schmitt is dwarfed by a massive boulder during his exploration of the Moon's surface in December 1972. NASA.

firm enough to support astronauts and their spacecraft. From Lunar Orbiter photographs, astronomers compiled their most detailed maps of the near and far sides of the Moon.

The Moon's far side was first glimpsed by the Soviet probe Luna 3 in October 1959. Its photographs were poor by modern standards, but they did at least reveal the main difference between the two hemispheres of the Moon: there are scarcely any mare areas on the Moon's far side. Instead, heavily cratered bright uplands dominate the scene. The reason for this asymmetry is that the Moon's crust is several kilometres thicker on the far side. Large lowland basins like Mare Imbrium exist on the far side of the Moon, but they have not been flooded by dark lava. Volcanic lavas from the Moon's interior found it easier to leak out through the thinner crust of the Earth-facing hemisphere. The most prominent dark area on the Moon's far side is not a true mare at all, but a deep crater called Tsiolkovsky, 240 km in diameter (almost as large as Sinus Iridum on the visible hemisphere).

Once the Lunar Orbiters had spied out potential landing sites, the stage was set for astronauts to follow. On July 20, 1969, the Apollo 11 lunar module called Eagle carried Neil Armstrong and Edwin Aldrin to the first manned lunar landing, in the southwestern Sea of Tranquillity. The two astronauts spent two hours exploring the Moon's surface, setting up experiments and collecting samples for geologists to study back on Earth. For the first time, humans had touched another world in space.

By the time Apollo 17 concluded the series of manned landings in December 1972, astronauts had brought back to Earth over 380 kg of Moon samples, most of which is stored in Houston, Texas. When divided into the overall cost of Apollo, one kilogram of Moon rocks is worth 100 million US dollars. In addition to the Apollo samples, three Soviet automatic lunar probes have returned to Earth with a few hundred grams of Moon soil.

What have we learned from these precious specimens? The most astounding fact about the Moon rocks is their immense age. The Apollo 11 samples, for instance, proved to be 3700 million years old,

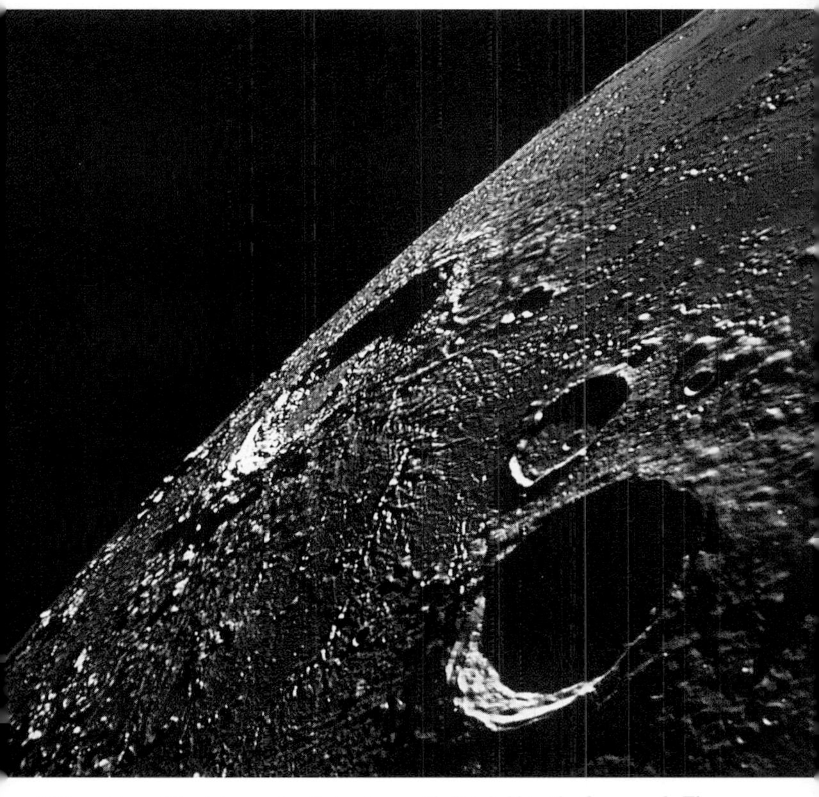

Oblique view of the lunar crater Copernicus, with Reinhold in the foreground. The surface of Oceanus Procellarum is splashed with ejecta thrown out by the meteorite impact that formed Copernicus about 1000 million years ago. NASA.

older than virtually any rocks on Earth – yet the site from which they came, Mare Tranquillitatis, is one of the youngest areas on the Moon. The youngest Moon rocks of all, found at the Apollo 12 site in Oceanus Procellarum, are 3200 million years old. As expected, the lunar maria turned out to be covered with lava flows similar in composition to volcanic basalt on Earth. They do not resemble a jagged, lumpy lava field on Earth because over the thousands of millions of years since the lavas were laid down, the sandblasting effect of micrometeorites has eroded the surface rocks to form a layer of soil several metres deep, called the *regolith*.

By contrast, the highlands, sampled by later Apollo missions, consist of a paler rock called anorthosite, rare on Earth. Rocks from the highlands proved older than those from the maria, mostly dating back 4000 million years or more. Their jumbled, fragmented nature bears witness to the violent bombardment from meteorites which the Moon suffered early in its history.

Despite scientists' hopes, the rich haul of Moon rocks did not conclusively answer the question of the Moon's origin. All three existing theories – that the Moon was a fragment of our planet, that it was once a separate body that was captured by the Earth's gravity, or that the Earth and Moon formed side by side, much as they are now – had drawbacks. Since the Apollo landings, a new and more promising theory has emerged in which a stray body the size of Mars hit the Earth, spraying debris into orbit around the Earth where it formed into the Moon.

That problem aside, we now have a much clearer picture of the rest of the Moon's history. Evidence from the rocks reveals that the Moon formed 4600 million years ago, at about the same time as the Earth and, presumably, the rest of the Solar System. Each body in the Solar System is believed to have been built up by an accumulation of many smaller bodies. This process, termed *accretion*, released sufficient heat to melt the Moon. A scum of less dense rock formed a primitive crust which was battered by the infall of debris left over from the formation of the Solar System. This bombardment carved out the Moon's cratered highlands and the mare basins. Other rocky bodies in the Solar System, notably Mercury, also bear the scars of this same mopping-up operation.

About 4000 million years ago the storm of debris abated. Then, slowly, molten lava began to seep out from inside the Moon, solidifying to form the dark lowland maria. Wrinkles of solidified lava are visible through binoculars and small telescopes when the maria are under low illumination. In particular, look out for the Serpentine Ridge in eastern Mare Serenitatis when the Moon is five to six days old, or six days past full. Mare Tranquillitatis and Mare Imbrium also have prominent wrinkle ridges. In fact, one 'crater' on Mare Tranquil-

Meandering near the foothills of the Apennine Mountains, Hadley Rille is a dried-up lava channel that was visited by the Apollo 15 astronauts in 1971. NASA.

litatis, called Lamont, consists of nothing more than low ridges of solidified lava. In places, particularly in western Oceanus Procellarum, a number of blister-like domes have been produced by the upwelling of molten lava. But volcanic cones like Vesuvius are absent on the Moon, evidently because the lunar lavas were too runny to build up into mountains.

By 1000 million years ago the volcanic outpourings had ceased, leaving the Moon cold and dead. Since then it has remained virtually unchanged, save for the arrival of an occasional meteorite to punch a new crater in its surface. For instance, the crater Copernicus was formed roughly 1000 million years ago; Tycho was blasted out about 300 million years ago.

Yet the Moon may not be completely inactive today. Observers have from time to time reported transient events, such as glows and obscurations, around the edges of the maria and in certain craters – Aristarchus seems to be a favourite spot for these *transient lunar phenomena* (TLPs). Most TLPs have been observed by amateur

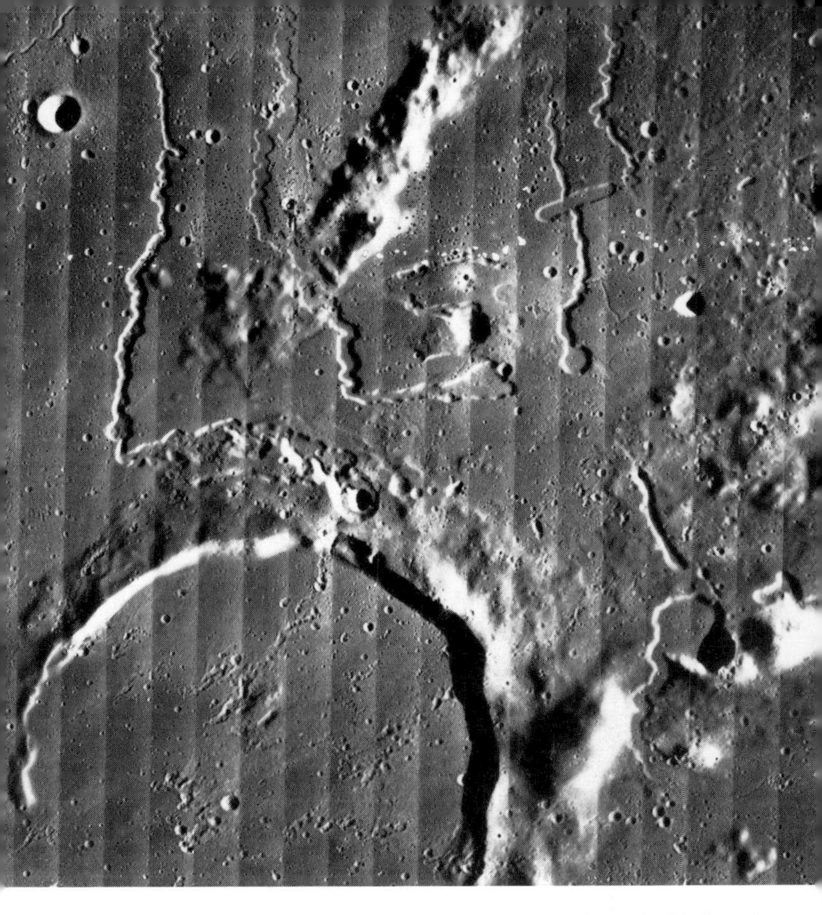

Rilles and faults on the surface of Oceanus Procellarum to the north of the horseshoe-shaped crater Prinz, photographed by Lunar Orbiter V. The picture appears banded because it was transmitted to Earth in strips. NASA.

astronomers. Their origin remains puzzling, but they are probably caused by the release of gas from the Moon's interior.

Other features to look out for when observing the Moon are grooves in the surface, known as *rilles*, apparently caused by faulting. For instance, the crater Hyginus near the Moon's centre lies in the middle of a long, rimless cleft along which smaller craters have been formed by subsidence. Another type of rille, called a sinuous rille, snakes over mare surfaces like a meandering river. Apollo 15 landed at the edge of one of these sinuous rilles, called Hadley Rille. Such winding rilles are not dried-up river beds but are probably collapsed tunnels through which underground lava streams once flowed.

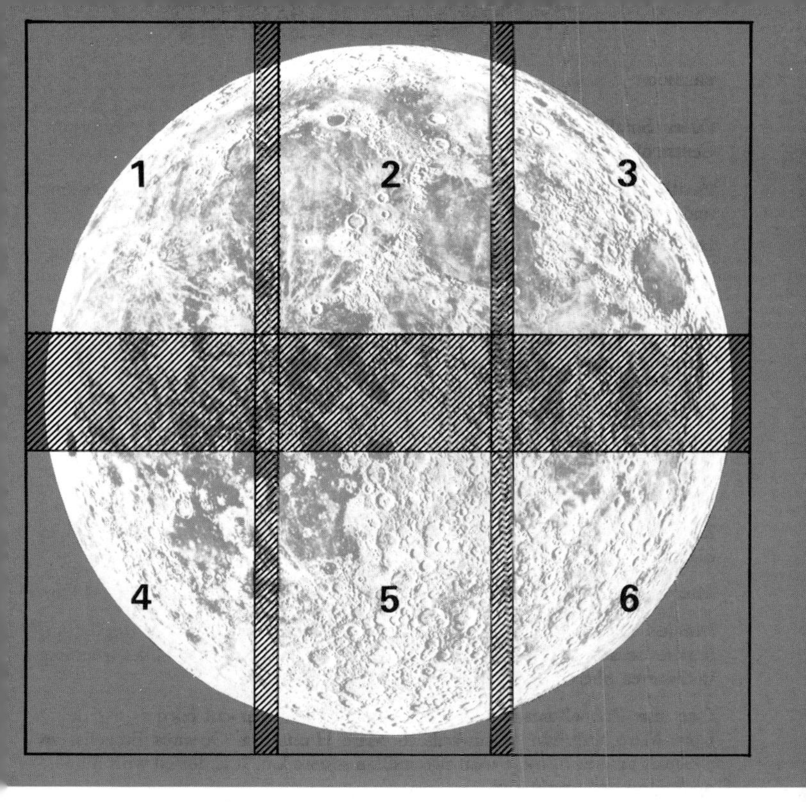

A note on orientation. These maps show the Moon in conventional orientation, with north at the top, as it appears to the naked eye and in binoculars. Through an astronomical telescope, south will be at the top, so telescope users must invert the maps. West is at the left, as on Earth, although in the rest of the sky west is in the opposite direction, towards the western horizon.

Moon Map 1

Aristarchus Brilliant young crater 45 km in diameter with multiple terracing on its inner walls. The brightest area on the Moon, and centre of a major ray system. Dark bands are visible on the inner walls under high illumination. Aristarchus is the location of numerous red glows known as transient lunar phenomena (TLPs), apparently caused by outgassing from the surface. Mountains to the north exhibit fantastic sculpturing.

Copernicus Diameter 93 km. One of the most magnificent craters on the Moon. Centre of a major ray system. Terraced walls and numerous hummocky central peaks. Rays from Copernicus are splashed for more than 600 km across the Mare Imbrium and Oceanus Procellarum.

Encke Low-walled crater 29 km in diameter covered by rays from nearby Kepler. Brilliant under high illumination.

Euler Small, sharp crater, 28 km across, in southwestern Mare Imbrium. Centre of a minor ray system.

Fauth Keyhole-shaped crater south of Copernicus, nearly 2 km deep. Striking under low illumination.

Harpalus Crater 40 km in diameter on Mare Frigoris, bright under high illumination. The smaller crater Foucault, 24 km in diameter, lies between it and Sinus Iridum.

Herodotus Companion to Aristarchus, similar in size (diameter 35 km) but different in structure – it has a dark, lava-flooded floor and it is not a ray centre. The W-shaped Schröter's Valley, 200 km long, starts at a craterlet outside the northern wall of Herodotus.

Hevelius Large, bright crater 118 km in diameter on western shore of Oceanus Procellarum, with clefts crossing the floor. Smaller, sharper crater Cavalerius (diameter 64 km) adjoins to the north.

Kepler Major ray centre in Oceanus Procellarum. Brilliant crater, 32 km in diameter, with central peak and heavily terraced walls.

Mairan Prominent crater in highlands west of Sinus Iridum. Diameter 41 km.

Marius Dark, flat-floored ring 41 km in diameter on Oceanus Procellarum, notable because it lies on a wrinkle ridge. There are many dome-like structures in this area where lava has bubbled up through the surface.

Oceanus Procellarum Vast dark plain with no clear-cut borders extending from Mare Imbrium southwards to Mare Humorum. Oceanus Procellarum occupies an area of more than two million square km. It is dotted with numerous craters and bright rays.

Prinz U-shaped formation 52 km in diameter, remains of a half-destroyed crater flooded by lava from Oceanus Procellarum. Notable because of sinuous rilles to the north.

Pythagoras Magnificent large crater 128 km in diameter with terraced walls rising nearly 5 km and prominent central peak, near northwestern limb of the Moon.

Reiner Sharp crater 30 km in diameter west of Kepler in Oceanus Procellarum. Particularly notable is a tadpole-shaped splash of brighter material on the dark plain to the west and north.

Reinhold Prominent crater 48 km in diameter southwest of Copernicus, with terraced walls. A smaller, lower ring to the northeast is Reinhold B.

Rümker Remarkable formation on northwestern Oceanus Procellarum, visible only under low illumination. Rümker is an irregular, lumpy dome 55 km wide.

Sinus Iridum Beautiful large (diameter 260 km) bay on the Mare Imbrium. Its seaward wall has been broken down by invading lava and is reduced to a few low wrinkle ridges. Its remaining walls, angular in shape, are known as the Jura Mountains, which are brilliantly illuminated by the morning Sun. A deep, 39-km-diameter crater on the northern rim is Bianchini.

Moon Map 2

Agrippa Oval-shaped crater 46 km in diameter with central peak, forming a neat pair with Godin.

Alpine Valley Flat-floored valley 150 km long through the lunar Alps, connecting Mare Imbrium with Mare Frigoris.

Anaxagoras Crater 51 km in diameter near the north pole of the Moon, centre of an extensive ray system.

Archimedes Distinctive, flooded ring 83 km in diameter in eastern Mare Imbrium, notable for its almost perfectly flat floor.

Aristillus Prominent crater 55 km in diameter in eastern Mare Imbrium with terraced walls, numerous surrounding ridges and a complex central peak 900 m high. Under high illumination it is seen to be surrounded by a faint ray system. Forms a pair with Autolycus to the south.

Aristoteles Magnificent, partially lava-flooded crater 87 km in diameter, touching the smaller crater Mitchell to the east. Numerous ridges radiate from its outer walls. Makes a pair with Eudoxus to the south.

Autolycus Prominent crater 39 km in diameter, south of Aristillus. High illumination shows it to be the centre of a faint ray system.

Cassini Flooded ring of unusual appearance with partially destroyed walls, 56 km in diameter. Contains a bowl-shaped crater, Cassini A, 17 km wide.

Eratosthenes Prominent, deep crater on the edge of Sinus Aestuum at the southern end of the Apennine Mountains. Terraced walls and craterlet on central peak. Diameter 58 km.

Eudoxus Rugged crater 67 km in diameter with small central peak and terraced walls. South of Aristoteles.

Godin Smaller but deeper companion to Agrippa; 35 km in diameter with central peak, bright walls and faint ray system.

Hyginus Rimless crater 10 km in diameter, centre of a cleft or rille 225 km long, visible in small telescopes, evidently formed by collapse of the surface. To the east is another cleft, the Ariadaeus Rille.

Lambert Crater 30 km in diameter in Mare Imbrium, with central craterlet. Situated on a wrinkle ridge. Under low illumination a larger 'ghost' ring, Lambert R, is seen to the south.

Linné Bright spot on Mare Serenitatis, best seen under high illumination. At its centre is a small, young crater 2.4 km in diameter.

Manilius Bright crater 39 km in diameter in Mare Vaporum, with terraced walls and central peak. Develops a ray system as illumination increases, as does its neighbour Menelaus (diameter 27 km) on the rim of Mare Serenitatis.

Mare Imbrium Enormous circular plain 1250 km in diameter, dominating this section. Bounded by the Alps, Caucasus, Apennine and Carpathian

Mountains but open on the southwest to Oceanus Procellarum. Mare Imbrium has a double structure: traces of a smaller inner ring are visible, marked by a few isolated mountains and wrinkle ridges. Note different shades of lava on its surface. Isolated mountains protruding from its dark floor include Pico, Piton, the Straight Range, the Teneriffe Mountains and the Spitzbergen Mountains.

Mare Serenitatis A major rounded lunar sea 680 × 580 km, bounded on the northwest by the Caucasus Mountains and on the southwest by the Haemus Mountains. A bright ray from Tycho crosses the dark lava plain, passing through the crater Bessel, 16 km in diameter. Under high illumination, Mare Serenitatis is seen to be rimmed with darker lava. Note also a major wrinkle ridge (the Serpentine Ridge) on the east side. Apollo 17 landed at the southeastern edge of Mare Serenitatis.

Plato Unmistakable large, dark-floored crater in the highlands north of Mare Imbrium, 100 km in diameter; prominent under all conditions of illumination. Tiny craterlets pockmark its flat floor. Temporary obscurations of the surface, presumed to be caused by outgassing, have been observed in this area. Landslips appear to have detached part of the inner western wall.

Pytheas Small (diameter 20 km) but deep and prominent crater on Mare Imbrium, rhomboidal in outline. Becomes brilliant under high illumination.

Stadius 'Ghost' ring to the east of Copernicus, outlined only by a few ridges and craterlets. Visible only under low illumination. Diameter 64 km.

Timocharis Bright crater 35 km in diameter on Mare Imbrium, with terraced walls and distinctive central crater. Faint ray centre.

Triesnecker Crater 26 km in diameter surrounded by a system of clefts.

Moon Map 3

Atlas Large crater, 87 km in diameter, with terraced walls and complex floor. A ruined ring abuts to the northwest. Atlas and Hercules form one of the many crater pairs in this region.

Burckhardt Complex elliptical (45 × 55 km) crater, appearing to overlap older formations on either side.

Bürg Prominent crater despite its moderate size (diameter 40 km). Central peak. Lies in the centre of Lacus Mortis. Note nearby rilles.

Cleomedes Large, irregular-shaped crater 126 km in diameter with partially flooded floor, to north of Mare Crisium. West wall is interrupted by 43-km-diameter crater Tralles.

Endymion Large, dark-floored enclosure 125 km in diameter, with walls up to 4900 m high.

Franklin Crater 56 km in diameter. Forms a pair with the 40-km-diameter Cepheus to the northwest.

Geminus Prominent crater 86 km in diameter with central peak. Bright rays emanate from two smaller craters nearby, Messala B and Geminus C.

Hercules Flat-floored enclosure 67 km in diameter containing sharp, bright bowl crater Hercules G.

Lemonnier Old and flooded crater with its western wall washed away by lava from Mare Serenitatis. Diameter 61 km.

Mare Crisium Unmistakable dark lowland plain ringed by high mountains like an oversized crater, 435 × 565 km. Main feature on its flat, lava-flooded floor is the crater Picard, 23 km across, with smaller crater Peirce to its north.

Mare Tranquillitatis An irregular-shaped lowland, 650 × 900 km. Wrinkle ridges attest to numerous lava flows over the plain. Main crater on the mare is the distorted Arago, on the western side, 26 km in diameter. Apollo 11 made the first manned lunar landing in the southwest corner of Mare Tranquillitatis.

Plinius Distinctive crater 43 km in diameter with complex central peak combined with a central craterlet. Stands between Mare Serenitatis and Mare Tranquillitatis, along with Dawes to the northeast, 18 km in diameter.

Posidonius Large (diameter 100 km) partially flooded and ruined crater on the northeast shore of Mare Serenitatis. Its floor contains a curving ridge, several rilles and a small bowl crater. Ruined structure Chacornac, 51 km in diameter, abuts to the southeast.

Proclus Small (28 km wide) but brilliant crater on the western edge of Mare Crisium. High illumination shows it is the centre of a fan-shaped ray system.

Taruntius Low-walled crater in northwestern Mare Fecunditatis, 56 km in diameter, with concentric inner ring. Crater Cameron (formerly Taruntius C) interrupts the northwestern wall. Centre of a faint ray system.

Thales Bright ray crater 32 km in diameter, northeast of Mare Frigoris.

Moon Map 4

Bullialdus Handsome crater in Mare Nubium with terraced walls and complex central peak. Diameter 59 km. Two smaller craters, Bullialdus A and B, form a chain extending south.

Flamsteed Small (diameter 21 km) crater on Oceanus Procellarum with much larger ring of eroded hills to its north, Flamsteed P.

Gassendi Large, 100-km-diameter, partially flooded ring on the northern border of Mare Humorum. Complex internal pattern of clefts, ridges and hillocks. A rich area for transient lunar phenomena (TLPs). Deeper crater Gassendi A interrupts the north wall, with the smaller Gassendi B lying further north.

Grimaldi Vast, 222-km-diameter, dark-floored enclosure at the western limb of the Moon with broad, crater-strewn walls. Nearer the limb is a smaller dark patch, marking the floor of Riccioli, 152 km in diameter.

Hainzel Curious keyhole-shaped crater, 92 × 66 km, evidently composed of three craters fused together.

Hippalus Lava-flooded bay 58 km in diameter on the shores of Mare Humorum, in an area with many rilles.

Lansberg Prominent crater 40 km in diameter with massive walls and a central peak, on Oceanus Procellarum.

Letronne Large bay, 119 km wide, on the south side of Oceanus Procellarum. Its seaward facing wall has evidently been washed away by invading dark lava.

Mare Humorum Rounded lowland plain 420 km across with irregular borders. Ringed by clefts and wrinkle ridges. On the south it invades the rings Doppelmayer and Lee, although Vitello escapes destruction. On the east is the ringed bay Hippalus, associated with much surface faulting.

Schickard Major dark-floored enclosure 227 km across. South of Schickard are the overlapping craters Nasmyth (diameter 77 km) and Phocylides (114 km). Adjoining it to the southwest is the extraordinary plateau Wargentin, 84 km wide, evidently a crater filled to the brim with solidified lava.

Schiller Curious footprint-shaped enclosure, 179 × 71 km.

Sirsalis and Sirsalis A Twin craters near a long cleft, Rima Sirsalis, which stretches towards Darwin.

Moon Map 5

Abulfeda Prominent smooth-floored crater with sculptured inner walls, 62 km in diameter. Apollo 16 landed in the highlands north of here in 1972.

Albategnius Large (diameter 136 km) walled enclosure with central peak. The prominent crater Klein, 44 km in diameter, breaks the southwest wall.

Aliacensis Prominent crater with irregular outline, 80 km in diameter. Forms a pair with Werner.

Alpetragius 3900-m-deep, bowl-shaped crater on outer slopes of Alphonsus with large central dome. Diameter 40 km.

Alphonsus Large enclosure 118 km in diameter with complex walls and a central ridge. Numerous craterlets and clefts cover the floor; several dark patches are visible under high illumination. Alphonsus is the site of reported obscurations believed to be caused by the release of gas from the surface.

Arzachel Magnificent crater 97 km in diameter with terraced walls and prominent central peak. To its east is a noticeable crater 32 km in diameter with central peak, like a smaller version of Alpetragius, called Parrot C.

Barocius Large formation southeast of Maurolycus, diameter 82 km. Its northeast wall is broken by Barocius B. To the southwest is Clairaut, 75 km in diameter, between Barocius and Cuvier.

Birt Sharp, bright crater 17 km in diameter on eastern Mare Nubium. Telescopes reveal a smaller crater (diameter 7 km), Birt A, on the east wall, and under low illumination a rille to the west.

Blancanus Crater 105 km in diameter, south of Clavius.

Clavius Magnificent walled plain 225 km in diameter. Note the distinctive arc of smaller craters across its convex floor. Its south wall is interrupted by the 50-km crater Rutherfurd, and its northeast wall by 52-km Porter.

Delambre Prominent crater 53 km in diameter with irregular interior, southwest of Mare Tranquillitatis.

Deslandres Huge, low and eroded formation 234 km in diameter, southeast of Mare Nubium. The partially ruined ring Lexell, diameter 63 km, opens onto its southern side, and the prominent crater Hell (diameter 33 km) lies on its western floor.

Fra Mauro Largest member, 94 km wide, of an old, eroded crater group north of Mare Nubium, also including Bonpland (diameter 60 km), Parry (46 km), and Guericke (58 km). Apollo 14 landed at Fra Mauro in 1971.

Heraclitus Curious elongated formation with central ridge, south of Stöfler. Its southern end is rounded off by the crater Heraclitus D. Between Heraclitus and Stöfler is the crater Licetus, diameter 75 km. Touching Heraclitus to the east is Cuvier, also 75 km in diameter.

Herschel Deep (3900 m) crater with elongated central peak, north of Ptolemaeus; diameter 41 km. Farther north is the slightly smaller Spörer, a partially filled ring.

Hipparchus Large, eroded enclosure 150 km in diameter north of Albategnius. Its central 'peak' is actually a small ruined crater. On its northeastern floor is Horrocks, 30 km in diameter and 2800 m deep. Between Hipparchus and Albategnius lies the crater Halley (diameter 35 km), east of which is Hind, 29 km wide and 2800 m deep.

Longomontanus Large walled plain in the rugged southern uplands of the Moon. Diameter 145 km. A ridge to the east forms a crescent-shaped enclosure known as Longomontanus Z.

Maginus Major walled plain 163 km in diameter north of Clavius. Convex floor with small peaks in centre. Southwest wall is interrupted by smaller crater, Maginus C.

Mare Nubium Irregular dark lowland plain, covered with numerous wrinkle ridges and ghost craters. On its southern shore is the flooded crater Pitatus, and in the southwest are the dark-floored pair Campanus and Mercator (48 km and 47 km in diameter). Its most celebrated feature is a 26-km-long fault on the eastern side called the Rupes Recta, or Straight Wall, between the craters Birt and Thebit. The Straight Wall appears to run north–south through the remains of an old flooded crater of which only the eastern half remains. At the southern end of the Straight Wall are the Stag's Horn Mountains, apparently the remains of a flooded crater.

Maurolycus Distinctive crater, 114 km wide, with twin central peak. Its walls rise to 5000 m. It partially obliterates a smaller, unnamed formation to the north.

Moretus Crater with a central peak in the jumbled uplands southeast of Clavius. Diameter 114 km. Closer to the south pole are the craters Short (diameter 71 km) and Newton (diameter 64 km), heavily foreshortened.

Pitatus Large, dark-floored ring 105 km across on the southern shore of Mare Nubium. Invading lava from Mare Nubium has partly destroyed the crater's walls and left only the vestige of a central peak. Note the rilles around its inner walls. A smaller, similarly flooded ring adjoining it to the east is Hesiodus.

Ptolemaeus Vast walled plain 153 km in diameter, hexagonal in shape. Its ancient floor is heavily pockmarked with smaller craters, the most prominent being Ptolemaeus A.

Purbach Battered but still prominent large crater, 118 km in diameter. Its floor contains ridges, and its north wall is interrupted by the oval crater Purbach G, while its south wall intrudes into Regiomontanus.

Regiomontanus Flooded crater, 126 × 110 km. Its central peak has a summit craterlet. Forms a pair with Purbach, both noticeably hexagonal in outline.

Scheiner Crater 110 km in diameter southwest of Clavius. The largest of three craterlets on its floor is Scheiner A.

Stöfler Large, flat-floored formation 137 km wide, west of Maurolycus. Its eastern wall is destroyed by the intrusion of several craters, the largest being Faraday, diameter 69 km. The southern wall of Faraday is disturbed by Faraday C, which itself intrudes into Stöfler P.

Thebit Fascinating triple crater on southeastern Mare Nubium. The main crater, 55 km in diameter, is broken by the 20-km Thebit A, which in turn is broken by the smaller Thebit L.

Tycho Magnificent crater in the Moon's southern uplands, 85 km in diameter. Prominent at all angles of illumination, and brilliant under high lighting. Massively terraced walls rising up to 4500 m, imposing central peak and rough floor. Tycho is the major ray crater on the Moon. Rays from Tycho extend for 1500 km or more in all directions. Note the dark 'collar' around Tycho under high illumination. Probably the youngest of the Moon's major features.

Walter Large crater 135 km in diameter, considerably modified by landslips on the inner walls and by several interior craterlets. Appears almost square in outline.

Werner Prominent crater 70 km in diameter with walls 4200 m high, notably sharper and more rounded than neighbouring craters. The floor of Werner is dotted with several hills.

Moon Map 6

Capella Prominent crater north of Mare Nectaris, 45 km in diameter, deformed by a surface fault. Large central peak. Isidorus, 39 km in diameter, adjoins it to the west.

Catharina One of a curving trio of craters around western Mare Nectaris. Diameter 97 km. A faint ring, Catharina P, covers much of its northern floor.

Cyrillus Crater 93 km in diameter, with complex terraced walls, multiple central peak and rugged floor. Overlapped by Theophilus.

Fracastorius Horseshoe-shaped bay 124 km in diameter on the southern shore of Mare Nectaris. Dark lava has breached its northern wall and flooded its interior.

Janssen Vast, irregular enclosure, 190 km in diameter, heavily bombarded. In the north it is interrupted by Fabricius, 78 km in diameter, which has a central peak. On the west wall is a smaller (diameter 34 km) sharp crater, Lockyer. Southeast of Janssen are the twin craters Steinheil and Watt (67 km and 66 km in diameter respectively). Farther north of Fabricius is Metius, diameter 88 km.

Langrenus Magnificent, bright walled plain on eastern Mare Fecunditatis with terraced walls, outer ridges and a complex central peak. Its diameter is 132 km. Langrenus is a ray centre. Northwest of it, on Mare Fecunditatis, is a trio of smaller craters called Langrenus F, B and K in order of decreasing size. South of Langrenus is the large, flooded formation Vendelinus, 147 km across.

Mädler Prominent crater (diameter 28 km) on northwest Mare Nectaris, with a central ridge.

Mare Fecunditatis Irregularly shaped dark lowland area, connecting with Mare Tranquillitatis. On its western border it invades several craters, notably Gutenberg (diameter 71 km) and Goclenius (diameter 60 km). Note the numerous clefts in this area.

Mare Nectaris Rounded lowland plain 400 km wide, bordered by several large craters, notably Theophilus, Cyrillus, Catharina and Fracastorius. An outer mountain ring, the Altai Scarp, surrounds Mare Nectaris.

Messier and Messier A Elliptical pair of craters on Mare Fecunditatis, prominent despite their small sizes of 9 × 11 km and 13 × 11 km. From the western member, Messier A, extend two bright rays. Both craters appear brilliant under high illumination.

Palitzsch A crater and valley on the eastern side of Petavius. The crater itself is at the southern end of the valley, which is 150 km long.

Petavius Magnificent walled enclosure 177 km in diameter. A prominent rille runs across the floor from the massive, complex central peak to the terraced walls, which appear double in parts. Ridges radiate from its outer walls. West of Petavius is Wrottesley, 57 km in diameter, and with a central peak. South of it is Hase, 83 km in diameter.

Piccolomini Beautiful crater on the Altai Scarp. Diameter 89 km, with a broad central peak and terraced walls.

Rheita Valley Crater chain northeast of Janssen and Fabricius. It can be traced for a total length of about 450 km. The crater Rheita itself, 70 km in diameter and with a small central peak, lies at the northern end of the valley.

Theophilus Imposing crater 100 km in diameter on the northwestern rim of Mare Nectaris, with a massive 2200-m central mountain. Terraced walls rise over 5000 m above the floor, with many external ridges.

Map 1

Northwest Section

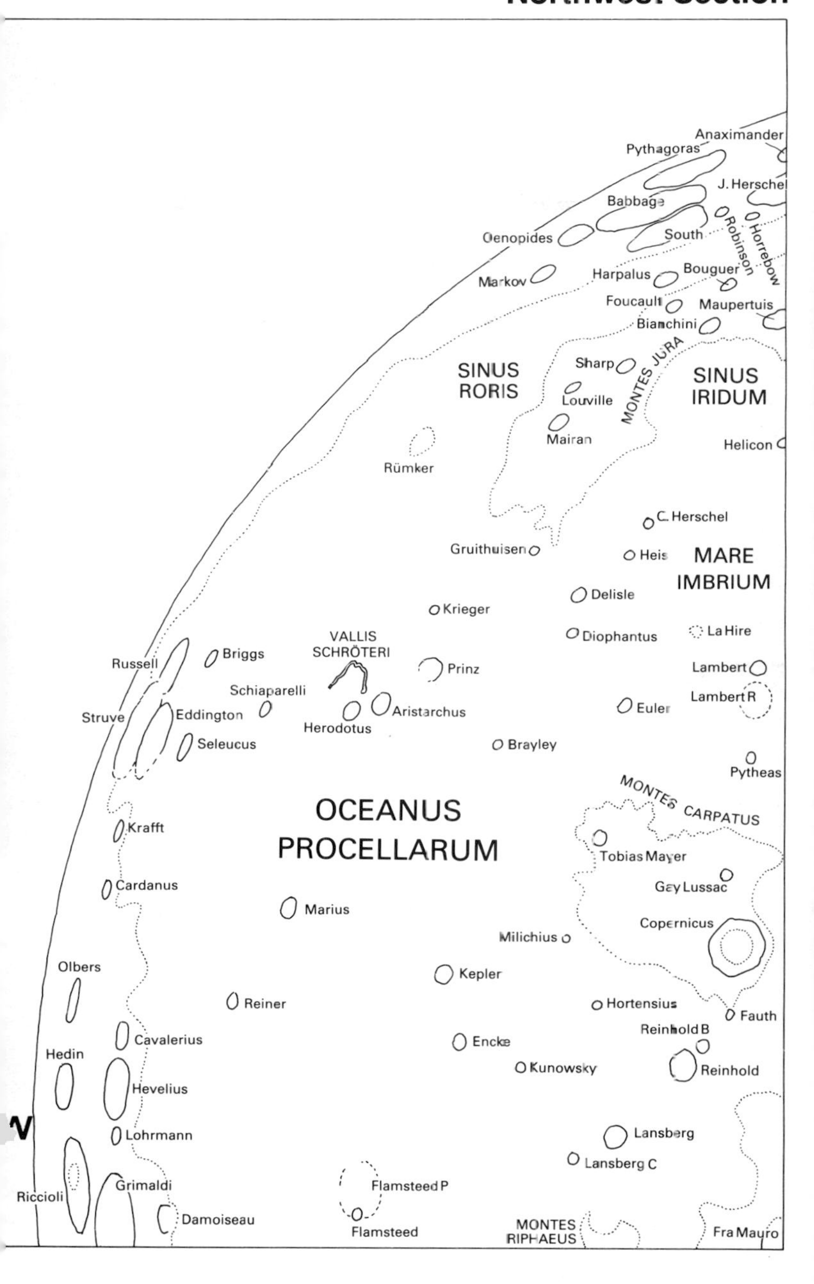

Map 2

North Central Section

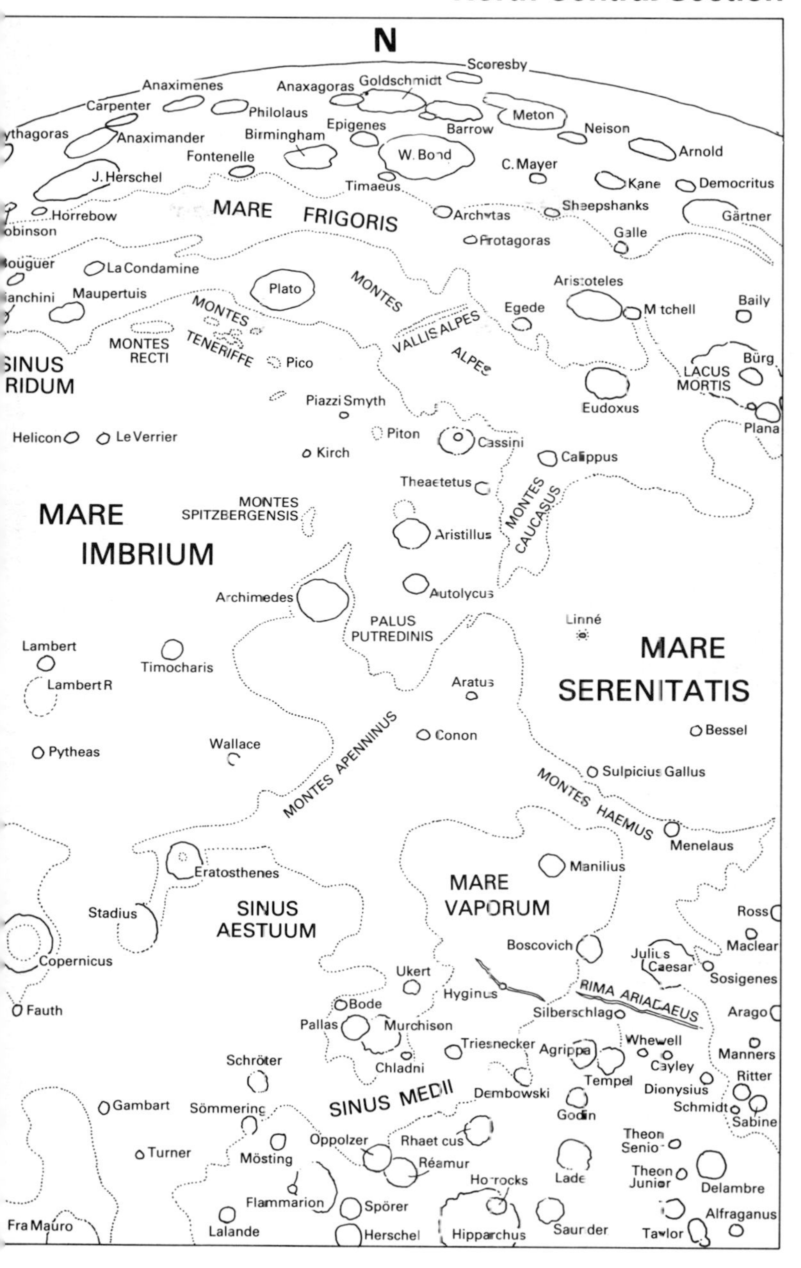

Map 3

Northeast Section

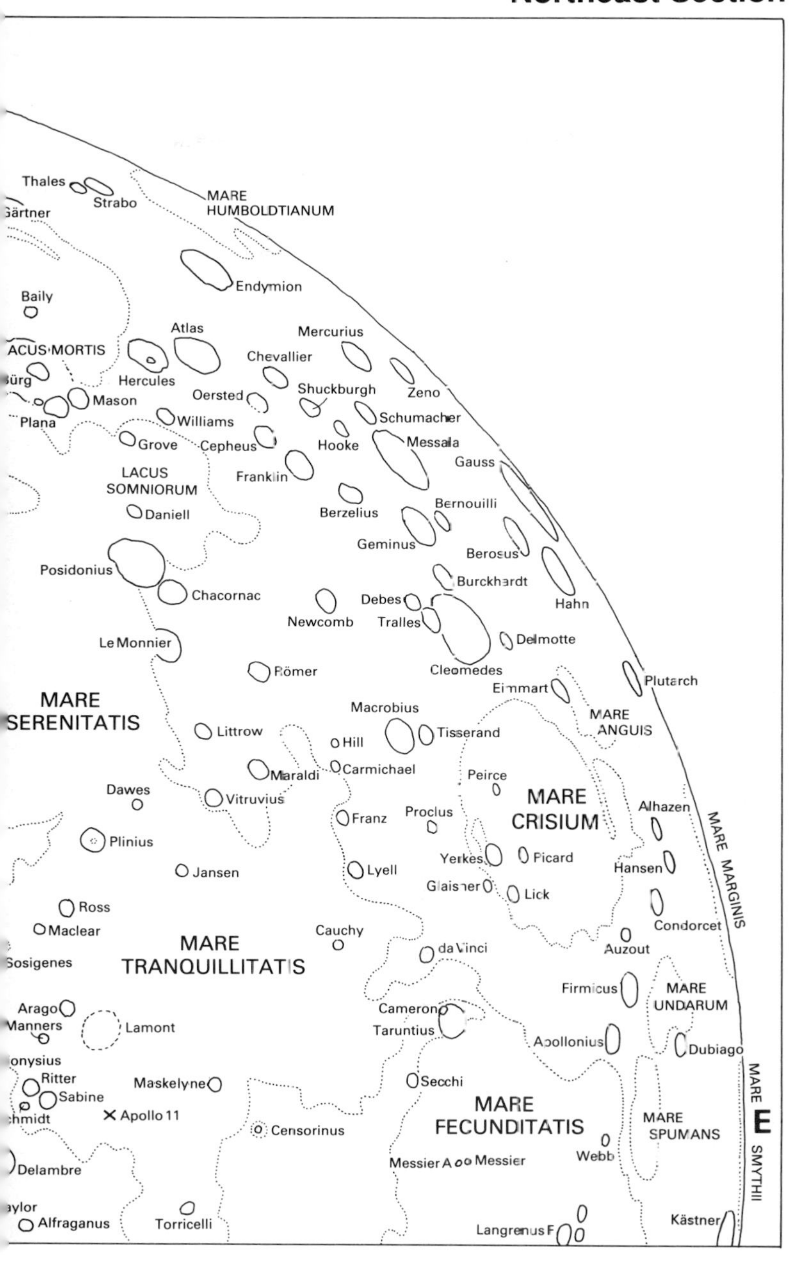

Map 4

Southwest Section

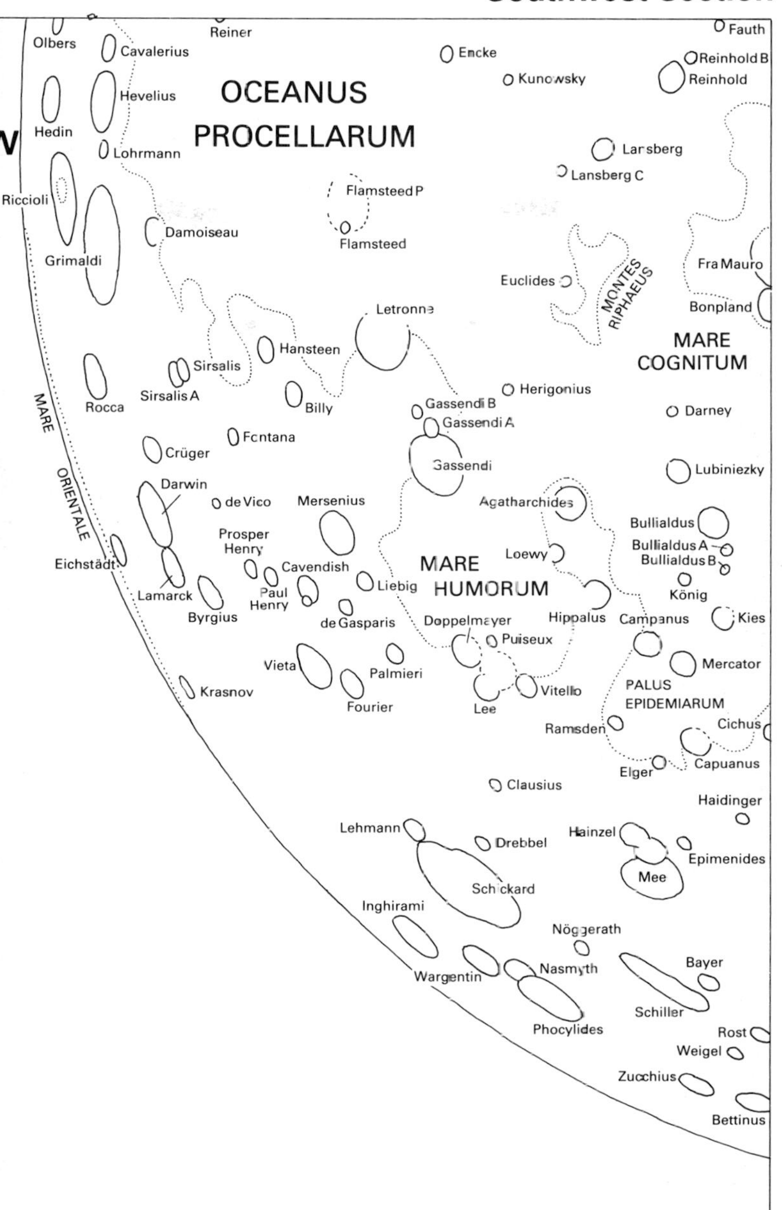

Map 5

South Central Section

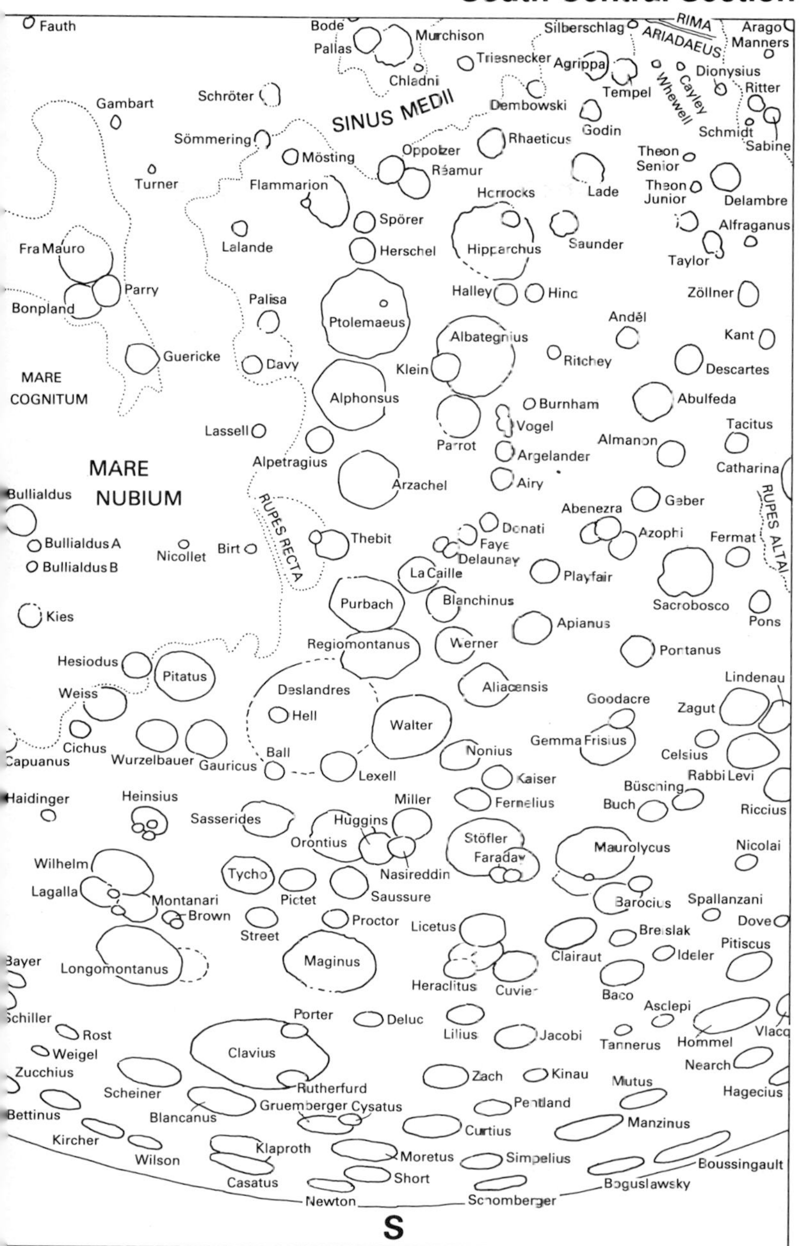

Map 6

Southeast Section

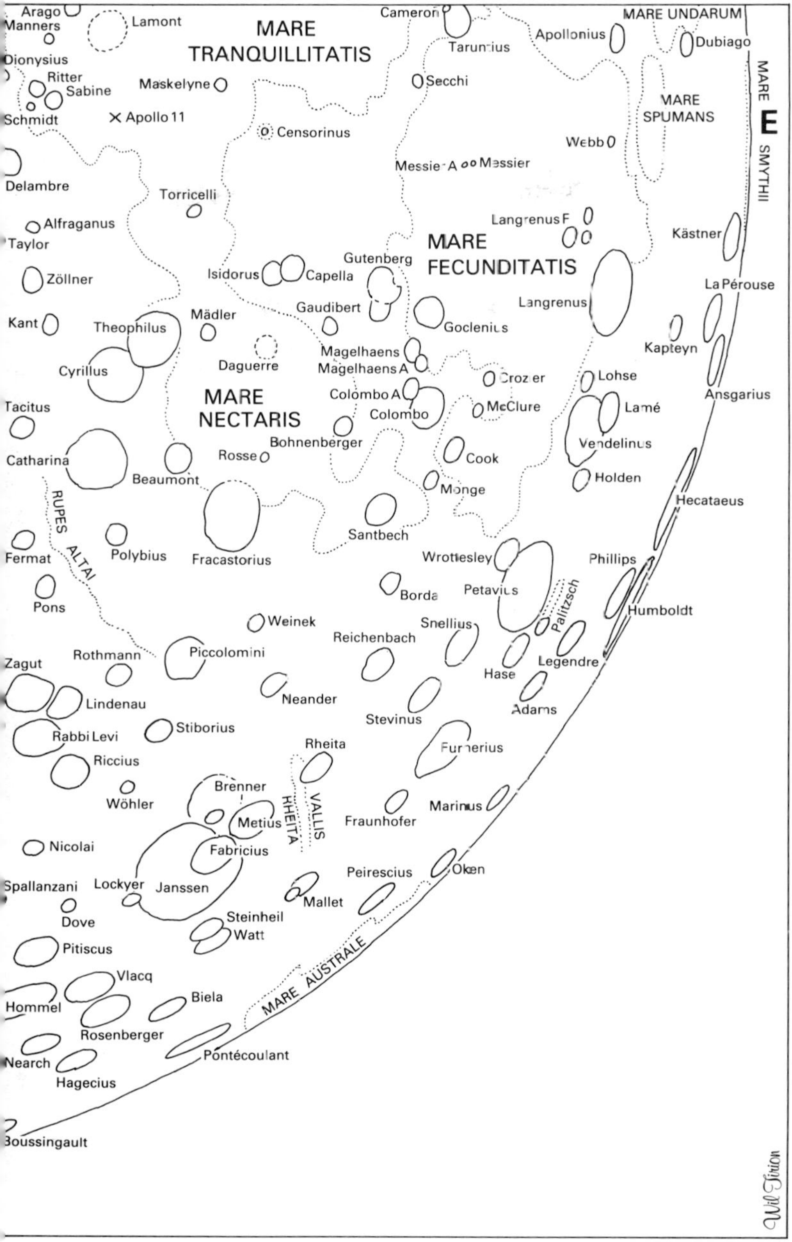

Arago
Manners
Lamont
MARE TRANQUILLITATIS
Cameron
MARE UNDARUM
Apollonius
Taruntius
Dubiago
Dionysius
Ritter
Sabine
Maskelyne
Secchi
MARE SPUMANS
MARE SMYTHII
E
Schmidt
Apollo 11
Censorinus
Webb
Messier A
Messier
Delambre
Torricelli
Langrenus F
Alfraganus
Taylor
MARE FECUNDITATIS
Kästner
Gutenberg
Isidorus
Capella
Zöllner
La Pérouse
Langrenus
Gaudibert
Mädler
Kant
Theophilus
Goclenius
Kapteyn
Magelhaens
Daguerre
Magelhaens A
Cyrillus
Crozier
Lohse
Colombo A
MARE NECTARIS
Ansgarius
Lamé
Tacitus
Colombo
McClure
Bohnenberger
Vendelinus
Cook
Catharina
Rosse
Beaumont
Monge
Holden
RUPES ALTAI
Hecataeus
Santbech
Fermat
Polybius
Fracastorius
Wrottesley
Phillips
Borda
Petavius
Pons
Palitzsch
Humboldt
Weinek
Snellius
Reichenbach
Zagut
Rothmann
Piccolomini
Legendre
Hase
Lindenau
Neander
Adams
Stevinus
Rabbi Levi
Stiborius
Rheita
Furnerius
Riccius
Brenner
VALLIS RHEITA
Wöhler
Marinus
Metius
Fraunhofer
Nicolai
Fabricius
Oken
Peirescius
Spallanzani
Lockyer
Janssen
Mallet
Dove
Steinheil
Watt
Pitiscus
MARE AUSTRALE
Vlacq
Biela
Hommel
Rosenberger
Pontécoulant
Nearch
Hagecius
Boussingault
Wil Tirion

Eclipses of the Sun and Moon

Occasionally the Sun, Moon and Earth line up exactly to cause an eclipse. When the Moon comes between the Sun and the Earth it blocks off the Sun's light from part of the Earth, causing an eclipse of the Sun. When the Moon is on the opposite side of the Earth to the Sun, the Earth can block off the Sun's light from the Moon, causing an eclipse of the Moon. If the Moon orbited in the same plane as the Earth's orbit around the Sun, then an eclipse would occur at each full or new Moon. But since the Moon's orbit is inclined at 5° to the orbit of the Earth, eclipses occur only on those occasions when the Moon is crossing the Earth's orbit at new or full Moon.

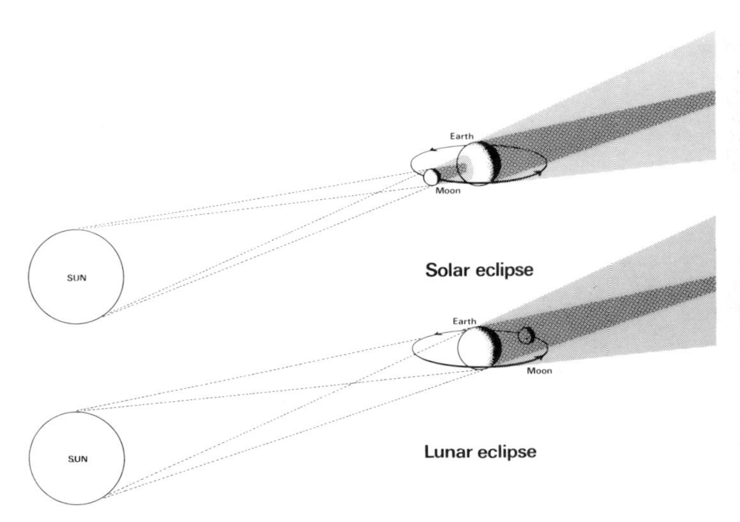

When the Moon passes in front of the Sun as seen from Earth, a solar eclipse occurs. When the Moon enters the Earth's shadow, a lunar eclipse occurs. Wil Tirion.

Each year there are at least two eclipses of the Sun, and a maximum of five, and from zero to three of the Moon, although not all of them will be visible from one place on Earth. The maximum number of eclipses possible in one year, both solar and lunar, is seven. An eclipse of the Moon is visible from wherever the Moon is above the horizon, but an eclipse of the Sun can be seen only from within the narrow band on which the Moon's shadow falls.

Scientifically, total eclipses of the Sun are by far the most important. At a total solar eclipse the Moon completely blots out the brilliant disk of the Sun, allowing astronomers to observe the Sun's faint outer halo of gas, the corona. (It is a fortunate coincidence that the Sun and Moon appear almost exactly the same size in the sky.) Astronomers travel across the globe for the few precious moments that totality affords. The longest that a total eclipse of the Sun can last is 7 minutes 30 seconds, but the usual duration is 2–3 minutes.

The band of totality is a mere 300 km across at its widest, but outside this band is a much wider area from which a partial eclipse of the Sun can be seen. Sometimes, when the Moon is at its farthest from us in its elliptical orbit, it is too small to cover the Sun's disk completely,

A complete ring of Baily's beads around the Sun at the annular eclipse of May 20, 1966, as seen from the Greek island of Lesvos. This 'diamond necklace' effect is caused by mountains around the rim of the Moon. John Mathers.

Above: *The 'diamond ring' effect, plus pink prominences visible against the pearly light of the Sun's inner corona, at the end of the total solar eclipse of July 11, 1991, viewed from Mexico.* Armagh Planetarium.

Left: *A crescent Sun. Over half the Sun's disk is blotted out by the Moon at the partial solar eclipse of February 25, 1971, seen through clouds.* Ian Ridpath.

and so a ring of bright sunlight remains visible around the dark outline of the Moon. Such an eclipse is termed an *annular* eclipse (from the Latin word *annulus* meaning a ring, not because they occur yearly).

Partial and annular eclipses are of curiosity value only; they have none of the scientific importance of a total eclipse. A solar eclipse starts at *first contact*, when the edge of the Moon begins its progress across the face of the Sun. Totality is still 1½ hours away. The partial phases of the eclipse can be observed by looking at the Sun through a dark filter, or by projecting the Sun's image through binoculars or telescope onto a white card. When you do this, compare the jet-black outline of the Moon with the umbra of sunspots. This will confirm that sunspots are not totally black; instead their colour appears somewhat brownish.

A suitable filter for looking at the Sun is heavily overexposed black-and-white photographic negative film (use two or three layers if necessary). You can look at the Sun quite safely through this, because the silver in the film absorbs both the light and the heat from the Sun. Colour film is not suitable as a filter because the colour dyes do not absorb infrared (heat) radiation; although colour film dims the light from the Sun sufficiently, the Sun's heat will still get through to damage your eyes. Another simple but safe way of observing an eclipse of the Sun is to make a pinhole in a piece of card and allow the Sun's light to pass through this hole onto a white surface. You are in fact using a pinhole camera to observe the Sun. The drawback with this arrangement is that the image so formed is small and faint.

Not until about 15 minutes before totality, when the Sun's disk is more than 80 per cent covered, does the sky start to get noticeably dark. An eerie half-light falls over the landscape; animals act as though night is falling. Totality itself rushes up as the last crescent sliver of sunlight is blotted out by the Moon. In the final seconds, chinks of sunlight peep between the mountains at the Moon's rugged edge, producing *Baily's beads*, named after the English astronomer Francis Baily who described them after the eclipse of 1836. Often one bead shines brighter than the others, producing the effect of a diamond ring.

Then, second contact: the Moon completely covers the Sun, and the pearly-coloured corona springs into view. Feather-like plumes and streamers of the corona extend outwards from the polar and equatorial regions of the Sun for several solar diameters. Pinkish-red prominences can be seen looping out from the Sun's chromosphere around the dark outline of the Moon. Bright stars are visible in the darkened sky. All too soon, the beautiful spectacle is over. The diamond ring flashes out at third contact, signalling the end of totality. At fourth contact, the Moon moves clear of the Sun. The eclipse is over.

Eclipses of the Moon are far less spectacular. The Moon takes several hours to move completely through the dark inner part of the Earth's shadow, the umbra. The outer part of the shadow, the penumbra, is so light that it produces little noticeable darkening of the Moon's surface.

A total eclipse of the Moon can last for up to 1¾ hours, but even when totally eclipsed the Moon seldom completely disappears. The reason for this is that light is bent into the Earth's shadow by the Earth's atmosphere, giving the eclipsed Moon a coppery-red colour. Dark eclipses occur when there are a lot of clouds and dust in the Earth's atmosphere, which block the light. Although of interest as a natural spectacle, a lunar eclipse is of little scientific importance.

Mercury

Mercury is a disappointing object for observers. Through even the largest telescopes it appears as nothing more than an almost featureless orange blob, less distinct than the Moon appears to the naked eye, that goes through phases as it orbits the Sun every 88 days. Most observers must therefore be content with simply catching a glimpse of this elusive object during one of its periodic appearances in the morning or evening sky.

Because Mercury is the closest planet to the Sun, it and the Sun never appear far apart in the sky. Circumstances dictate that there are two good times to look for Mercury. One is when it is setting after the Sun during evenings in spring (around March–April in the northern hemisphere or September–October in the southern hemisphere). The second is when it is a morning object rising before the Sun in the autumn (September–October in the northern hemisphere, March–April in the southern hemisphere).

An additional complication is that its orbit is markedly elliptical, ranging from 46 to 70 million km from the Sun, so that even on these occasions there are times when Mercury is easier to see than at others. Even when Mercury is best placed, a low, clear horizon will be needed to see it; binoculars help to pick the planet out of the twilight glow, for Mercury can never be seen against a truly dark sky. With all these complications, it is little wonder that many town dwellers have never seen the planet. Nevertheless it is worth looking for, because at its brightest it can shine at magnitude –1 or more.

The difficulty in observing Mercury led to a long-standing mistake concerning the time it takes to rotate on its axis. Towards the end of the 19th century, the Italian astronomer Giovanni Schiaparelli proposed, after a long series of observations, that the planet spins on its axis in 88 days, the same time as it takes to orbit the Sun. It would therefore keep one face turned permanently towards the Sun, as the Moon does to the Earth. In the 1920s, the Greek-born astronomer Eugène Antoniadi compiled a map showing smudgy markings on the surface of Mercury based on an assumed 88-day rotation period. This map seemed to settle the matter once and for all.

Then, in 1965, came a surprise. At Arecibo Radio Observatory, astronomers Rolf Dyce and Gordon Pettengill bounced radio waves off the surface of Mercury. From the change in frequency of the reflected radio waves, they deduced that Mercury spins once every 59 days, two-thirds of the time it takes to orbit the Sun. Therefore the Sun does rise and set on Mercury, but very slowly. For the Sun to go once around the sky as seen from the surface of the planet – say, from one noon to the next – takes 176 Earth days, during which time Mercury orbits the Sun twice, spinning three times on its axis.

Not the crescent Moon, but the planet Mercury photographed by the space probe Mariner 10 in March 1974. The bright ray crater just above the centre is called Kuiper. The largest craters are about 200 km across. Jet Propulsion Laboratory.

In the skies of Mercury the Sun appears two and a half times as large as it does from Earth. The Sun's intense heat roasts the surface rocks of Mercury to over 400°C at noon on the equator, hot enough to melt tin and lead. Without an atmosphere to hold in the heat, the planet's surface cools to a frigid –170°C during the long night. The daytime side of Mercury is continually blasted by lethal doses of high-energy solar radiation.

Astronomers had long assumed that Mercury resembled our Moon in appearance. For one thing, it is only 50 per cent larger than our Moon, with a diameter of 4880 km. The only planet in the Solar System smaller than that is Pluto. But it took the space probe Mariner

10 in 1974 to show how remarkably similar Mercury and the Moon appear. As Mariner 10 flew past Mercury, its cameras photographed a surface heavily pockmarked with craters of all sizes, similar to the lunar highlands. Mariner 10 photographed little more than a third of Mercury's surface, yet, if that portion is typical of the rest of the planet, it is enough to tell us much of the previously unknown story of Mercury.

Craters on Mercury look almost identical to their lunar counterparts. There are deep, young craters, eroded ancient craters, craters with terraced walls, central peaks and bright rays. Many of the features on Mercury have been named after artists, composers and writers, thereby breaking the near-monopoly that astronomers' names previously had in the naming of surface features of Solar System bodies. For instance, we now find on Mercury memorials to Bach, Mozart, Van Gogh and Chekhov.

Sometimes it is difficult to distinguish at a glance between a picture of Mercury and one of the Moon. Almost certainly, the craters on both bodies have been formed in the same way, by the impact of large meteorites early in the history of the Solar System. One noticeable difference on Mercury is that material ejected from the craters has not travelled as far as on the Moon, because Mercury's surface gravity is stronger – over twice that of the Moon, but still only 38 per cent of the Earth's. Another result of the higher gravity of Mercury is that craters tend to be shallower for a given diameter than on the Moon.

Mercury has cliffs hundreds of kilometres long and 2–4 km high, unlike anything on the Moon. These are believed to have been caused by a shrinking of the planet as its core cooled early in its history, leading to compression and faulting in the crustal rocks. The surface rocks of Mercury are actually slightly darker in colour than those of the Moon, reflecting a mere 6 per cent of the sunlight hitting them, compared with 7 per cent for the Moon; Mercury in fact has the darkest surface of any planet in the Solar System.

Between many of the large craters on Mercury are areas of ancient crust peppered only with small craters. These areas, termed the intercrater plains, have no real counterpart on the Moon. They clearly predate the large craters, and may be examples of the original crust of Mercury. By contrast, the Moon's original crust has been totally churned up by impacts so that none of it remains unaltered. Alternatively, the original crust of Mercury may have been churned up like that of the Moon, but was subsequently smoothed out again by volcanic activity. Only study of actual rocks from the planet will resolve this dispute. Other areas of particular interest for first-hand study are deep craters near the poles whose interiors are permanently shaded from the Sun, thereby preserving in deep freeze any gases that have seeped from the planet over its history.

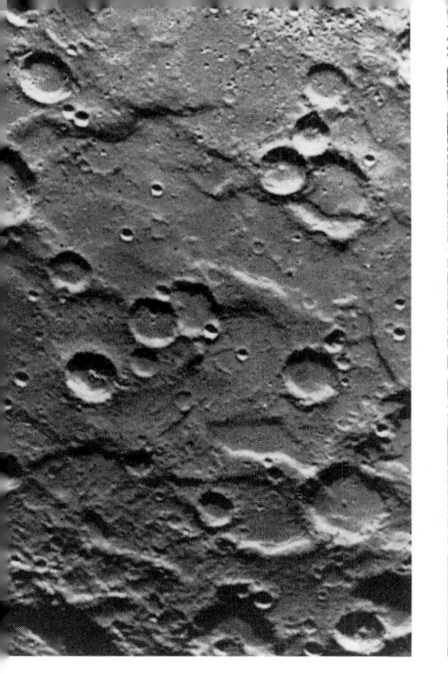

Mercury from Mariner 10. Left: *Plains between the craters in the region of the south pole are traversed by ridges and scarp slopes.* Right: *Bright rays similar to those radiating from craters on the Moon cross the terrain of Mercury. The largest crater seen here is 100 km in diameter.* Jet Propulsion Laboratory.

Of all the features seen by Mariner 10, the most prominent is an enormous bull's-eye structure, partly hidden by shadow, named the Caloris Basin. Its diameter is 1400 km, similar to Mare Imbrium on the Moon; it was presumably formed by the impact of an asteroid after most of the surface had already been cratered. The Caloris Basin contains several concentric rings of mountains, and is surrounded by a number of radial ridges and grooves. Most importantly from a geological point of view, its interior, and much of the low-lying land around it, has been flooded by lava. Geological activity died out on Mercury over 3000 million years ago, as it did on the Moon. Since then, little has changed except for the random arrival of a stray meteorite.

Despite its outward resemblance to the Moon, inwardly Mercury is believed to be more like the Earth. Mercury has a relatively large mass for its small diameter, which implies that it has a large iron core three-quarters of the planet's diameter. A core that size would be as big as the Moon. The existence of an iron core was directly confirmed when Mariner 10 measured a magnetic field around the planet, albeit with only 1 per cent of the strength of the Earth's magnetic field – but that is still far stronger than the magnetic fields of Venus and Mars. In its own way, Mercury turns out to be a fascinating world containing many clues about the origin and development of the Solar System.

Venus

Many people have seen Venus without realizing it. The planet appears as the brilliant evening or morning 'star', the most prominent object in the twilight, unmistakably outshining every genuine star with its cold white light. So striking is Venus at its best that it is frequently reported as a hovering UFO.

Venus orbits the Sun every 225 days at a distance of 108 million km; it can pass within 40 million km of the Earth, closer than any other planet. In size (only 650 km less than the Earth's diameter) it is almost a twin of the Earth, at 12,100 km in diameter. But the brilliance of Venus in the sky is due not just to its proximity and size. The main reason is its cloak of unbroken clouds that reflect 76 per cent of the light hitting them. Although making Venus so prominent, the existence of these clouds prevented astronomers from seeing its surface.

Through a telescope, Venus appears like a white billiard ball that goes through phases similar to the Moon's as it orbits the Sun. As with the Moon, one complete cycle of phases (the synodic period) takes longer than one orbit (the sidereal period) because the Earth and Venus are moving relative to each other in their respective orbits around the Sun; for Venus, the synodic period is 584 days. When it is a crescent, Venus is close enough to the Earth for its phase to be picked out in modest binoculars; some people even claim to have seen Venus as a crescent with the naked eye. The planet appears most brilliant when 28 per cent of its disk is illuminated as seen from Earth, this being the most favourable combination of distance and phase. Venus can reach a maximum magnitude of –4.7, nearly seven times brighter than the next most prominent planet, Jupiter. Being so bright, Venus is best observed against a twilight sky to reduce the dazzle.

A simple pair of binoculars, or a small telescope, will show the crescent Venus as well as in this photograph. Unbroken clouds veil the surface of the planet from the view of even the largest telescopes. Hale Observatories.

Only the vaguest markings can be made out in the clouds of Venus with Earth-based telescopes; some dusky shadings are visible on the disk, and often the clouds appear to be brighter at the poles, with a surrounding darker collar. The terminator (the edge of the illuminated portion) can appear irregular, due not so much to differences in the height of the clouds as to differences in brightness. These effects are caused by the corkscrew circulation of clouds around Venus from its equator to its pole, as was realized from space-probe photographs.

Unable to see the planet's surface because of the enveloping clouds, astronomers could only guess at the rotation period of Venus – and they guessed incorrectly. As with Mercury, radar observations provided the surprising truth. It turns out that Venus rotates on its axis from east to west, the opposite direction from the Earth and other planets, and it does so very slowly: once every 243 days, longer than the 225 days it takes to orbit the Sun. Its clouds, though, rotate every four days, also retrograde (east to west), a result of high-speed winds in the upper atmosphere.

Before the days of space probes, theories about the nature of the planet's surface abounded. Since Venus was so similar in size to the Earth, it was tempting to speculate that conditions there might be Earth-like. One charming notion was that Venus resembled our planet as it had been in Carboniferous times, with steaming jungles and even dinosaurs. Some astronomers proposed that the planet was entirely covered with water, while others imagined it to be a world of deserts. None of the theories came close to anticipating the uniquely hostile conditions on Venus.

Radio astronomers provided the first clue in the late 1950s when they detected radio wave emissions from the planet, which implied that it was very hot (even hotter than boiling water). The deserts of the Earth are mildly warm by comparison. These readings, doubted at the time, were confirmed in 1962 by the American probe Mariner 2 which scanned the planet as it flew past at a distance of 35,000 km.

Conditions on Venus were experienced directly for the first time by a Soviet probe, Venus 4, when it parachuted into the atmosphere in October 1967. It found that Venus's atmosphere is made almost entirely of unbreathable carbon dioxide gas, but the probe was destroyed by the intense heat and crushing pressure long before it reached the surface. Venus 7 was the first probe to land intact on the surface, on December 15, 1970. It registered a temperature of 475°C and an atmospheric pressure 90 times that on Earth. Venus 7 landed on the night side of the planet; in 1972 Venus 8 landed on the day side, finding conditions there to be identical. The dense atmosphere of Venus traps heat like a blanket, keeping the temperature constant over the entire planet. Under such pressure cooker conditions, the atmosphere of Venus behaves more like a liquid than a gas. To explore

The clouds of Venus circulate once around the planet in 4 days, spiralling from equator to pole as they do so. The cloud patterns are visible only in ultraviolet light, as in this photograph from the Pioneer Venus orbiter. NASA.

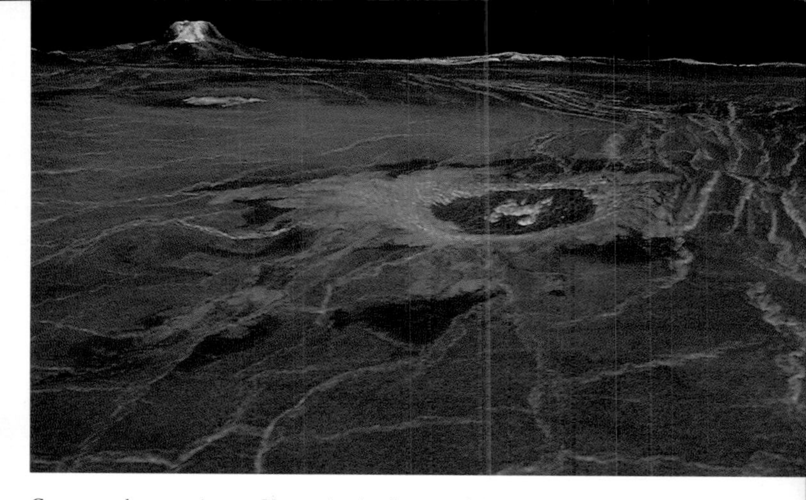

Craters and mountains on Venus: in the foreground is an impact crater named Cunitz, 50 km across, and on the horizon is the volcanic mountain Gula Mons, 3 km high, in this view reconstructed from Magellan radar imagery. The vertical scale is exaggerated, making the terrain seem rougher than it really is. NASA.

Venus is like trying to explore a scalding hot ocean, and it requires a refrigerated spacecraft reinforced like a submarine.

Why should Venus be so hot – hotter even than the Sunward face of Mercury, despite the fact that its clouds reflect over three-quarters of the incoming sunlight? The answer lies with what is termed the greenhouse effect. About 1 per cent of the incoming sunlight penetrates to the planet's surface, according to space probes, so that it is as gloomy there as on a heavily overcast day on Earth. That incoming sunlight is absorbed by the surface and is reradiated at longer wavelengths, in the infrared. Although the carbon dioxide of the atmosphere is transparent to visible light, it traps infrared; since infrared is heat energy, the temperature of the atmosphere rises.

It turns out that Venus and the Earth have similar amounts of carbon dioxide, but on Earth most of it is locked away in rocks such as limestone. Whereas the amounts of carbon dioxide are similar, Venus has far less water than the Earth. Whatever water it originally possessed has long since evaporated and been lost to space. Only a trace of water vapour remains, but that is sufficient to boost the effect of the carbon dioxide in causing the greenhouse effect of the atmosphere of Venus.

A final contribution to the greenhouse effect is provided by the clouds of Venus. These are made not of water vapour, as are the clouds of Earth, but of sulphuric acid of 80 per cent concentration, stronger than in a car battery. Sulphuric acid, too, absorbs infrared. Taken together, these three factors of carbon dioxide, water vapour and sulphuric acid turn Venus into a perfect suntrap. The clouds add to the nastiness of Venus in another way: from them descend showers of corrosive sulphuric acid rain. Despite its heavenly name, Venus turns out to be an incarnation of Hell.

In December 1978 a flotilla of five American space probes called Pioneer plummeted into the atmosphere of Venus. They found that the uppermost layer of sulphuric acid clouds, the one which observers view through telescopes, lies about 65 km above the planet's surface and is a few kilometres thick. Around 58 km altitude is a thin haze layer apparently consisting of sulphuric acid particles, which give the clouds their yellow tinge. The densest cloud layer of all is at about 50 km altitude, and it is from this layer that the rain of sulphuric acid droplets falls. Below the clouds, the gloom is broken by flashes of lightning, while thunder reverberates in the atmosphere.

Although the clouds of Venus mask its surface from view, astronomers have nevertheless been able to map the planet's features by using radio waves, which penetrate the clouds. Radar observations from Earth during the 1970s revealed some features, but the first complete map of the planet was made by a radar device aboard a Pioneer spacecraft that went into orbit around Venus in December 1978, a companion to the probes that entered the atmosphere.

Venus is mostly rolling plains, but there are three main continental areas. One, called Ishtar Terra, the size of the United States, has a mountain range which towers 12 km above the mean surface level, higher than Mount Everest on Earth. The largest continental area of all, Aphrodite Terra, the size of South America, is cut by a system of rift valleys that extends for thousands of kilometres.

Spectacular detail was revealed by the Magellan probe that went into orbit around Venus in August 1990. Magellan's radar 'eye' spotted impact craters ranging in size from over 100 km across down to 3 km wide, thus demonstrating that large meteorites can get through the dense atmosphere unhindered. Most excitingly of all, there were volcanic mountains with fresh-looking lava flows on their flanks. Most impressive was Maat Mons, 8½ km high, the second-highest peak on the planet, which lies near the equator in Aphrodite Terra; it is surrounded by lava flows estimated to be no more than ten years old. Clearly, Venus is still an active planet, and its highlands were formed by volcanic action.

Photographs from Soviet lander probes sitting on the surface of Venus show a rocky wasteland bathed in a sulphurous orange glow. Chemical analyses by these probes confirm that the surface rocks of Venus are similar in composition to volcanic basalts on Earth, as would be expected from the volcanic activity on the planet. Venus is a tantalizing vision of an Earth that might have been – and a terrifying demonstration of what the Earth itself might have become had it been born closer to the Sun.

Mars

Mars is distinguishable by its intense reddish-orange hue, stronger than the colour of any star and the cause of its association with the god of War. At its best Mars can shine at magnitude –2.8, rivalling Jupiter. But the most favourable appearances of the planet are few and far between, the main reason being the marked ellipticity of its orbit, which takes it from 206 to 249 million km from the Sun (average distance 228 million km). If the Earth passes Mars when Mars is closest to the Sun, only about 55 million km separates the two bodies and astronomers get their best views of the red planet. But when Mars is farthest from the Sun, 100 million km separate it from Earth, and it appears unimpressive even in powerful telescopes. The closest approaches of Mars occur at intervals of about 15 years, so astronomers do not waste their chances of observing Mars at its best.

Mars has a diameter of 6790 km, just over half that of the Earth. Its day is just over half an hour longer than our own – 24 hours 37 minutes – but its year is nearly twice as long as ours, 687 Earth days. Its orbit lies outside the Earth's, so Mars can never appear as a crescent. But at times it displays a distinctly gibbous phase, like the Moon when it is a couple of days from full.

Binoculars show Mars as nothing more than an orange dot of light; a telescope is needed to bring into view the main features of the planet. Among the most obvious are the white polar caps, which stand out in stark contrast to its ochre-coloured deserts. Occasionally, violent winds whip up dust storms in the thin atmosphere, obscuring the dusky surface markings. Frustratingly for observers, the worst dust storms tend to occur when Mars is at its closest to the Sun, thus ruining the best observing opportunities. The most prominent dark surface marking, a large triangular area named Syrtis Major, was first noted by the Dutchman Christiaan Huygens in 1659. Syrtis Major, along with the polar caps, should be visible in a modest amateur telescope.

The temptation to assume similarities between Mars and the Earth led early astronomers astray. The dark areas, which range in colour from brown to grey-green, were termed seas and lakes in the belief that they really were filled with water, while the orange areas were named after places on Earth – there is an Arabia, Libya, Syria and Sinai on Mars. Towards the end of the 19th century astronomers realized that there were no oceans on Mars after all, but this opened the way for a much more intriguing explanation for the dark areas: that they were covered with primitive vegetation, such as moss or lichen. In support of this view, observers noted that when the polar caps melted in the Martian summer, the surface markings became larger and darker; this was interpreted as being due to the vegetation growing in the milder, wetter conditions.

The most extreme proponent of the life-on-Mars idea was an American astronomer, Percival Lowell. His ideas were inspired by the discovery made in 1877 by an Italian, Giovanni Schiaparelli, of apparent long, straight lines criss-crossing the planet's surface. Schiaparelli called these *canali*, a word which in Italian means 'channels'; but inevitably it was translated as 'canals', implying that they were artificial, although Schiaparelli himself kept an open mind about their true nature.

For Lowell, there was no doubt: the canals were evidence of an advanced civilization on Mars. His beliefs sparked off a whole generation of science fiction, including the famous *War of the Worlds* by H.G. Wells. Lowell set up his own observatory at Flagstaff, Arizona,

Mars photographed by the Hubble Space Telescope in December 1990. The dark tongue-shaped marking is Syrtis Major, a volcanic shield. At top right, hazy clouds cover the planet's north pole. NASA.

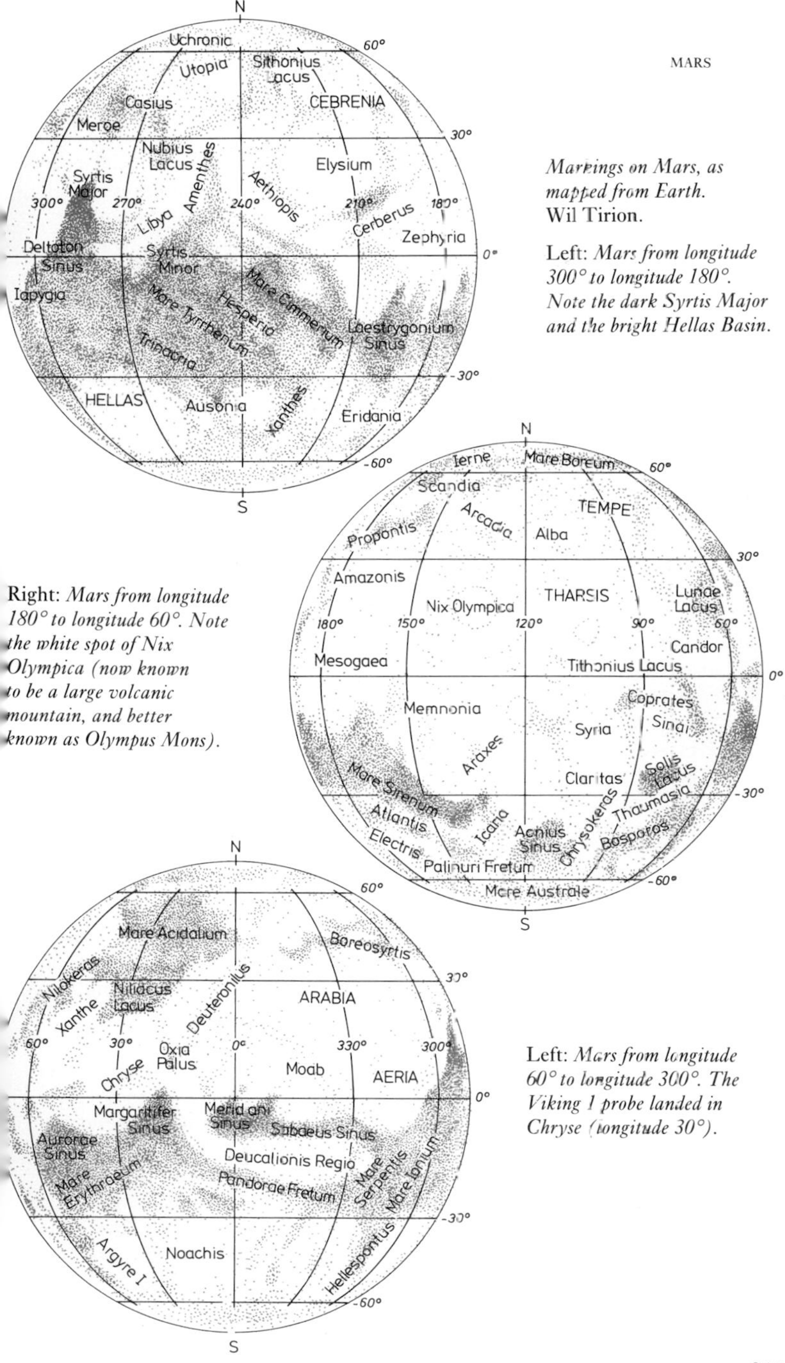

Markings on Mars, as mapped from Earth.
Wil Tirion.

Left: *Mars from longitude 300° to longitude 180°. Note the dark Syrtis Major and the bright Hellas Basin.*

Right: *Mars from longitude 180° to longitude 60°. Note the white spot of Nix Olympica (now known to be a large volcanic mountain, and better known as Olympus Mons).*

Left: *Mars from longitude 60° to longitude 300°. The Viking 1 probe landed in Chryse (longitude 30°).*

in 1894 specifically to study Mars. There, he produced fanciful maps of the canal network and wrote books such as *Mars as the Abode of Life*, published in 1908, in which he expounded his theory of a Martian civilization clinging to existence on an arid planet, reliant on the canals to bring meltwater from the polar caps to irrigate crops at the equator.

Most other astronomers failed to see the canals, or in their place could detect only broad, irregular smudges. After Lowell's death in 1916 a few devoted followers kept the canal theory going, but the idea of Martian civilization was doomed to oblivion in the light of new knowledge about conditions on the planet.

By the 1950s it was clear that the atmosphere on Mars was far too thin for a human to breathe. Under so thin an atmosphere, temperatures would be frigid and dangerous amounts of ultraviolet radiation would penetrate to the ground. To make matters worse, no oxygen could be detected in the planet's atmosphere, although carbon dioxide was known to be present. No advanced life form could exist under such conditions, although hardy vegetation was still not ruled out.

Our knowledge of Mars took a major step forward in July 1965, when the American space probe Mariner 4 sent back the first close-up photographs as it flew past Mars at a distance of 10,000 km. Its most important revelation was that there are craters on Mars, looking similar to those on the Moon, only somewhat more eroded because of the effects of the planet's atmosphere. Actually, the discovery of craters on Mars was not unprecedented. The great American observer Edward E. Barnard saw signs of them in 1892 with the 91-cm (36-inch) refracting telescope at Lick Observatory, as did John E. Mellish in 1915 using the Yerkes Observatory's 102-cm (40-inch) refractor, but neither man published his observations for fear that they would be disbelieved.

Unfortunately, a lunar-like Mars seemed an unpromising abode for life of even the lowliest form. The image of Mars as a dead world, in both the geological and biological sense, was reinforced in 1969 when Mariners 6 and 7 photographed more craters, apparently of impact origin. The first full survey of the entire planet came in 1971–2, when Mariner 9 went into orbit around Mars. It revealed major formations that previous Mariners had, by ill luck, completely missed.

For a start, there was a chain of three volcanic mountains atop a highland area known as the Tharsis Bulge; they are named Arsia Mons, Pavonis Mons and Ascraeus Mons. ('Mons' means mountain). Other types of feature on Mars and the names given to them include lowland plains (Planitia); highland plateaus (Planum); valleys (Vallis); canyons (Chasma); and eroded craters (Patera). Northwest of the chain of Tharsis volcanoes is an even bigger volcanic mountain, Olympus Mons, originally seen from Earth as a white ring and then known as Nix Olympica ('the snows of Olympus'). This whole area is noted for the frequent appearance of white clouds, often described as being

Olympus Mons, an immense volcanic mountain, blisters the surface of Mars in this mosaic of photographs from the Mariner 9 space probe. Ancient lava flows spread out around it. Olympus Mons is visible to observers on Earth as a white ring, long known by the name Nix Olympica. Jet Propulsion Laboratory

shaped like a W. The existence of mountains here explains the preference of clouds for this region.

Olympus Mons, 600 km wide and 26 km high, is larger even than the volcanic islands of Hawaii on Earth; it is often given the title of 'largest volcano in the Solar System', although there may be larger ones on Venus. Whatever the case, Mars certainly has impressive volcanoes.

Another dramatic and unexpected feature was an immense system of faults, up to 600 km wide and 7 km deep, extending east from the Tharsis Bulge. This massive rift valley, now called the Mariner Valley, not merely dwarfs the Earth's Grand Canyon, but at 4000 km long it could span the entire United States. The Mariner Valley corresponds to a broad, fuzzy canal seen from Earth, called Coprates; it is visibile from Earth mainly because dark dust collects along its floor. Coprates is one of the few 'canals' that has any corresponding surface feature on Mars. A few other canals, notably one called Cerberus, do coincide with dark streaks on the Martian surface, caused by darker dust or rock; but these are all broad and irregular markings, unlike the thin, straight canals that Lowell drew. Close comparison of Lowell's canal maps with space probe pictures reveals few correlations at all. There seems no real explanation for the canal network that Lowell and his followers drew, other than the fallibility of human observers straining at, or past, the limits of visibility.

Although the Lowellian canals were not in evidence, Mariner 9 found convincing signs that water had once flowed on Mars. Sinuous channels, looking like dried-up river beds, snaked across parts of the planet's surface. Some of the lowlands appeared to have once been inundated by flash floods. Liquid water cannot exist on Mars today because the atmospheric pressure is too low. The existence of ancient watercourses implied that in the past the atmosphere was denser – and a denser atmosphere would keep the planet warmer. Perhaps the volcanoes of Mars gushed out enough gas to change the climate temporarily. If that were so, life might have had a chance to arise on Mars after all, and microscopic organisms such as bacteria might still cling to existence among the red sands of the planet. The only way to find out was to go and look.

In 1976 two American space probes called Viking were sent to look for life on Mars. Each probe came in two halves: a lander and an orbiter. The Viking 1 lander touched down on a lowland plain called Chryse in the northern hemisphere, over which water appeared to have flowed during the wet times on Mars. The other lander descended on

The Mariner Valley, a huge fault system stretching for 4000 km. It is shown here in its entirety on a mosaic of Viking orbiter pictures. US Geological Survey.

The rusty-red surface of Mars photographed by the Viking 2 lander. Part of the lander itself is visible in the foreground. The largest rocks are about a metre across. Fine dust suspended in the atmosphere makes the sky pink. NASA.

the opposite side of the planet in an area known as Utopia, which is encroached upon by the outer fringes of the north polar cap during winter.

The landers carried colour cameras as well as instruments to analyse the soil and the atmosphere. Each Viking lander looked around, and found itself on a rock-strewn desert without any visible signs of life – no plants, insects or animal tracks. A mechanical arm reached out to pick up samples of the soil and tip them into an on-board biological laboratory which set to work in search of Martian microorganisms.

To the disappointment of many, neither Viking found life in the soil of Mars, despite exhaustive tests. That is not to say that Mars is necessarily totally sterile; maybe there is life of a kind that would not respond to the Viking experiments, or possibly it is concentrated in certain oases that the Vikings missed. But the negative Viking results do make it extremely unlikely that there is, or ever has been, anything alive on Mars.

A lifeless Mars is not too surprising in the light of readings from Viking's other instruments, which revealed just how hostile conditions on the planet really are. The Viking 1 lander recorded a maximum summer afternoon air temperature of –29°C, whereas farther north at the Viking 2 landing site temperatures dropped well below –100°C as winter approached and the carbon dioxide in the atmosphere began to freeze, forming patches of white frost on the surface. Atmospheric pressure was a mere 7.5 millibars (750 pascals), equivalent to the pressure at a height of 35 km above the Earth's surface.

Although Mars was a disappointment for the biologists, there was plenty to interest the geologists. As expected, the rusty redness of Mars is caused by large amounts of iron oxide in the surface rocks. Mars may be the richest source of iron ore in the Solar System. Even the sky is pink, a result of fine particles of dust suspended in the atmosphere.

While the landers studied Mars from the surface, the orbiters continued the surveying job begun by Mariner 9. Mars turns out to be a planet of two differing halves. The northern hemisphere of the planet is the lower and smoother of the two, flooded by lavas from the volcanoes of Mars. The southern hemisphere is higher and has been

Phobos and Deimos, the moons of Mars, are thought to be captured asteroids. Here at bottom (Phobos on the right) they are compared with asteroid Gaspra, top, all at the same scale and under similar lighting conditions. Gaspra was photographed by the Galileo probe, and Phobos and Deimos by the Viking orbiter. NASA.

heavily cratered by meteorite impacts, giving it a broad similarity to the lunar highlands. This was the area that the first Mariners photographed. In the southern hemisphere of Mars are two large impact basins, Hellas and Argyre. Argyre, 1000 km in diameter, is about the same size as the Moon's Mare Imbrium, while Hellas is even larger, equivalent in size to Oceanus Procellarum. Unlike the lunar maria, however, Argyre and Hellas appear to be filled with light-coloured dust.

It comes as a surprise to learn that Mars is not short of water, although most of that water is in frozen form, in the polar caps and in a subsurface permafrost layer. This offers another possible explanation for the ancient water channels of Mars: the melting of subsurface ice by volcanic heating. The polar caps of Mars are made of water ice a few metres thick, augmented each winter by carbon dioxide. This freezes out of the atmosphere to produce a thin sprinkling of frost that can extend more than halfway to the equator. During each Martian year the atmospheric pressure changes by about 20 per cent as carbon dioxide evaporates from one polar cap with the coming of spring, migrates to the opposite hemisphere and then freezes out again as winter arrives at the other pole.

There is one final puzzle from the pre-space-probe era of observation that remains to be explained: what caused the seasonal changes in the dark areas if there is no vegetation on Mars? Wind-blown dust provides the answer. Mariner 9 and the Viking orbiters recorded many examples of changes in surface markings caused by light and dark dust being blown by seasonal winds. Syrtis Major, for example, is a sloping area of darker rock which periodically becomes partly covered by paler dust and is later swept clear again. Dust blown by the winds on Mars, which can reach 200 km per hour, is also a powerful erosive agent, sandblasting the surface features of the planet.

No discussion of Mars would be complete without reference to its two tiny moons, Phobos and Deimos. They were discovered in 1877 by Asaph Hall using the 66-cm (26-inch) refractor at the US Naval Observatory in Washington DC, and both are beyond the range of normal amateur telescopes. Space probe photographs show that both are cratered lumps of rock shaped like lumpy potatoes. Phobos, larger and closer to Mars, measures about 27 × 19 km; Deimos is about 15 × 11 km.

Phobos is remarkable in that it orbits Mars three times a day. It also lies closer than any other moon to its parent – a mere 6000 km above the surface of Mars. Probably they are asteroids which strayed too close to Mars and were captured by the planet's gravity. They will make a fascinating sight in the sky for the first astronauts to set foot on Mars, the planet of red, frozen deserts which was so nearly suitable for life of its own.

Jupiter

Jupiter is the king of the planets, and the most fascinating of all to study with small instruments. Humble binoculars reveal the planet's cream-coloured disk and four main moons, called the Galilean satellites after Galileo, who discovered them in 1609. Some people with exceptionally acute vision claim to be able to see the Galilean satellites with the naked eye.

A small telescope brings into view some of the details of Jupiter's disk: dark belts of cloud parallel to the equator and an eye-shaped marking in the southern hemisphere known as the Great Red Spot, first observed in 1665. Careful study of these features reveals that Jupiter's rotation period varies with latitude, from 9 hours 50 minutes at the equator (the fastest rotation of any planet in the Solar System) to 9 hours 55 minutes at higher latitudes. In addition, the Great Red Spot drifts slightly in relation to its surroundings. These effects demonstrate that the visible surface of Jupiter is not solid. We are looking at clouds, constantly seething and swirling, changing in colour and shape. Jupiter never appears the same twice. That is its attraction.

Jupiter photographed by the Hubble Space Telescope in May 1991, showing multi-coloured bands of cloud. The Great Red Spot is at lower right. NASA.

Jupiter's Great Red Spot stares like a Cyclopean eye from this Voyager 2 photograph taken in July 1979. Other clouds exhibit fantastic scalloping as they sweep around the Red Spot and a white oval to the south of it. NASA.

Jupiter is easy to spot with the naked eye, reaching a maximum of magnitude –2.9 when at its closest to Earth, 590 million km away. And it manages to outshine every star except Sirius, even when at its most distant from us. Its brightness results from its highly reflective clouds and its imposing size, the largest planet in the Solar System. Examination of its outline gives added confirmation that it is not a solid planet: Jupiter has a bulging midriff. The diameter at its equator is 142,800 km, but from pole to pole it measures 133,500 km. A line of 11 Earths would be needed to equal Jupiter's equatorial width. And even though Jupiter is made of gas, its bulk is such that its mass is two and a half times as much as all the other planets put together.

Jupiter orbits the Sun every 11.9 years at an average distance of 778 million km, over five times the distance of the Earth from the Sun. It is well placed for observation from Earth every 13 months. Because Jupiter's cloud features are so impermanent and mobile, it is impossible to give more than a generalized description of the planet's appearance. Its disk is crossed by alternating bright zones, caused by ascending gas, and dark belts where the gases descend. Frozen ammonia crystals form high, cold clouds in the bright zones; the darker belts are lower and warmer ('warm' is only a relative term, for the average temperature of the cloud tops is –150°C). The colours of the belts can vary from yellow and brown to orange, red or even purple as a result of complex chemicals in the atmosphere of Jupiter.

High-speed winds of up to 400 km per hour whip the edges of the zones and belts into turbulent eddies, giving them a scalloped appearance. The weather on Jupiter is unpredictable. Dark and light spots can suddenly erupt in the clouds, lasting for weeks or even decades before fading away. One of the main roles of amateur observers is to track these storms as they erupt and move around the planet.

Of all the markings on Jupiter, the most famous, and by far the most permanent, is the Great Red Spot. It certainly is great: 14,000 km wide and 40,000 km long, enough to swallow three Earths. But it is not always red; most often it is pinkish, and sometimes it can fade to a colourless grey. Its colour is believed to be due either to red phosphorus or sulphur. By good fortune, the spot was particularly prominent when the Voyager 1 and 2 spacecraft reached the planet in 1979. Even now its nature is not fully understood, but it appears to be an upwardly spiralling column of gas similar to a hurricane on Earth, its top spreading out about 8 km above the surrounding cloud deck. Jupiter's other, smaller spots are believed to be similar swirling storm systems. As if to emphasize the storminess of the planet, the Voyager probes photographed massive flashes of lightning on Jupiter's night side.

The key to Jupiter's meteorology is the fact that it gives off twice as much heat as it receives from the Sun. The planet was hot when it formed, and still retains some of that heat today. This internal store of heat drives the complex cloud systems of Jupiter, keeping the Great Red Spot and its smaller relatives alive for far longer than any storms can persist on Earth.

Interestingly, Jupiter has virtually the same chemical composition as the Sun: mostly hydrogen and helium. There is thought to be a rocky core about twice the size of the Earth at the centre of Jupiter, but no space probe could ever land on it. Beneath the wispy high-altitude clouds of frozen ammonia are complex chemicals that give the dark belts their colour. Deeper still, temperatures are similar to those on Earth; here are clouds of water vapour. About 1000 km below the visible cloud tops, temperatures and pressures have increased to the point where hydrogen is compressed into a liquid. The liquid hydrogen seas of Jupiter are about 20,000 km deep. Below, under the crushing pressure of 3 million Earth atmospheres, hydrogen is compressed into a superdense state with the properties of a metal; hence it is known as metallic hydrogen. Convection within the hot metallic hydrogen interior of Jupiter is thought to be responsible for the planet's intense magnetic field, 10 times stronger than the Earth's, extending 100 times Jupiter's radius into space. If the magnetic field around Jupiter were visible to the naked eye, it would appear over twice the size of the full Moon.

Jupiter possesses a fascinating collection of satellites, like a mini Solar System. We have already mentioned the four largest, known as the Galilean satellites. With the simplest optical aid they can be seen performing a merry dance around Jupiter – changing position from night to night, sometimes out of sight behind the planet, sometimes transiting across its face and sometimes being eclipsed in its shadow.

The closest of the Galilean satellites to Jupiter is Io, 3600 km in diameter (slightly larger than our own Moon), orbiting every 42½ hours. Io is the most volcanically active body in the Solar System. The Voyager 1 probe in 1979 photographed eight volcanoes erupting simultaneously on it. Hundreds of other volcanic vents were visible, though not actually erupting. Those volcanoes erupt not lava but liquid sulphur which solidifies to form the red, orange and yellow of Io's surface.

What keeps Io molten remains an open question. According to one theory, Io is caught in a gravitational tug of war between Jupiter and the other Galilean satellites; their opposing pulls release tidal energy that melts Io's interior. But the amount of energy generated by this mechanism may not be sufficient, and the answer may instead lie with powerful electrical currents that flow through Io as it orbits within Jupiter's magnetic field, heating the moon like an electric bar fire.

Io recycles its interior onto its surface, endlessly turning itself inside out. Some of the sulphur escapes and showers onto the innermost moon of Jupiter, Amalthea, giving it an orange coating. Amalthea is an irregularly shaped lump of rock only about 200 km in diameter, too faint to be seen in amateur telescopes.

Within the orbit of Amalthea the Voyager probes discovered a faint ring of dust, a mere 50,000 km above Jupiter's cloud tops. This tenuous ring of Jupiter is believed to result from the break-up of one or more tiny moons, two of which were discovered by the Voyagers, orbiting at the ring's outer edge.

Moving outwards past Io we come to Europa, smallest of the Galilean satellites, with a diameter of 3100 km. Europa is encased in a white shell of ice, veined with fine cracks. Underneath its ice crust, Europa is believed to have a rocky interior. Next in line from Jupiter is Ganymede, the largest and brightest of the satellites; what's more, at 5260 km in diameter it is the largest moon in the Solar System, larger even than the planet Mercury. Ganymede and the fourth of the Galilean satellites, Callisto, 4800 km in diameter, are both balls of rock and ice, rather like giant muddy snowballs Callisto is saturated with impact craters, the largest 300 km in diameter and called Valhalla; it is similar to the large basins on the Moon and Mercury, surrounded by wave-like ridges. Ganymede is also cratered by impacts, but less

A volcano called Loki erupts at the limb of Jupiter's moon Io, sending clouds of sulphur 100 km into space; photographed by Voyager 1 in March 1979. NASA.

heavily than Callisto. It also exhibits a strange grooving on its surface. Bright patches on Ganymede and Callisto mark the places where recent impacts have exposed fresh ice.

The rest of Jupiter's total of at least 16 moons are small and insignificant. Several of these – particularly the outer four, which move in highly elliptical, retrograde orbits – may be passing bodies that were captured by Jupiter's gravity.

Jupiter's volcanic moon Io.

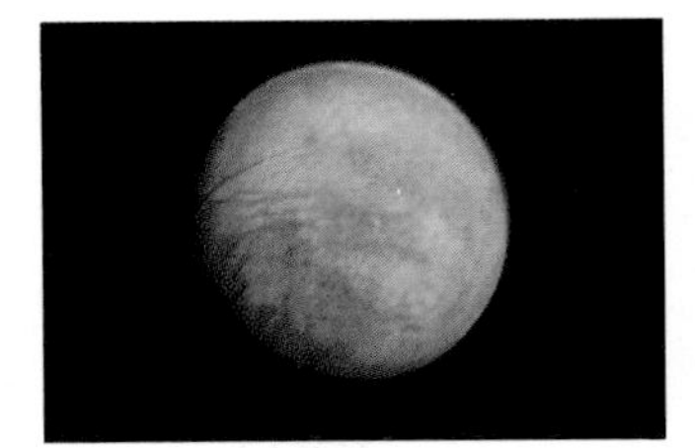

Europa has a cracked, icy surface.

Ganymede, largest moon of Jupiter.

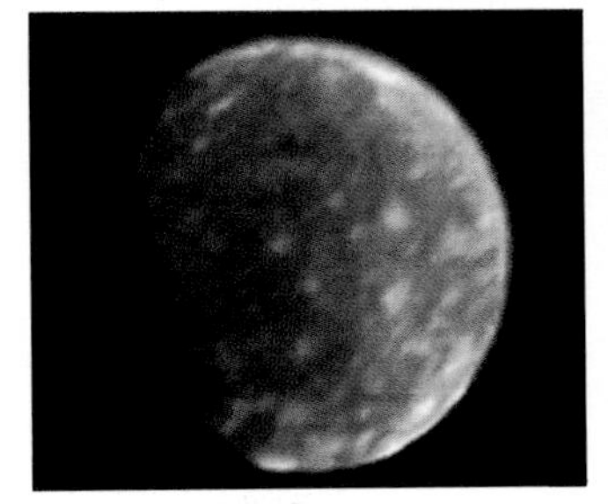

Callisto is dark and cratered. NASA.

Saturn

Bright rings girdle Saturn's equator, making it the most beautiful of the planets. Those distinctive rings can be spotted clearly in a small telescope; good binoculars, mounted steadily, will show the small disk of the planet elongated into an ellipse by the rings. Binoculars should also pick out Saturn's largest moon, Titan, which orbits the planet every 16 days.

Saturn's rings can reflect more light than the body of the planet itself, so that at its best Saturn appears of magnitude –0.3. Without the rings, Saturn would be no brighter than magnitude 0.7. Strangely enough, from time to time Saturn can appear to be without rings. The reason is that the planet's axis is tilted at 29° to the vertical. As Saturn orbits the Sun, the rings are sometimes tipped towards us, while at other times the rings are presented edge-on. So thin are the rings that when edge on (as happens about every 15 years) they disappear from view in even the largest telescopes on Earth. The brightness of Saturn therefore depends not only on its distance from Earth, but also on the aspect of the rings.

Serenely beautiful Saturn and its rings, photographed by the Hubble Space Telescope in August 1990, showing subtle cloud belts and a dark polar hood. In addition to the broad Cassini Division in the rings, note the narrower Encke Division near the rings' outer edge. The shadow of Saturn's globe can be seen on the rings. NASA.

The ringed planet orbits the Sun every 29½ years at an average distance of 1430 million km, 9½ times farther than the Earth is from the Sun. In many ways, Saturn is a smaller brother of Jupiter. Its equatorial diameter of 120,000 km is second only to that of Jupiter; its rotation period of 10¼ hours is second-fastest to that of Jupiter; and Saturn is another planet made mostly of hydrogen and helium gas.

In one way, though, it is unique among the planets: its average density is less than that of water. This remarkable fact comes about because the planet's mass is less than a third of Jupiter's mass, so its gravity is less and hence its central regions are not compressed as densely. There probably is a rocky central core, but the region surrounding the core, in which hydrogen is compressed into a liquid metallic form, extends out to only half the planet's radius, against three-quarters of the radius in the case of Jupiter. That is not sufficient to compensate for the low density of Saturn's outer layers, so the planet's overall density is a mere 70 per cent of that of water.

Saturn's low density is apparent from its outline, which is even more squashed than that of Jupiter. Saturn's pole-to-pole diameter is 108,000 km, fully 10 per cent less than its equatorial diameter; with Jupiter the difference is 6 per cent.

Through a telescope, Saturn appears as a tranquil, ochre-coloured disk, darker at the poles and with some dusky horizontal bands. Serious study of Saturn requires a telescopes of at least 200 mm aperture, for there are none of the swirling, multicoloured storm clouds that make Jupiter's globe so interesting in a small telescope, and no equivalent of the Great Red Spot. However, every 30 years or so a large white spot erupts in the planet's northern hemisphere, when the planet's north pole is tilted at its maximum towards the Sun. The white spots are evidently storm clouds produced by solar heating. Such an outbreak was last seen in 1990, and lasted a few months.

White spots apart, this is not to say that Saturn lacks activity in its atmosphere; it is just that the cloud patterns are usually concealed by high-altitude haze. Cameras aboard the probes Voyager 1 and 2 which reached Saturn in 1980 and 1981 recorded low-contrast cloud swirls resembling those on Jupiter. The meteorology of both planets should be similar, for they both have internal sources of heat. Saturn radiates two and a half times as much heat as it receives from the Sun, a legacy of its birth. Tracking of cloud systems revealed that gales of up to 1800 km per hour blow on Saturn, four times faster than on Jupiter.

Inevitably, the attention of the space probes focused on the glorious rings. As seen from Earth, the rings look like a continuous disk encircling the planet, but appearances are deceptive. The Dutchman Christiaan Huygens in 1655 was the first to realize that the rings are not solid, but consist of a swarm of tiny particles orbiting Saturn. The central part of the rings, known as Ring B, is the widest and brightest.

It is separated from the outer, fainter Ring A by a 3000-km-wide gap known as Cassini's Division. Extending inwards from Ring B towards the planet is the faintest ring of all, the transparent Ring C, also known as the crepe ring. A narrow gap called Encke's Division can be seen in Ring A, while observers looking through large telescopes under good conditions reported apparent ripples within the rings, a sign that the density of ring material varies from place to place.

But even the best telescopic views did not prepare astronomers for the astounding wealth of detail that was revealed by space probes. Under the close scrutiny of the Voyagers' cameras, the rings broke up into thousands of narrow ringlets and gaps, like the ridges and grooves of a gramophone record. Some ringlets were not perfectly circular, but were elliptical in shape. Even Cassini's Division was not empty, but contained thread-like ringlets. A new outer ring, called the F Ring, appeared to consist of twisted strands like a rope.

In some places, fine dust overlays the rings, apparently supported by electromagnetic forces in Saturn's magnetosphere, producing darker features known as spokes. Spoke-like features have been reported from time to time by ground-based observers, but it took the Voyager pictures to establish their reality.

The particles that make up Saturn's rings range in size from tiny specks to lumps the size of a house or larger. Their composition is mostly frozen water, possibly mixed with dust, resembling loosely compacted snowballs. The rings are most likely material that was prevented from forming into a moon by the overpowering force of Saturn's gravity. Alternatively, they may be the remains of a former moon that strayed too close to the planet and broke up.

Saturn's rings are extraordinarily thin in relation to their 270,000 km diameter. Voyager observations showed that the rings are no more than 100 m thick. On the same ratio of thickness to diameter, a gramophone record would be 5 km in diameter.

The Voyagers discovered several moons of Saturn too small to be seen from Earth, bringing the total of known Saturnian moons to 18, with others suspected. One of these tiny moons, Pan, actually orbits in Encke's Division in Ring A. Another, Atlas, patrols the outer edge of Ring A. Two other moons, Prometheus and Pandora, orbit either side of the F Ring, shepherding its particles.

Slightly farther out from Saturn, Janus and Epimetheus move along the same orbit, which initially confused astronomers who first spotted them from Earth in 1966. Orbit-sharing is common around Saturn. Tethys, a satellite visible from Earth, has two small siblings, Telesto and Calypso, moving on the same path. Dione, visible from Earth, shares its orbit with tiny Helene, another Voyager discovery.

The gravitational effect of some of these satellites helps produce the gaps in Saturn's rings. For instance, the gravitational tug of Mimas

In close-up, Saturn's rings break up into thousands of ringlets like the ridges and grooves of a gramophone record. Here the colours have been artificially enhanced by computer. Voyager 2 photograph, taken on August 17, 1981. NASA.

pulls particles out of the Cassini Division. Mimas itself, like most of the moons of Saturn, is a dirty snowball of frozen water and rock. It sports a remarkable crater 135 km in diameter, larger than Copernicus on the Moon, fully a third of its own 390 km diameter. The impact that caused this grotesque feature must have nearly shattered Mimas.

Iapetus, Saturn's outermost moon but one, is another strange body: one side is five times darker than the other. This harlequin effect is probably caused by dust knocked off Saturn's outermost moon, Phoebe, which is the darkest of Saturn's moons. The dark dust from Phoebe falls inwards to be swept up by Iapetus, coating its leading face, while bright ice on its trailing side remains exposed.

Saturn's largest moon, Titan, 5150 km in diameter, deserves to be considered as a planet in its own right. Bigger than Mercury (but slightly smaller than Jupiter's main moon Ganymede), it is the only moon to have a substantial atmosphere – in fact, the atmospheric pressure on the surface of Titan is 50 per cent greater than at sea level on Earth. Titan's atmosphere is 90 per cent nitrogen, with methane making up most of the rest. Clouds of orange smog top the atmosphere, obscuring the surface from the prying eyes of space probes. Despite that thick atmosphere, Titan's surface temperature is very low, about –180°C. Perhaps a rain of liquid methane falls from the rust-coloured sky into seas of methane on Titan's surface. Titan has been described as a deep-freeze version of the Earth. It will certainly be a fascinating spot for a space probe of the future to land.

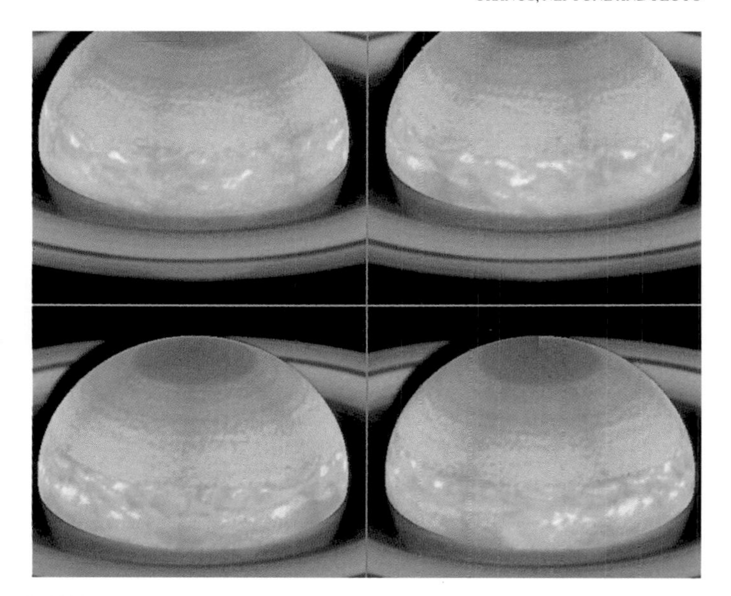

In 1990 a huge white spot erupted in the clouds of Saturn, just north of its equator, spreading ribbons of cirrus around the planet. This sequence was taken as Saturn made one complete rotation, showing how the white clouds encircled it. NASA.

Uranus, Neptune and Pluto

Uranus should in theory be visible to the naked eye: at its brightest it is of magnitude 5.5. But it is so insignificant that it was never noticed by ancient astronomers, for whom the Solar System stopped at Saturn. Uranus was not discovered until March 13, 1781, when William Herschel spotted it through his telescope during a systematic survey of the skies. Once its position is known, Uranus can easily be followed in binoculars as it moves against the background stars. There is a certain fascination in seeing the blue-green disk of this distant planet through a telescope, but even in large telescopes Uranus displays no detail.

Uranus is one of the four 'gas giants' of the outer Solar System, the others being Jupiter, Saturn and Neptune. Uranus is 51,000 km in diameter at the equator, less than half that of Saturn but four times larger than that of the Earth. It lies 2900 million km from the Sun, 19 times farther than the Earth. Seasons would pass slowly on Uranus, for it takes 84 years to complete one orbit, but the seasons would be extreme, for Uranus appears to have been knocked over onto its side:

the planet's axial tilt is 98°, meaning that its axis of rotation lies almos in the plane of its orbit. Every 42 years, therefore, one of the poles of Uranus is pointing towards the Sun, while its opposite pole is in darkness for decades. In between times its equatorial region faces sunwards. During each 84-year orbit of Uranus the Sun can appear overhead at every latitude, something that never happens on any other planet. No one knows why Uranus should be lying on its side in this unique way; perhaps it suffered a collision with another large body long ago.

Uranus is celebrated for another reason: it was the second planet discovered to have rings. In 1977, astronomers watched as Uranus passed in front of a star. Unexpectedly, they noticed that the star winked on and off a number of times both before and after it was obscured by the disk of Uranus. From this observation, and others like it which followed, astronomers deduced that Uranus is encircled by nine rings, so faint they are invisible by direct observation from Earth.

The rings are narrow, 10 to 100 km wide, separated by gaps of between 300 km and over 2000 km. The rings lie 15,000 to 25,000 km above the cloud tops of Uranus, and are probably caused by the break-up of a former moon, or moons, that strayed too deeply into the planet's gravitational clutches. In addition to the rings, Uranus has five moons visible from Earth: Miranda, Ariel, Umbriel, Titania and Oberon, ranging in size from 500 to 1500 km in diameter. Ten small moons, and two additional faint rings, were discovered by the Voyager 2 space probe when it flew past Uranus in January 1986. Disappointingly, the gaseous surface of the planet itself was bland and greenish, with scarcely any cloud features. The moons, like the rings, all move around Uranus's crazily tilted equator.

In structure, Uranus is believed to have a rocky core surrounded by a mantle of ice, topped by an atmosphere of hydrogen and helium mixed with a fair proportion of methane, which gives the greenish colour. Neptune, the next-farthest planet from the Sun, is a twin of Uranus. It is slightly smaller – diameter 49,000 km – but shows the same featureless greenish disk through a telescope. Of magnitude 7.8 at brightest, it is completely invisible to the naked eye, but can be followed in binoculars when one knows where to look.

Unlike the accidental discovery of Uranus, the existence of Neptune was predicted in advance of its discovery. Astronomers found that Uranus was not keeping to its expected course, and one suggested reason was that it was being pulled by the gravity of an as yet unseen planet. In England, mathematician John Couch Adams calculated the new planet's position in 1845, and the following year Urbain Le Verrier of France came up with a similar conclusion. Neptune was found close to the predicted position by astronomers at the Berlin Observatory on September 23, 1846.

Neptune crawls around the Sun in 165 years at an average distance of 4500 million km, 30 times farther away than the Earth. Being so distant from the Sun, it is a very cold and dark planet. Neptune has a cloud feature equivalent to Jupiter's Great Red Spot, called the Great Dark Spot, revealed by the cameras of Voyager 2 when it reached Neptune in August 1989.

But the planet's main distinction is its two curious outer satellites, Triton and Nereid. Triton, the largest of Neptune's eight known moons, is in a retrograde (east to west) orbit. Tidal forces from Neptune mean that Triton's orbit is gradually shrinking, so that the moon will spiral closer to the planet until it is broken up in 10–100 million years' time. A shattered Triton will form a far more substantial set of rings around Neptune than the existing thin, faint ones, for it is 2700 km in diameter, over three-quarters the size of our own Moon. The outer satellite of Neptune, Nereid, is much smaller, only about 300 km across, but it has an exceptionally elliptical orbit that swings it between 1.4 and 9.7 million km from Neptune every 360 days. Something seems to have disturbed the satellite system of Neptune; exactly what happened remains one of the mysteries of the outer Solar System.

Pluto is the odd man out among the planets – indeed, many astronomers doubt whether it deserves to be called a planet at all. It is by far the smallest planet. Pluto's diameter of 2300 km is less than that of our own Moon, making it smaller even than Neptune's largest satellite, Triton; in fact, the icy surface of Triton, photographed by Voyager 2, probably looks much like that of Pluto. Triton and Pluto may both have wandered the outer reaches of the Solar System before Triton was captured by Neptune; the strange orbits of Triton and Nereid around Neptune could therefore be the consequences of that disruptive event.

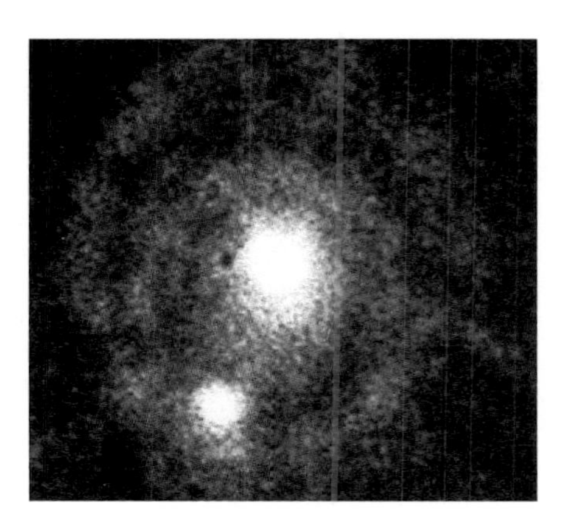

Pluto and its moon Charon, photographed by the Hubble Space Telescope.
Even with the power of the Space Telescope, Pluto is so small and distant that it appears as nothing more than a dot of light.
NASA/ESA.

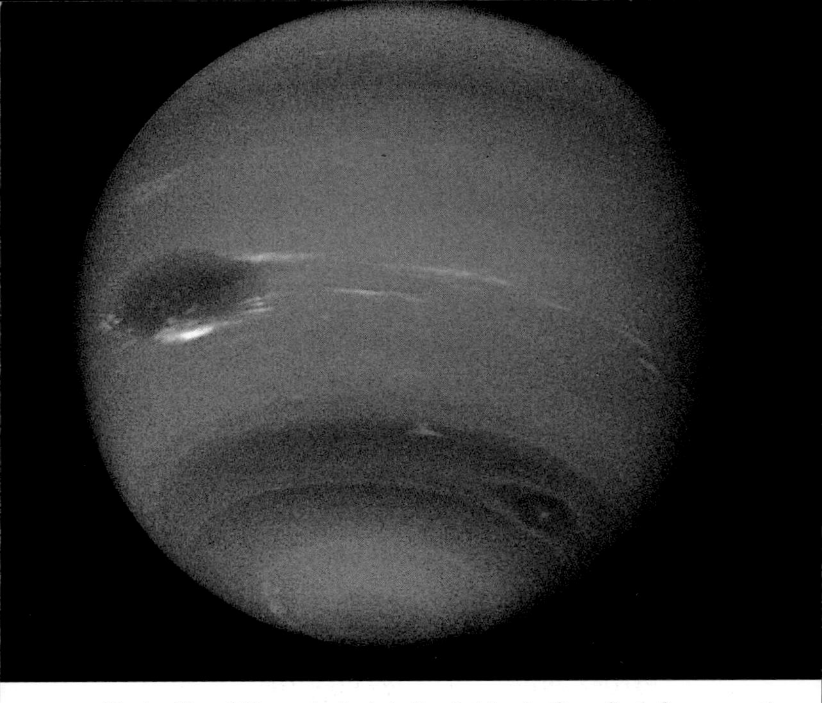

The icy blue of Neptune's clouds is disturbed by the Great Dark Spot, an anti-cyclone the size of the Earth, fringed with white methane cirrus. This photograph was taken by Voyager 2 in August 1989. NASA.

Certainly, Pluto has the oddest orbit of any planet: it crosses the orbit of Neptune, so that at times Neptune temporarily becomes the outermost planet of the Solar System, as is the case between January 1979 and March 1999. Pluto's average distance from the Sun is 5900 million km.

Pluto is not alone in its 250-year orbit around the Sun. In 1978 astronomers discovered that Pluto has a moon, called Charon, fully half Pluto's diameter. Charon orbits its parent every 6.4 days, the same time that the planet takes to spin on its axis. Charon therefore hangs over one spot on the surface of Pluto, visible permanently from one hemisphere of the planet but invisible from the other.

Being so faint and insignificant, Pluto was not discovered until 1930. For decades prior to that, various astronomers had tried to predict where a planet beyond Neptune might lie, but without success. In the end, Pluto was found by Clyde Tombaugh at the Lowell Observatory in Arizona as the result of a deliberate round-the-sky photographic search for new planets. Tombaugh's search failed to detect any sign of a tenth planet lurking beyond Pluto, as have all subsequent searches. Probably all that exists beyond Pluto is a swarm of icy comets.

Comets and Meteors

Comets are insubstantial bodies, loosely knit collections of frozen gas and dust that loop through the Solar System on elongated orbits, returning to the Sun at intervals ranging from a few years to many thousands of years. A cloud of thousands of millions of comets, named the Oort Cloud, is believed to exist on the dim outer edges of the Solar System, about a light year from the Sun. The gravitational influence of passing stars nudges comets from this cloud into new orbits that bring them towards the Sun, where they become visible to us as ghostly, glowing apparitions. A closer swarm of comets, termed the Kuiper Belt, lies just beyond the orbit of Pluto.

When far from the Sun, a comet shines only by reflecting sunlight. At that stage it is small – only a few kilometres across – and faint. As a comet approaches the Sun it warms up, turning the ice into gas. Under the influence of the Sun's radiance the gases of the comet begin to fluoresce, in similar fashion to the gas in a neon tube, thereby considerably increasing the comet's brightness. Gas and dust released from the warming comet produce a halo or *coma* 100,000 km or so in diameter. At the centre of the coma is the nucleus, only a few kilometres in diameter and the only solid part of the comet, consisting of a dirty snowball of ice, dust and perhaps some rock. It would take over a thousand million comets to equal the mass of the Earth.

Not all comets have tails, but many do. One part of the tail consists of gas blown away from the comet's head by the solar wind of atomic particles streaming from the Sun. The other part of the tail is made up of dust particles liberated from the head by the evaporating gases. Comet tails always point away from the Sun. A comet's tail can extend for 100 million km or more, yet despite its glorious appearance it is less

Halley's Comet pursues an elliptical path around the Sun, steeply inclined to the orbits of the planets. Its 76-year orbit takes it from between the orbits of Venus and Mercury out to a greater distance than Neptune, and back again. Wil Tirion.

Halley's Comet on April 15, 1986, shortly after its closest approach to Earth. Its gas tail extends to the upper right, while the broad dust tail fans out to the left, appearing foreshortened because it is pointing away from Earth. At bottom left is the galaxy NGC 5128 (Centaurus A). National Optical Astronomy Observatories.

dense than a laboratory vacuum. The tail of a comet gives it the appearance of rapid motion across the sky, but actually its motion against the stars is noticeable only by comparing its position from night to night.

Two dozen or more comets may be visible through a telescope each year, although only occasionally do any of them become bright enough to be prominent to the naked eye. The comets that are seen each year are a mixture of known comets returning to the Sun and completely new discoveries. About a thousand comets have well-known orbits, and more are being discovered all the time. Dedicated amateur astronomers sweep the skies to discover new comets; each new comet is given its discoverer's name.

Many comets that stray into the inner regions of the Solar System have their orbits altered by the gravity of the planets so that they never again recede far from the Sun. Most of these *periodic comets*, which have orbital periods shorter than 200 years, are now thought to come from the Kuiper Belt rather than the Oort Cloud. The comet of shortest known period is Encke's Comet, which orbits the Sun every 3.3 years. It is so old that it has lost most of its gas and dust, and is too faint to see with the naked eye.

The most famous comet of all is Halley's Comet, named after the English astronomer Edmond Halley, who calculated its orbit in 1705. Halley's Comet returns every 76 years or so, and last appeared in 1985–6. Its orbit takes it from 88 million km from the Sun (between the orbits of Mercury and Venus) out to 5300 million km (beyond Neptune).

The dust lost from a comet disperses into space. The Earth and other planets are continually sweeping up cometary dust. When a particle of cometary dust comes whizzing into the atmosphere, it burns up by friction at a height of about 100 km, producing a streak of light known as a shooting star or meteor. The whole event is over in less than a second. On any clear night, a few meteors are visible each hour as particles of dust dash at random to their deaths in the atmosphere.

Such random meteors are termed *sporadic*. Occasionally, though, the Earth crosses the orbit of a comet and encounters a dense swarm of particles. This gives rise to a so-called meteor shower, in which

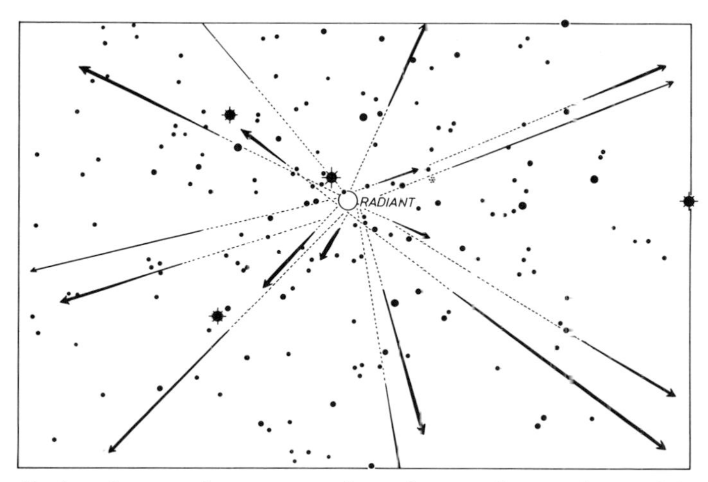

Members of a meteor shower appear to diverge from a small area of sky termed the radiant. This diagram shows the radiant of the Lyrid meteors, which lies in the constellation Lyra, near the bright star Vega. Wil Tirion.

meteors may be visible coming from one direction in the sky at the rate of dozens per hour. The area of sky from which the meteors seem to come is known as the *radiant.*

A meteor shower is named for the constellation in which the radiant lies. For instance, the Perseids, an abundant shower of bright meteors which the Earth encounters each August, seem to radiate from Perseus; the Geminids from Gemini; and so on. One historical oddity concerns the Quadrantids, which come from an area in Boötes that was once part of the now-defunct constellation of Quadrans, the quadrant.

The strength of a meteor shower is measured by its zenithal hourly rate (ZHR), which is the number of meteors that an individual observer would see if the radiant were directly overhead. Since the radiant is seldom, if ever, at the zenith, the number of meteors actually seen per hour will be less than the theoretical ZHR. In addition, bright moonlight will wash out the fainter meteors, again reducing the observed ZHR.

Amateur astronomers make valuable observations of meteor showers, counting the number of meteors visible and estimating their brightness. Typical meteors are of magnitude 2 or 3, but the most spectacular are brighter than the brightest stars, and the occasional brilliant meteor, termed a fireball, can cast shadows. Some meteors seem to split up as they fall, and some leave trains of glowing gas which fade in a few seconds. The table shows the main meteor showers visible each year. The ZHR is only a guide, and can vary considerably from year to year.

One extreme case is the Leonids, normally a modest shower, which bursts into life at 33-year intervals when its parent comet, Tempel–Tuttle, returns to perihelion. An intense storm of Leonids was seen over the United States in 1966, and astronomers will be on the lookout for a repeat performance in and around 1999.

Major Meteor Showers

Shower	*Limits of activity*	*Date of maximum*	*Maximum rate (ZHR)*
Quadrantids	January 1–6	January 3–4	100
Lyrids	April 19–25	April 21–22	10
Eta Aquarids	May 1–10	May 5–6	35
Delta Aquarids	July 15–August 15	July 28–29	20
Perseids	July 23–August 20	August 12–13	75
Orionids	October 16–27	October 22	25
Taurids	October 20–November 30	November 4	10
Leonids	November 15–20	November 17–18	10
Geminids	December 7–15	December 13–14	75

Asteroids and Meteorites

Between Mars and Jupiter orbits a belt of rubble known as the asteroids or minor planets. The asteroids were unknown until 1801, when the Italian astronomer Giuseppe Piazzi discovered Ceres, the largest of them, although astronomers had previously speculated that an unknown planet might exist in the suspiciously wide gap that separates Mars and Jupiter.

However, even if every asteroid were rolled together (over 5000 of them are currently known, and there must be thousands of others too small to be seen from Earth), they would make a body less than half the size of the Moon. The asteroids are not, as has been suggested, the remnants of a former planet that was disrupted; they are merely the leftovers from the formation of the major planets.

Ceres itself is 1000 km in diameter and takes 4.6 years to orbit the Sun. Although Ceres is the largest asteroid, it is made of dark rock and so is not the brightest asteroid. That honour goes to Vesta, which is composed of paler rock; at times it is just visible to the naked eye. Vesta is the third-largest asteroid, 500 km in diameter. The second-largest asteroid is Pallas, diameter 550 km. These asteroids, and several others, are bright enough to be followed in binoculars.

In 1991 the space probe Galileo, while on its way to Jupiter, flew past the asteroid Gaspra. Galileo's photographs showed that Gaspra is an irregularly shaped, rocky body about 17 km long, and pitted with impact craters. Gaspra resembles the two moons of Mars, Phobos and Deimos, thus strengthening the suspicion that those moons are captured asteroids. (For a photograph, see page 348.)

Meteor Crater: the impact of a giant meteorite in the Arizona desert 25,000 years ago gouged out this massive scar, over 1 km across and 200 m deep. Most of the iron meteorite was vaporized in the heat of the impact. Ian Ridpath.

Ninety-five per cent of asteroids orbit in the main belt between Mars and Jupiter, but there are some notable exceptions. One group of asteroids, known as the Trojans, moves along the same orbit as Jupiter. Most significant from our point of view are the so-called Earth-crossing asteroids, which have orbits that take them across the path of the Earth; these are also known as Apollo asteroids after the first of their type, discovered in 1932. Some of these objects may be the nuclei of extinct comets. Famous Apollo asteroids include Hermes and Icarus, both of which have made close approaches to the Earth. Asteroids of this type must have hit the Earth in the past, and others may do so in future, with devastating effect.

Objects that do reach the surface of the Earth are termed meteorites. One particularly large meteorite, estimated to have weighed a quarter of a million tonnes, crashed into the Arizona desert about 25,000 years ago, digging out the now-famous Meteor Crater, 1.2 km in diameter. (Strictly, it should be called Meteorite Crater, but the old name is too well entrenched to be changed now.) Most of the meteorite was destroyed on impact, but enough fragments remain scattered around the crater to show that the meteorite was made of iron.

Several thousand meteorites are estimated to land on Earth each year, but only about a dozen are actually picked up; the rest fall in uninhabited areas or into the sea. Most meteorites observed to fall are of the stony kind, and if they are not picked up immediately they soon weather away – at least, they do under normal conditions. But in Antarctica, scientists have found large numbers of ancient meteorites, including many rare types, preserved in pristine condition by the natural deep-freeze of the ice cap. Another rich source is the deserts of Earth, where the arid conditions preserve the meteorites.

Stony meteorites are usually termed *chondrites*, because they contain mineral-rich blobs known as chondrules. The small percentage of stones without chondrules are called *achondrites*. The most interesting class of stony meteorite are the carbonaceous chondrites; these have a high carbon content, and are believed to be among the most primitive rocks in existence, virtually unchanged since the formation of the Solar System. Most meteorites are believed to be chips off asteroids, but the carbonaceous chondrites may be fragments from the heads of comets.

A small intermediate group of meteorites is the stony-irons, which consist of about 50 per cent iron and 50 per cent rock. But the other main group of meteorites apart from the chondrites is the irons. These contain 90 per cent iron and about 10 per cent nickel. The largest known meteorite is iron, and weighs about 60 tonnes. It must have fallen quite gently, for it did not dig a crater and did not break up. It lies where it fell in ancient times, near Grootfontein in Namibia. By contrast, the largest stony meteorite weighs 1.7 tonnes, and is one of a group of meteorites that fell near the city of Jilin, China, in 1976.

Astronomical Instruments and Observing

Binoculars and telescopes serve two main purposes: they collect more light than the human eye, and they magnify objects. In astronomy, the first of these two aspects is paramount. Astronomers seek large telescopes not because they magnify more, but because their increased light-gathering power brings fainter objects into view and allows finer details to be distinguished.

Telescopes come in two main types: refractors, which use a main lens (known as the object glass or objective) to collect light, and reflectors, which collect light with a mirror. Astronomical telescopes have interchangeable eyepieces to give different magnifications. Binoculars are a modified form of refractor in which the light path is

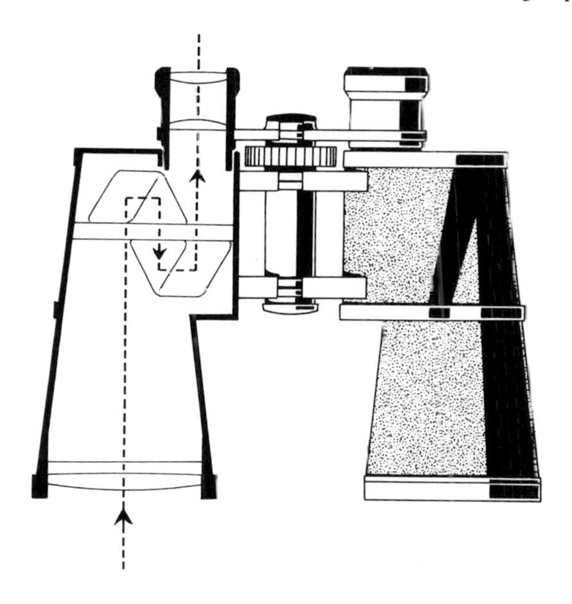

In binoculars the path of the light is folded by prisms. Wil Tirion.

folded by prisms to make them more compact. Telescopes that combine lenses and mirrors are becoming increasingly popular; these are known as catadioptric systems, and are best thought of as a modified form of reflector.

The first optical equipment that most astronomers own is a pair of binoculars. Indeed, they are virtually indispensable for any observer, even those who own a substantial telescope, for a binocular sweep can pick up objects for the telescope to home in on. Binoculars bear markings such as the following: 8 × 30, 8 × 40, 7 × 50, 10 × 50. In each case, the first figure gives the magnification, and the second figure is the aperture (in millimetres) of the front lenses.

Any of the above-mentioned combinations of magnification and aperture are suitable for astronomy. Binoculars with powers greater than about ×10 are difficult to hold steady, for the image jumps around with each tremble of the hand; instead, they need to be mounted on a stand. Remember also that the higher the magnifying power for a given aperture, the fainter the image and the narrower the field of view. Binoculars of modest power give breathtaking wide-angle views of the heavens that telescopes cannot match.

Binoculars have the advantage that they are relatively inexpensive. A small telescope – one with an aperture of 50–60 mm (2–2.4 inches) – can cost several times more than a pair of binoculars, with little gain in light grasp. What telescopes do offer is higher magnification and a tripod mounting to support the instrument. Unfortunately, in many mass-produced telescopes the mounting is not as steady as might be wished, and the image frequently vibrates for several seconds each time the telescope is moved. As a selling point, some small telescopes offer eyepieces with impressively high powers of over ×200. But such high magnifications applied to a small telescope produce images so faint that little can be seen, and are therefore a complete waste. As a good general rule, the maximum usable magnification on a telescope is ×20 for each 10 mm of aperture (×50 per inch). These pitfalls aside, there are many serviceable small telescopes available which will bring

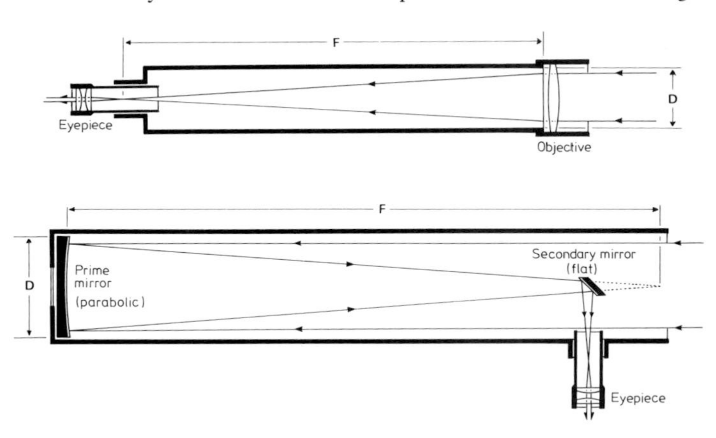

Top: *The path of light rays through a refracting telescope.* Below: *A reflecting telescope of Newtonian design, the most common type used by amateurs.* D *is the telescope's aperture,* F *the focal length.* Wil Tirion.

into view a wide selection of the celestial sights mentioned in this book.

For more ambitious observers, a refractor of 75 mm (3 inches) aperture is the minimum required to begin serious work. Telescopes greater than 75 mm in aperture are usually reflectors, because size for size they are cheaper to make than refractors. Popular sizes are reflectors of 150 mm (6 inch) and 220 mm (8½ inch) aperture, which should bring into view virtually all the targets for observation listed in this book. As for cost, remember that a telescope is a precision optical instrument, and so you must expect to pay at least as much as you would for any similar instrument such as a good camera.

Reflectors used by amateurs are usually built to the design invented in 1668 by Isaac Newton. In the Newtonian reflector, light collected by the concave main mirror is bounced back up the tube to a smaller secondary mirror which diverts the light into an eyepiece at the side of the tube. Inevitably, the secondary mirror blocks some of the incoming light from the main mirror, but this shadowing effect of the secondary is not significant and does not adversely affect the image. Another reflector design is the Cassegrain, in which the light is reflected back from the secondary through a hole bored in the main mirror. Large professional telescopes frequently use the Cassegrain design.

In *catadioptric* systems, which combine lenses and mirrors, the incoming light passes through a thin glass plate at the front of the telescope before falling onto the main mirror and being reflected as in a Cassegrain. The advantage of this design is that the overall length of the telescope can be made very much shorter than in a conventional reflector. The consequent saving in weight and space, allied with increased portability, compensates for the higher cost of a catadioptric telescope.

A telescope requires a mounting on which it can be supported and swivelled to point at various places in the sky. The quality of the mounting is just as important as the quality of the optics, for even the best telescope cannot be expected to show much if it is shaking around all the time, or if you have difficulty steering it to track objects as the Earth rotates. Sturdiness and smoothness of movement are all-important in a telescope mounting.

The simplest form of mounting is the *altazimuth*. This has two axes, one horizontal and one vertical, which allow the telescope to be pivoted up and down (in altitude) and to pan from side to side (in azimuth). In an altazimuth mount, the telescope is usually held in a fork on top of a tripod. The very smallest refractors have tripods so short that they have to be placed on a table. Such instruments are little more than toys and they are quite unsuitable for astronomy, where it must be remembered that the observer is looking upwards most of the time. A number of refractors get over this problem by providing prisms known as star diagonals to fit at the eyepiece end. These bend the light so that the

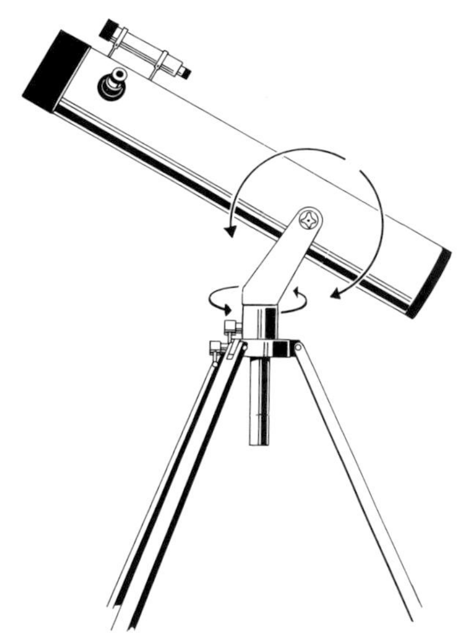

An altazimuth mounting, the simplest form of support for a telescope. The telescope is held in a cradle which allows it to rock up and down and swivel from side to side. Wil Tirion.

observer can look down into the eyepiece, which is certainly more convenient. For Newtonian reflectors, tall tripods are not necessary, since the eyepiece is near the top of the tube, and so is normally quite accessible from a standing position.

Convenient additions to an altazimuth mount are small knobs, known as slow motion controls, which can be turned to move the telescope slightly in each axis. This is useful both for centring an object in the field of view, and for tracking it as the Earth spins. When high powers are in use, the Earth's rotation can carry an object out of the telescope's field of view in a remarkably short time.

A popular version of the altazimuth for large Newtonian reflectors is the Dobsonian mount, named after the American amateur John Dobson who devised it. In a Dobsonian, the telescope tube is made of lightweight material so that its centre of gravity lies near the mirror end. The tube is supported at its lower (mirror) end by a wooden box with a Formica base that swings around in azimuth on pads made of Teflon (the slippery plastic material used in non-stick frying pans). Formica glides smoothly on Teflon, making a simple yet steady bearing. The tube pivots up and down in altitude on a shaft supported by more Teflon pads. Despite its deceptive simplicity, the Dobsonian has proved a very effective mounting. Cheapness and portability have ensured its widespread adoption.

Best of all for tracking objects is an *equatorial* mount. This has a main axis that is aligned parallel to the Earth's axis. This main axis is known as the polar axis because it points to the north celestial pole (or, in the southern hemisphere, the south celestial pole). It governs the telescope's movement in right ascension. The telescope pivots in declination on a second axis at right angles to the first, called (logically enough) the declination axis. Equatorial mounts are usually fitted with a small motor that slowly turns the equatorial axis at exactly the same speed as the Earth rotates. Once the telescope is pointed at a celestial object and the drive motor is running, the object will remain fixed firmly in the field of view for as long as the observer wishes. Experience soon demonstrates the desirability of a stationary image if you are trying to split a faint double star or make a drawing of a planet.

A few words need to be said about what you can expect to see with telescopes of different sizes, and why. Anyone looking through an astronomical telescope for the first time is usually surprised to find that the image is upside down. There is a simple practical reason for this: to turn the image the right way up, an extra lens would have to be inserted in the eyepiece. Every time light passes through a lens (or is reflected off a mirror), some light is lost. Loss of light in an astronomical telescope is undesirable, and in any case the additional lens is

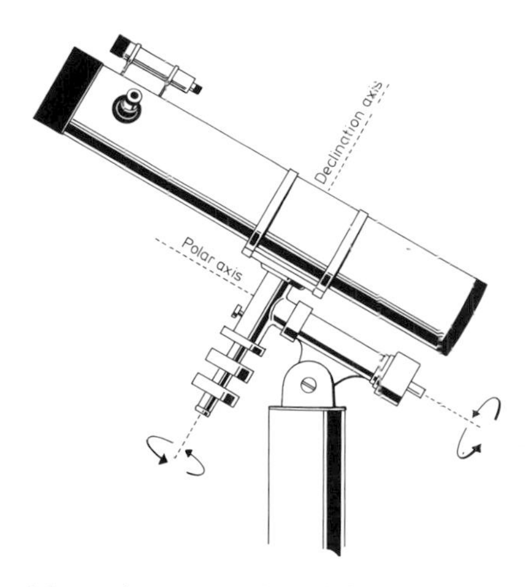

In an equatorial mounting, one axis (termed the polar axis) is set up so that it is parallel to the Earth's axis and hence points towards the celestial pole; the other axis (the declination axis) is at right angles to it. Wil Tirion.

The 4.2-metre William Herschel reflector on La Palma in the Canary Islands is one of the world's most powerful telescopes. In this photograph, the telescope can be seen inside its dome. During the exposure the dome was turned, with its slit open, to give a transparent effect. Royal Greenwich Observatory.

unnecessary. For astronomical purposes it does not really matter which way up the image appears, so the extra lens is left out, and the image remains inverted. Some telescopes come equipped with so-called terrestrial eyepieces which turn the image the right way up for everyday viewing.

When you make an observation you should record the date, the time, the instrument, the magnification and the viewing conditions. Unlike binoculars and spyglass telescopes, which have fixed eyepieces, the eyepieces in astronomical telescopes are interchangeable, offering a range of magnifications to suit the object under study. For instance, a star cluster or galaxy will require low magnification; planets are best viewed with medium magnification; and to split a close double star you will need your highest powers.

The magnification of an eyepiece depends on both its focal length and the focal length of the telescope. To work out the magnification of an eyepiece in a particular telescope requires a little straightforward arithmetic. You simply divide the focal length of the telescope by the focal length of the eyepiece. The result tells you the power of that eyepiece. It is easy to see that, the shorter the focal length, the higher the magnification. Eyepieces have their focal length marked on them.

Manufacturers often describe their telescopes in terms of focal ratio, such as *f*/6 or *f*/8. The focal ratio is the focal length of the lens or mirror divided by its aperture. If you don't know the focal length of your telescope you can easily find out by multiplying its aperture by the focal ratio. For instance, a 100-mm telescope with a focal ratio of *f*/6 has a focal length of 600 mm; if its focal ratio is *f*/8, its focal length is longer, at 800 mm. For a 150-mm telescope of *f*/6 or *f*/8, the focal length would be 900 mm or 1200 mm respectively.

Now, assume that you have an eyepiece of focal length 20 mm. In a telescope of 600 mm focal length it will give a magnification of 600 divided by 20, which is ×30. That is quite a low power for astronomical purposes. In a telescope of 1200 mm focal length, the same eyepiece will give a magnification of ×60. An eyepiece of half the focal length, 10 mm, will give twice the magnification. Note that the aperture of the telescope is irrelevant in these calculations; the focal length is the sole figure governing magnification.

Where the telescope's aperture becomes crucial is in the matters of light grasp and resolution of detail. All other things being equal, a larger aperture shows fainter stars and finer detail than a smaller aperture, but exactly how faint the stars are and how fine the detail is depends on atmospheric conditions, optical quality and the observer's eyesight. Practical experience shows that the faintest stars likely to be visible through amateur telescopes of various apertures are as follows:

Aperture	*Limiting magnitude*
50 mm (2 inches)	11.2
60 mm (2.4 inches)	11.6
75 mm (3 inches)	12.1
100 mm (4 inches)	12.7
150 mm (6 inches)	13.6
220 mm (8.5 inches)	14.4

Incidentally, your eyes need time to become accustomed to the dark before you can hope to see the faintest objects; when you go out at night from a bright room allow at least 10 minutes for your eyes to become dark adapted. One useful trick when trying to glimpse faint objects is to use *averted vision* – that is, to look to the side of the object under study so that its light falls on the outer, more sensitive part of the retina.

The resolution, or resolving power, of a telescope is expressed in seconds of arc ("). One second of arc is a very small quantity, equivalent to the size that a coin would appear to be from several kilometres away. A telescope's resolution governs the detail that one can see on the Moon or planets, and the closeness of double stars that can be separated. Here are the theoretical limits of resolution for telescopes of various apertures:

Aperture	*Limit of resolution (seconds of arc)*
50 mm (2 inches)	2.3
60 mm (2.4 inches)	1.9
75 mm (3 inches)	1.5
100 mm (4 inches)	1.1
150 mm (6 inches)	0.8
220 mm (8.5 inches)	0.5

Under exceptional conditions, the limits of magnitude and resolution tabulated above may be surpassed; but in many cases, particularly from cities, they will not be achieved.

Finally, we come to the atmosphere itself. Professional observatories are sited on high mountain tops to get above as much of the atmosphere as possible, but most amateurs are stuck with conditions in their own back yard, all too often including urban haze and glare from streetlights. There are two aspects of the atmosphere to take into account: its clarity, and its steadiness. A good index of atmospheric clarity is the magnitude of the faintest stars that can be seen overhead by the naked eye. The zenithal limiting magnitude should always be noted when meteor observing, because it affects the hourly rates. Atmospheric clarity needs to be at its best when looking for faint objects, particularly nebulae, galaxies and comets.

On the other hand, the steadiness of the atmosphere, known as the seeing, is paramount for planetary and double star observation. Turbulence in the atmosphere produces a boiling effect of the telescopic image, drastically reducing the resolution. Hot air rising from neighbours' houses can be a particularly annoying localized form of turbulence. Ironically, on a crystal-clear night after a rainstorm the seeing can be particularly atrocious, whereas slightly misty nights, when the clarity is low, can be the steadiest.

Seeing is estimated on a five-point scale: 1, perfect; 2, good; 3, moderate; 4, poor; 5, very bad. If a close double star cannot be clearly split on a night of indifferent seeing, examine it again on a better night.

Astrophotography

A camera can be used as an astronomical instrument. Photographs of the sky can be taken with any camera capable of time exposures. That usually means a single lens reflex (SLR) camera since most modern automatic cameras do not have a time-exposure facility. Depending on the length of the exposure, it is possible to photograph constellations, graceful curving star trails, groupings of planets and perhaps the occasional meteor.

Firstly, the camera must be loaded with fast film. Whether it is colour or black-and-white doesn't matter. Black-and-white has the advantage that it is more economical and can be easily processed at home, but colour transparencies are the most convenient for beginners. Colour negative film is not so good, as it requires special printing. Very fast colour slide films, rated ISO 1000 or faster, are available and these can produce beautiful results.

It is important to use the widest aperture on the camera lens. Camera apertures are quoted in terms of *f*/stops. Most camera lenses will open up to *f*/2.8 and many even wider, up to *f*/1.4. Cameras often have interchangeable lenses. A wide-angle lens will photograph more sky than a standard lens, but the images will be smaller and the constellation shapes will appear distorted towards the edges. Finally, the camera focus must be set to infinity.

A useful accessory is a cable release, which enables the shutter to be opened and closed without shaking the camera. For time exposure the camera needs to be on the setting marked 'B' (this, incidentally, stands for 'bulb', a legacy of the days when cameras were operated by air bulbs). On the B setting, the button must remain depressed to keep the shutter open, so the cable release must be locked for the length of the time exposure. Some cameras have a 'T' setting (for 'time'); on this, one press of the button opens the shutter, and a second press closes it.

The camera must be firmly mounted for the duration of the exposure. If there is no suitable tripod, the camera can be wedged firmly on the ground by pieces of wood, bricks or stones. One good way of avoiding camera shake is to hold a piece of card over the lens while the shutter is opened. Once all vibrations have died down, the card can be moved away. The card must be put back in front of the lens before the shutter is closed at the end of the exposure. When the camera is pointing in the rough direction of the area to be photographed (stars are so faint they will probably not be visible in the viewfinder), you are ready to shoot.

The exposure used depends on what you are trying to achieve, and on the darkness of the sky. A good way to start is by photographing star trails, by leaving the shutter open for a length of time. As the stars move across the sky with the rotation of the Earth, they leave trails of light on the film. When you point the camera in the direction of the pole, the star trails will come out noticeably curved. If an exposure of the equatorial region of the sky is taken, the star trails will appear straight. Colour film will register an attractive range of star colours.

In an area with lots of streetlights, you should restrict the exposure to five minutes, or else the sky brightness will build up on the film and fog the image. Under darker skies, exposure of half an hour or longer will be possible without fogging. The only way to find out for sure is by experimenting.

If the exposure is kept short – 15 to 20 seconds – the Earth will not have rotated far enough to produce noticeable trails, and the constellations will be recorded as they appear to the naked eye. Stars down to 6th magnitude will be recorded. Photographs taken with short exposures in twilight, or with moonlight illuminating the surrounding landscape, can be particularly attractive. From time to time, several planets form an interesting grouping which is worth capturing with a short exposure. Always take a succession of photographs with different exposures to be sure of getting the best result. Use of a telephoto lens can sometimes add drama to a scene – for instance, a partially eclipsed Moon rising over a distant skyline. But you must remember that the effect of the Earth's rotation shows up more quickly through a telephoto, so exposures should be restricted to no more than a few seconds to avoid trailing of the images.

When a bright meteor shower such as the Perseids or Geminids is due, try a series of long exposures with a wide-angle lens in the hope of catching some. Meteors dart across the sky so quickly that only the brightest will register on your film, and many of them will in any case fall outside the camera's field of view. But the excitement of photographing the brilliant streak of a meteor more than compensates for the disappointment of the other frames that show nothing but star trails.

If a camera is placed on an equatorial mount it can be guided to prevent star trails from forming. Guided exposures of only a few minutes' duration will capture stars far fainter than those visible to the naked eye. Most exciting of all is to use a camera to photograph what can be seen through a telescope. An SLR camera is essential for this. The camera lens must be removed completely and the camera body attached to the eyepiece mount (adaptors are available for this). In effect, the entire telescope is acting like a super telephoto lens.

With such a system, craters on the Moon, the belts of Jupiter and the rings of Saturn can be photographed. Exposures range from a fraction of a second for the Moon to a second or more for the planets. Even longer exposures, of several minutes' duration, can reveal delicate details of faint nebulae and galaxies. But that is only for advanced astrophotographers, and the techniques involved are outside the scope of a basic book such as this.

Index

All named stars of mag. 2.0 or brighter are included in the Index. Stars such as η (eta) Carinae will be found under the relevant constellations. Page numbers in bold type refer to photographs or illustrations.

Biographical Notes

Ian Ridpath is an English amateur astronomer, and an author and broadcaster on astronomy and space. He is editor of *Norton's 2000.0*, a new edition of the world-famous *Norton's Star Atlas and Reference Handbook*, and the author of *Star Tales*, about constellation mythology. Wil Tirion is a Dutch amateur astronomer and graphic artist. In 1981 he published *Sky Atlas 2000.0*, which was immediately hailed as the finest atlas of its kind. Since then he has published the more detailed *Uranometria 2000.0*. Ian Ridpath and Wil Tirion are also authors of the Collins *Gem Guide to the Night Sky*, and *The Monthly Sky Guide*.